Proceedings of the 4th International Symposium in Environmental Science and Industrial Ecology

Yuanqing Zhu
Editor

Proceedings of the 4th International Symposium in Environmental Science and Industrial Ecology

Springer

Editor
Yuanqing Zhu
College of Power and Energy Engineering
Harbin Engineering University
Harbin, Heilongjiang, China

ISBN 978-981-96-1577-3 ISBN 978-981-96-1578-0 (eBook)
https://doi.org/10.1007/978-981-96-1578-0

This Springer imprint is published by the registered company Springer Nature Singapore Pte Ltd.
The registered company address is: 152 Beach Road, #21-01/04 Gateway East, Singapore 189721, Singapore

Preface

This book presents selected papers from the 4th International Symposium in Environmental Science and Industrial Ecology (ISESIE 2024), held as a hybrid conference on August 23–24, 2024, in Bangkok, Thailand. The papers include contributions from researchers, scientists, academics, and practitioners covering topics in the fields of environmental science and ecology, as well as industrial ecology and resource management. The book serves as a valuable reference for students, researchers, and professionals studying the broad areas of environmental science and industrial ecology.

Harbin, China

Yuanqing Zhu

Contents

The Study on the Migration Law of Ba^{2+} in the Plastic Concrete Isolation Barrier in a Chemical Plant

Longjin Jiao, Bowen Bai, Peng Ge, Haoqing Xu, Aizhao Zhou, and Shaowen Hou

Abstract The use of plastic concrete barriers to prevent pollutant migration is a common technical approach. By integrating soil–water characteristic curve parameters and retardation factors obtained from experiments. This study thoroughly examines the influences of convection, dispersion, diffusion, and adsorption on pollutant transport. The impact of different permeability coefficients of plastic concrete barriers on the retardation of pollutants is evaluated. A two-dimensional simulation model integrating groundwater flow and pollutant movement was developed for a chemically contaminated site using HYDRUS-2D software. Numerical methods were applied to explore how pollutants migrate through groundwater and to evaluate how effectively the separation wall controls this movement. The results indicate that for an isolation barrier with an initial 5000 mg/L Ba^{2+} concentration, a depth of 48 m, and with a 1 m thickness, the failure criterion is based on the Ba^{2+} pollution levels outlined in the "Environmental Quality Standard for Groundwater": With a permeability coefficient of 1×10^{-7} cm/s, the barrier fails after 4284 days (11.73 years). In contrast, with a permeability coefficient of 1×10^{-8} cm/s, the failure time increases to 53,848 days (147.53 years).

Keywords Hydrus-2D · Barium ion migration · Plastic concrete · Isolation barrier · Breakdown time

1 Introduction

Plastic concrete cutoff walls are widely used in hydraulic dams due to their low elastic modulus, high strain capacity, and low permeability, achieving excellent results [1]. Since the 1980s, China has researched and applied plastic concrete in major projects,

L. Jiao · P. Ge
The No. 3 Brigade of Jiangsu Geology and Mineral Resources Bureau, Zhenjiang Jiangsu, China

B. Bai · H. Xu (✉) · A. Zhou · S. Hou
Jiangsu University of Science and Technology, Zhenjiang Jiangsu, China
e-mail: hank1nxu@just.edu.cn

Y. Zhu (ed.), *Proceedings of the 4th International Symposium in Environmental Science and Industrial Ecology*, https://doi.org/10.1007/978-981-96-1578-0_1

including the Three Gorges Dam [2]. With the development of China's economy, the process of industrialization and urbanization is constantly accelerating, and the soil pollution problem is more and more serious [3–5]. Among them, the soil pollution of industrial sites is particularly serious. To prevent pollutants from industrial sites from contaminating groundwater and causing environmental problems, vertical isolation barriers are commonly installed around these sites to block pollutant migration. Among these, plastic concrete vertical isolation barriers are frequently used because they can adapt to soil deformation and effectively prevent the migration and diffusion of groundwater and pollutants [6–8].

HYDRUS, developed in the Czech Republic, is a software tool for simulating the movement of water, heat, and various solutes in two-dimensional and three-dimensional variably saturated porous media. It includes various modules for tasks such as analyzing slope stability, studying seepage, designing constructed wetlands, assessing geochemistry, and evaluating colloid-facilitated transport. HYDRUS is widely used in fields such as hydraulics and environmental science due to its versatility and comprehensive capabilities [9, 10].

This study employed the HYDRUS-2D simulation program to create a 2D model for assessing water seepage and contaminant transport. This model was used to evaluate the effectiveness of cutoff walls in preventing seepage and containing pollutants. Parameters for various soil layers in the region were determined based on geotechnical investigation reports, experimental data, and literature. These parameters include the geometric dimensions, permeability coefficients, dispersion, diffusion coefficients, and retardation factors for each soil layer and the cutoff wall. A two-dimensional model for contaminant migration was developed to simulate the temporal and spatial changes in pollutant concentration with the isolation barrier installed. This model specifically evaluates the barrier's effectiveness in blocking Ba^{2+}.

2 Project Overview

A waste disposal site is located in a valley within a chemical plant area, surrounded by mountains on the east, south, and west sides, with an open valley to the north. The site generally inclines from a higher elevation in the south to a lower elevation in the north and belongs to a mid-low mountain erosion–deposition geomorphological unit. The surrounding mountain slopes range from 30° to 50°. The soil layers at the site, arranged from the surface downwards, consist of waste residue, silty clay, highly weathered slate, moderately weathered slate, slightly weathered slate, and unweathered slate. Several gullies on the mountain above the disposal area converge near the valley opening.

Investigations revealed that groundwater in the area contains multiple heavy metals exceeding permissible limits, with barium ions being the most widespread contaminant. To prevent environmental pollution from the migration of barium ions from the waste residue, a plastic concrete isolation barrier was installed on the northern side of the site. This barrier extends at least 1.0 m into the unweathered rock layer with a low permeability coefficient, thereby blocking the migration pathways of contaminants and ensuring the environmental safety of the surrounding area.

3 Mathematical Modelling

3.1 Establishment and Solution of Mathematical Model

F1 Flow Pattern of Water. The movement of soil water is mainly regulated by the Richards equation [11, 12].

$$\frac{\partial \theta}{\partial t} = \frac{\partial}{\partial x}\left[D\frac{\partial \theta}{\partial x}\right] + \frac{\partial}{\partial z}\left[D\frac{\partial \theta}{\partial z}\right] + \frac{\partial K}{\partial z} \tag{1}$$

where, θ is the volumetric water content, expressed in cubic meters per cubic meter (m^3/m^3); t denotes time in seconds (s); The variables x and z denote the horizontal and vertical distances, respectively, measured in meters (m). D represents the dispersion coefficient, which includes molecular diffusion and hydrodynamic dispersion, and is measured in square meters per second (m^2/s). K indicates the unsaturated permeability coefficient, expressed in centimeters per second (cm/s).

Concerning the soil hydraulic characteristics and K as mentioned in the preceding equation, van Genuchten equation [13, 14] was used to modify them. This equation offers a versatile representation of soil hydrodynamics, especially when the soil water is close to saturation. The equation is given by:

$$\theta = \begin{cases} \theta_r + \frac{\theta_s - \theta_r}{\left[1+|\alpha h|^n\right]^{1-1/n}}, & h < 0 \\ \theta_s, & h \geq 0 \end{cases} \tag{2}$$

$$S_e = \frac{\theta - \theta_r}{\theta_s - \theta_r} \tag{3}$$

$$K = K_s S_e^l \left[1 - \left(1 - S_e^{n/n-1}\right)^{1-1/n}\right]^2 \tag{4}$$

where: θ is the effective saturation of the soil; θs is the saturated water content of the soil, measured in cubic meters per cubic meter (m^3/m^3); θr is the residual water content of the soil, also in cubic meters per cubic meter (m^3/m^3); Ks is the saturated permeability, in centimeters per second (cm/s); n represents the correlation coefficient for the soil–water characteristic curve; and l is the gap connectivity coefficient, typically set to 0.5 and dimensionless.

Model for Solute Transport. The movement of solutes within the soil is primarily controlled by the convection–dispersion model [15, 16].

$$R_d \frac{\partial C}{\partial t} = D \frac{\partial^2 C}{\partial x_i^2} - v_i \frac{\partial C}{\partial x_i} \tag{5}$$

$$R_d = 1 + \frac{\rho_d}{n} K_d \tag{6}$$

where: C denotes the concentration of the solute (mg/L); R_d is a dimensionless blocking factor; n represents porosity; and K_d represents the linear adsorption partition coefficient under isothermal conditions (mg/kg).

4 Modeling of Contaminant Migration at the Site

4.1 Computational Model

Using the data from survey points in the geological survey report, this paper interpolates each soil layer to create a reference cross-section and create a finite element numerical model to analyze pollutant transport at the site, illustrated in Fig. 1. Through numerical modeling, we forecast the extended behavior of barium ions over time within the site layers and evaluate the performance of plastic concrete vertical isolation barriers across different geological scenarios.

4.2 Seepage Initial and Boundary Conditions

The simulation considers the most adverse scenario, where the water level within the contaminated site is elevated compared to the surrounding area. Specifically, there is a high water level inside the barrier and a lower level outside. According to the geological survey report, the water level inside the barrier is set at -5 m underground, while the level outside is − 20 m underground. Water head boundaries on the model's sides adjust in a linear fashion with depth.

Starting Conditions

$$H(x, y, t)|_{t=0} = H_0(x, y), (x, y) \in G \tag{7}$$

where, *G* is the research domain.

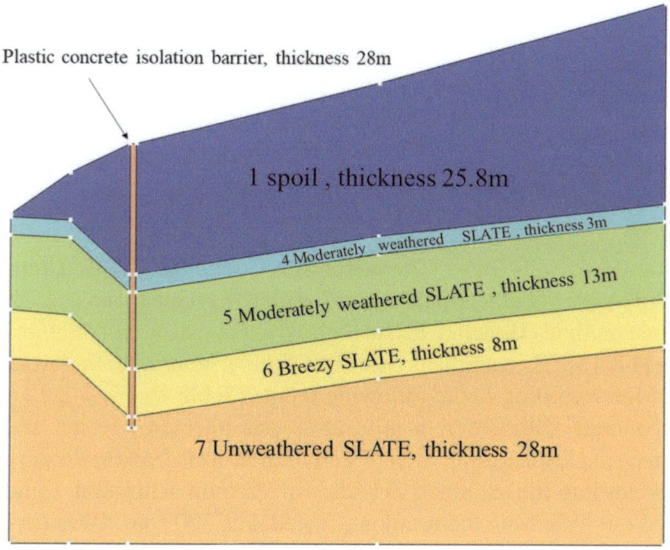

Fig. 1 FEM model

Boundary Conditions

$$H(x,y,t)\big|_{左} = H_1(x,y,t),(x,y) \in \text{left boundary}, t > 0 \tag{8}$$

$$H(x,y,t)\big|_{右} = H_2(x,y,t),(x,y) \in \text{right boundary}, t > 0 \tag{9}$$

4.3 *Starting and Boundary Conditions for Pollutant Transport*

Based on the preliminary design, the highest concentration of sulfate ions at the contaminated site can reach 5000 mg/L, which is used as the starting concentration for the simulation. The boundary for solute transport is set to a constant concentration.

Starting Conditions

$$c(x, y, t)|_{t=0} = c_0(x, y), (x, y) \in G \tag{10}$$

Boundary Conditions

$$H(x,y,t)\big|_{左} = 0,(x,y) \in \text{left boundary}, t > 0 \tag{11}$$

$$H(x,y,t)\big|_{右} = c_2(x,y,t),(x,y) \in \text{right boundary}, t > 0 \tag{12}$$

4.4 Design Conditions

Parameters of the Soil–water Characteristic Curve for Plastic Concrete. In this study, the soil–water characteristic curve for plastic concrete was determined using the centrifuge method (Hitachi CR21N high-speed constant temperature refrigerated centrifuge) [17, 18]. The sample is made of GS [17], bentonite, Fujian standard sand and water. Mix according to the following recipe (Table 1).

Plastic concrete with different mix ratios was injected into the centrifuge ring knife, and the ring knife sample was placed in the curing box for 28 days. The cured sample was put into the saturated cylinder for vacuum saturation. Centrifugal tests were carried out by setting the rotational speed n as 300 rpm, 1000 rpm, 1600 rpm, 25,000 rpm, 3500 rpm, 4500 rpm, 6000 rpm, 7500 rpm, 8500 rpm and 9500 rpm respectively, and each speed was run for 48 h. By substituting the test data into the calculation, the water head value at different rotational speeds can be obtained. The quality of the ring tool at different speeds is obtained, as shown in Table 2.

The above data were calculated and fitted [19], and soil–water characteristic curves of four different formulations were obtained. P1 was taken as an example (Fig. 2).

Soil–water characteristic curve parameters of different formulations were obtained, as detailed in Table 3.

Based on the above experimental results, the values of α, n, θ_r and θ_S of plastic concrete are 0.008, 1.09, 0.005, and 0.47, respectively. The permeability coefficient of plastic concrete is roughly between 10^{-7} and 10^{-8} cm/s. In order to simulate the different conditions of barium ions in plastic concrete barriers under the condition of high permeability and low permeability, the permeability coefficients of 10^{-7} cm/s and 1×10^{-8} cm/s were selected in this paper.

Parameters of the Soil–Water Characteristic Curve for Different Soil Layers. In this paper, mixed fill, strongly weathered andesite, moderately differentiated andesite

Table 1 Plastic concrete formula sheet

Formula	P1	P2	P3	P4
Water	180.5	190	171	199.5
Bentonite	20	20	20	20
GS	75	75	75	75
Grit	600	600	600	600
Water cement ratio	1.9	2	1.8	2.1
Bentonite content (%)	21.10	21.10	21.10	21.10

Table 2 Ring tool quality table at different speeds

	Number	1	2	3	4
	Saturated undisturbed soil	193.3	193.1	193.87	196.26
	Baked undisturbed soil	147.08	144.82	147.67	148.47
1	300 rpm	266.23	266.09	266.81	269.33
2	1000 rpm	265.32	264.17	265.51	268.49
3	1600 rpm	262.28	260.92	262.28	265.55
4	2500 rpm	259.83	259.01	260.13	263.47
5	3500 rpm	257.46	256.67	257.88	261.47
6	4500 rpm	255.8	255.34	256.36	260.14
7	6000 rpm	254.47	253.88	254.73	258.63
8	7500 rpm	252.89	252.41	253.34	257.33
9	8500 rpm	252.05	251.52	252.37	256.42
10	9500 rpm	252.02	251.37	252.17	256.11

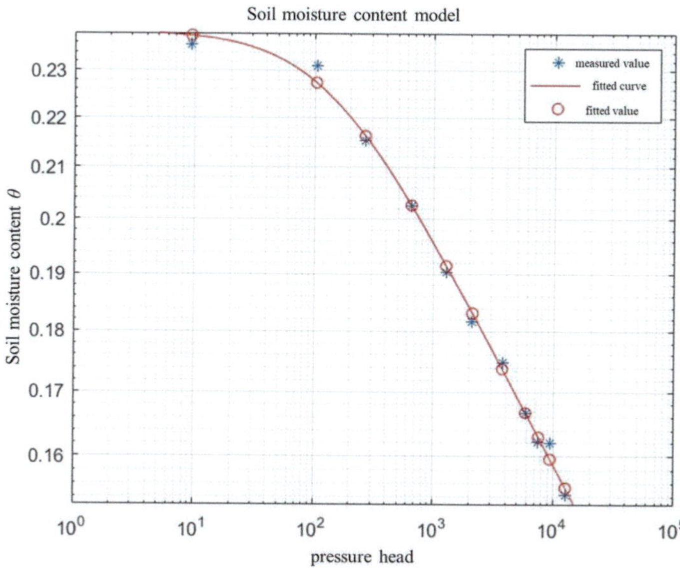

Fig. 2 P1 soil–water characteristic curve

Table 3 Soil–water characteristic curve parameters

	α	n	θ_r	θ_s
P1	0.0074	1.09	0.01145	0.46
P2	0.0097	1.08	0.00226	0.48
P3	0.0084	1.09	0.0005	0.46
P4	0.0046	1.096	0.0003	0.47

and cutoff wall are equivalent to soil. Drawing from the design and construction guidelines for anti-pollution barriers in landfills and contaminated sites [20], in conjunction with the geological survey report, the VG model parameters for the mixed fill were established, strongly weathered andesite, and moderately differentiated andesite were established (Table 4).

Dispersion, Diffusion Coefficient. Referring to the data on miscellaneous fill and various weathered rock layers provided in "Design and Construction of Anti-pollution Barriers in Landfills and Contaminated Sites" [20], relevant parameters are determined, as detailed in Table 5.

Retarding Factor. Batch test is often used to study the adsorption properties of adsorbent on adsorbent particles at constant temperature [21, 22]. In this paper, Batch test was used, taking Ba^{2+} as the research object, and the adsorption curves of plastic

Table 4 Common soil VG model parameters

Soil	Voidage (%)	α	θ_s	Saturation permeability coefficient	θ_r	n
1 Spoil	45	0.078	0.45	5.1×10^{-3}	0.036	1.56
4 Moderately weathered SLATE	30	0.001	0.3	2×10^{-5}	0.005	1.2
5 Moderately weathered SLATE	27	0.001	0.27	6.1×10^{-5}	0.005	1.15
6 Breezy SLATE	25	0.001	0.25	4.4×10^{-6}	0.005	1.12
7 Unweathered SLATE	20	0.001	0.2	6.71×10^{-8}	0.005	1.1

Table 5 Table summarizing the main parameters for each soil layer

Parameters	1 Spoil	4 Moderately weathered SLATE	5 Moderately weathered SLATE
Dry density (g/cm^3)	1.3	2.2	2.3
Lateral dispersion (m)	0.1	0.1	0.1
Longitudinal dispersion (m)	0.01	0.01	0.01
D (10 m^2/s)	8	2	1.8
Parameters	6 Breezy SLATE	7 Unweathered SLATE	8 Isolating barrier
Dry density (g/cm^3)	2.4	2.5	1.4
Lateral dispersion/m	0.1	0.1	0.1
Longitudinal dispersion/m	0.01	0.01	0.01
D (10 m^2/s)	1.6	1.5	2.6

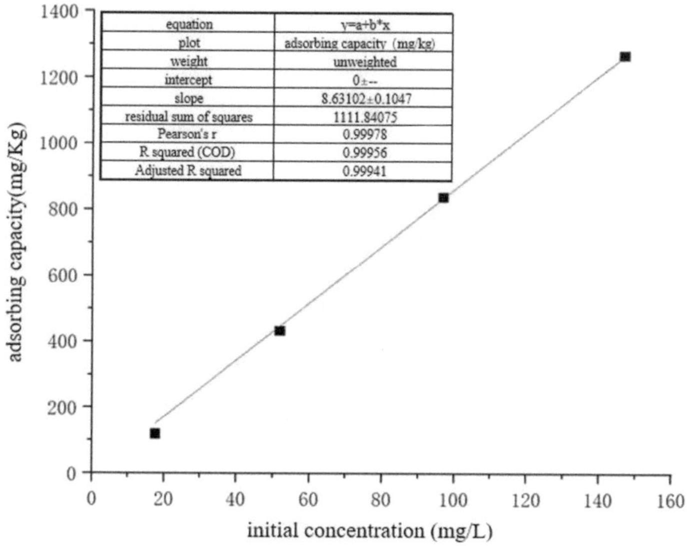

Fig. 3 Isothermal adsorption curve of P1 sample powder for Ba^{2+}

concrete powder with different formulations on heavy metal Ba were obtained, and then the blocking factor of plastic concrete barrier on Ba^{2+} was obtained.

The dried P1 and P2 samples were cooled and ground into powder. The powder was mixed with a solution of different initial Ba^{2+} concentrations in a centrifuge tube at a soil–water ratio of 1:10 (g/mL). The centrifugal tube was placed in a thermostatic oscillator and oscillated at 140 rpm and 25 °C for 24 h. The powder is separated from the contaminated liquid by centrifugal method. The clarified solution in the middle and upper part of the centrifugal tube was sampled and measured, and the amount of pollutants adsorbed by the soil material per unit mass was calculated, and then the adsorption curve of plastic concrete for heavy metals was obtained, as shown in Fig. 3.

The adsorption of Ba^{2+} by P1 plastic sample powder is consistent with Henry linear isothermal adsorption, and the distribution coefficient K_{dp} for Ba^{2+} is 8.631L/kg. According to Eq. (6), the blocking factor of P1 sample was 28.45. Considering that the retarding factor obtained by Batch test is greater than that in actual seepage [23, 24], the retarding factor of plastic concrete in seepage state is set to 15.

5 Analysis of Simulation Results of Site Pollutant Migration

5.1 Analysis of Pollutant Migration Results

Barrier With Permeability Coefficient of 1 x 10^{-7} cm/s

From Figs. 4, 5 and 6, it can be seen that Ba^{2+} migrated from the surface direction of the site under the drive of water flow from the year 0 to the year 50. When pollutants are obstructed by the barrier, they accumulate within it, leading to increased local concentrations. Simultaneously, pollutants are transported by water into the surrounding rock formations. During the period from the 50th to the 100th year, pollutants persist in moving towards the barrier, with some eventually penetrating through it.

Barrier With Permeability Coefficient of 1 x 10^{-8} cm/s

As can be seen from Figs. 7, 8 and 9, Ba^{2+} migrated from the site in the upward direction driven by water flow in the year 0 to 100. The isolation barrier effectively blocks the pollutants, preventing any from passing through.

Comparing the calculation results for different permeability coefficients reveals that the permeability of the barrier significantly impacts its effectiveness in isolating pollutants.

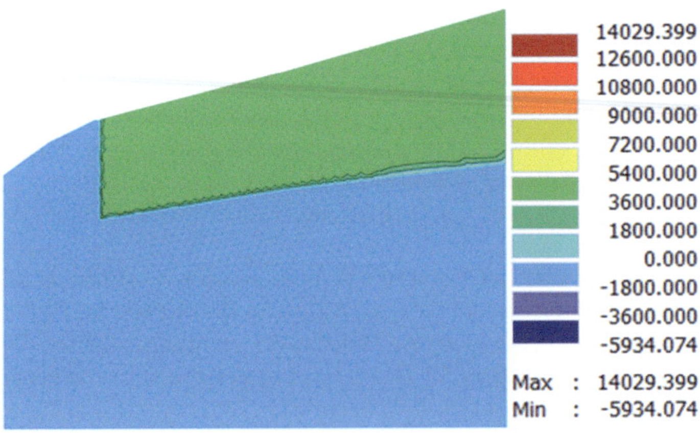

Fig. 4 Migration at year 0 for barrier with permeability 1×10^{-7} cm/s

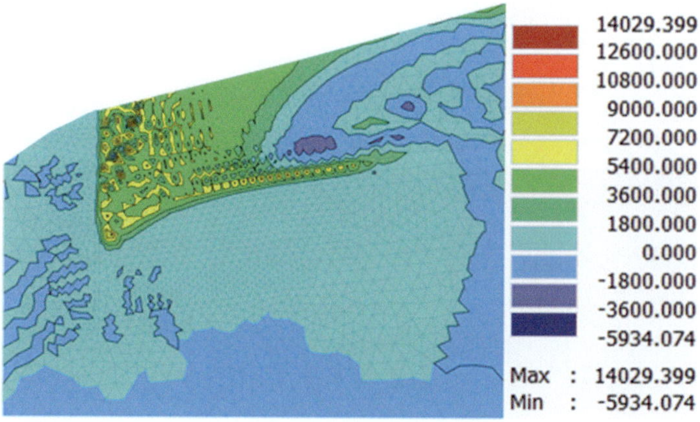

Fig. 5 Migration at year 50 for barrier with permeability 1×10^{-7} cm/s

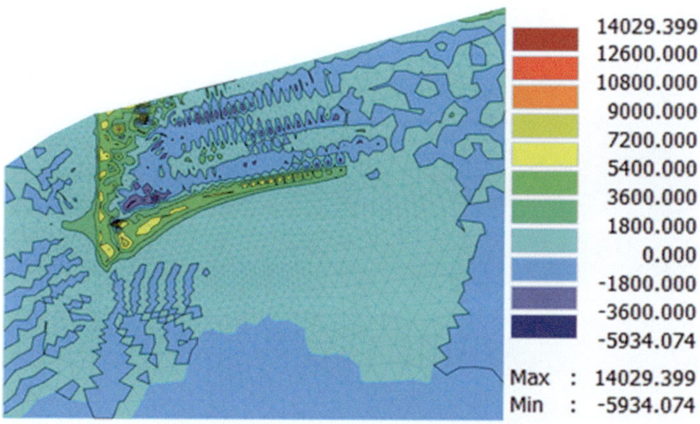

Fig. 6 Migration at year 100 for barrier with permeability 1×10^{-7} cm/s

5.2 Analysis of Observation Point Results

In the simulation process, observation points were set at 0 m, 4 m, 8 m, 12 m, 16 m, 20 m, 24 m, 28 m, 32 m, 36 m, 40 m and 44 m respectively, and the ion concentration diagrams of different observation points at different times were obtained, as shown in Figs. 10 and 11.

The breakdown curve of Ba^{2+} is obtained from the data of observation points, as shown below (Fig. 12).

The breakdown of the anti-fouling barrier by leachate occurs when the leachate infiltrates through the barrier via prolonged processes of infiltration, migration, dispersion, and adsorption, leading to the barrier's failure. This failure signifies the loss of the barrier's effectiveness in safeguarding environmental safety, marking the

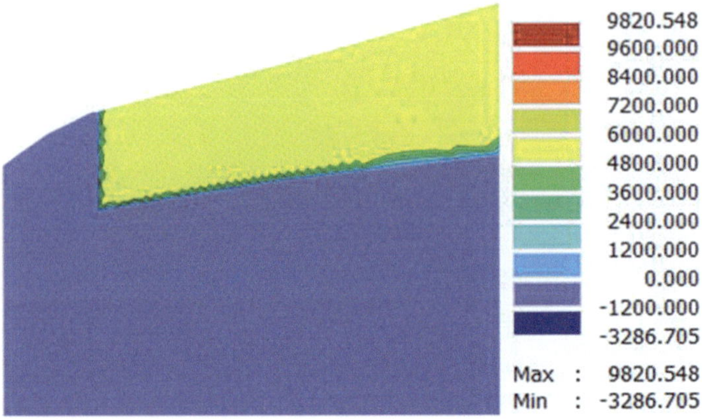

Fig. 7 Migration at year 0 for barrier with permeability 1×10^{-8} cm/s

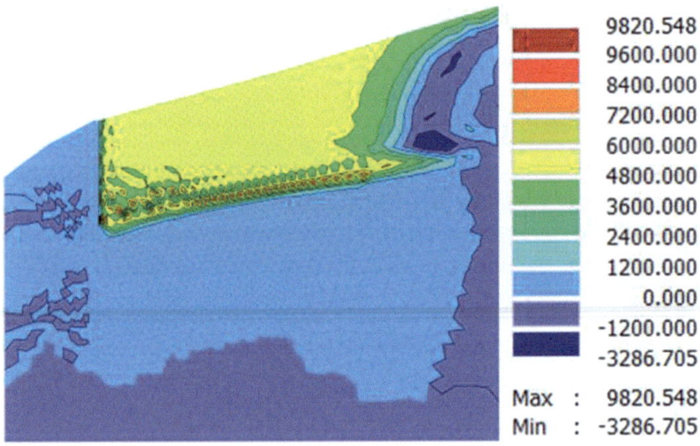

Fig. 8 Migration at year 50 for barrier with permeability 1×10^{-8} cm/s

point at which the leachate begins to harm the environment. In this paper, The breakdown standard was based on the contamination threshold specified in the Groundwater Environmental Quality Standards [25]. According to relevant guidelines, the concentration of Ba^{2+} that causes contamination is set at 85 mg/L. As shown in the figure, With a barrier permeability coefficient of 1×10^{-7} cm/s, the failure occurs in 4284 days (11.73 years). In contrast, For a permeability coefficient of 1×10^{-8} cm/s, the failure time extends to 53,848 days (147.53 years).

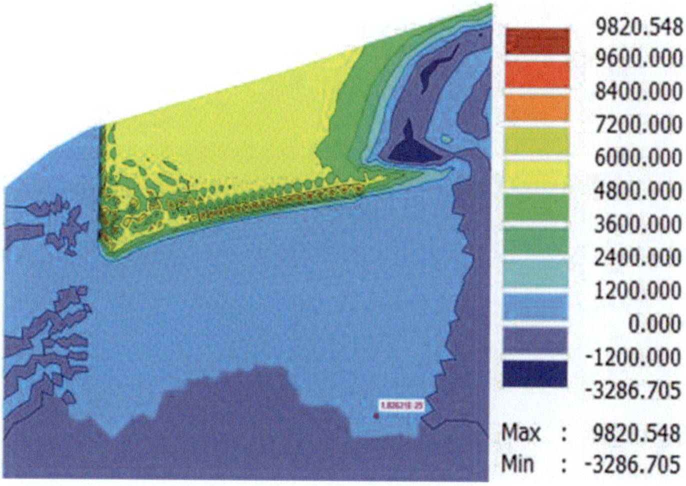

Fig. 9 Migration at year 100 for barrier with permeability 1×10^{-8} cm/s

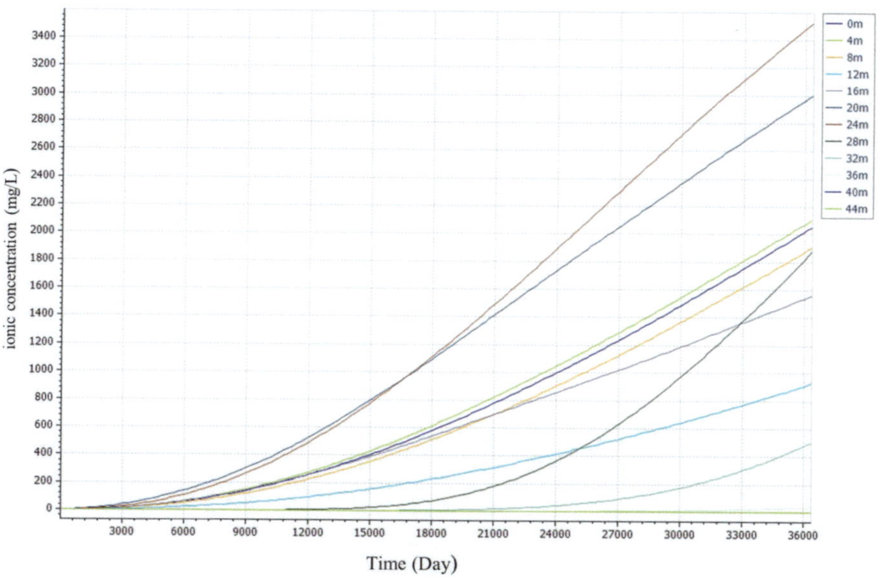

Fig. 10 Permeability coefficient plot of Ba^{2+} concentration plot at the observation point 1×10^{-7} cm/s

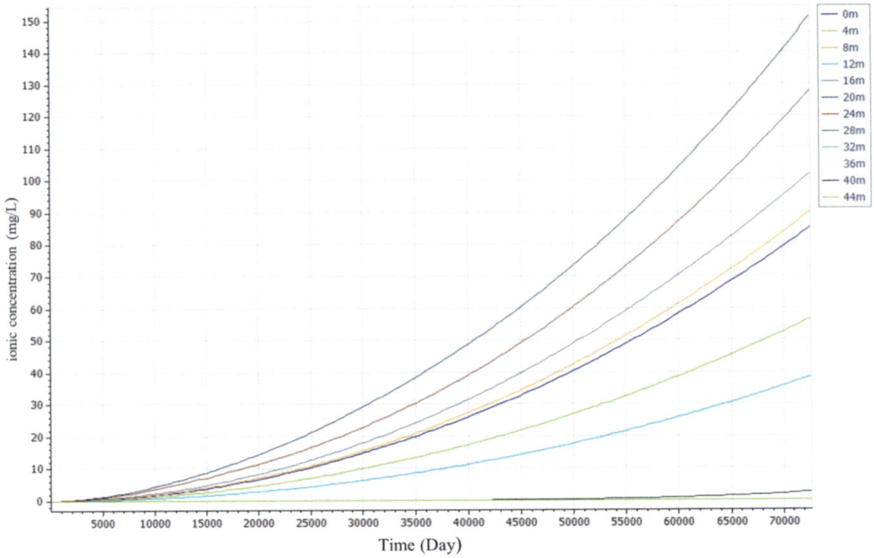

Fig. 11 Permeability coefficient plot of Ba^{2+} concentration plot at the observation point 1×10^{-8} cm/s

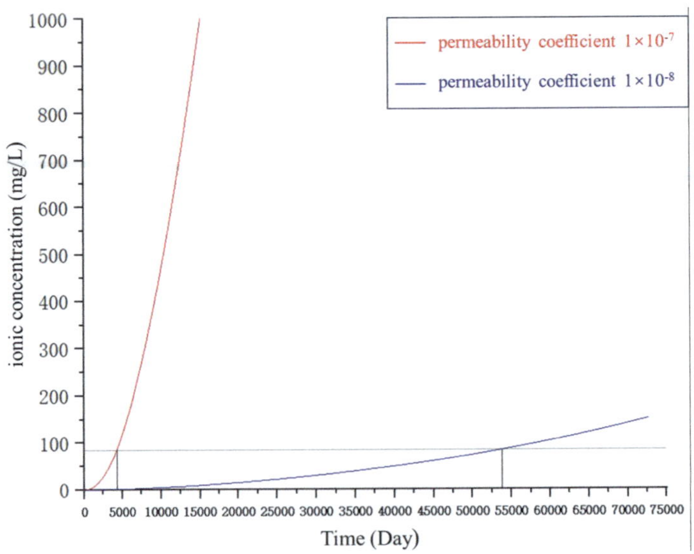

Fig. 12 Ba^{2+} breakdown diagram

6 Conclusion

1. The values of soil–water characteristic curve parameters α, n, θ_r and θ_S of plastic concrete are roughly 0.008, 1.09, 0.005, and 0.47.
2. Through Batch test, the retarding factor values of Ba^{2+} and plastic concrete without water head were approximately 28.45.
3. The permeability of the barrier significantly impacts its performance. With a barrier permeability coefficient of 1×10^{-7} cm/s, the failure occurs in 4284 days (11.73 years). In comparison, For a permeability coefficient of 1×10^{-8} cm/s, the failure time extends to 53,848 days (147.53 years).

References

1. Li, W.L.: A review of plastic concrete cutoff wall technology. Des. Hydraulic Hydro-Power Eng. **1995**(03), 54–59 (1995)
2. Li, J.Z., Yan, J.J., Yang, H.Q.: Application research of plastic concrete in the three Gorges project. J. Hydroelectr. Eng. **28**(01), 159–164 (2009)
3. Chen, M.F.: Review of heavy metal remediation technologies for soil and groundwater in industrially polluted sites in China. Bull. Chin. Acad. Sci. **29**(03), 327–335 (2014)
4. Song, Y., Wei, L., Wang, H.J.: Current status and selection strategies of heavy metal contaminated soil remediation technologies in China. Environ. Prot. **42**(09), 32–36 (2014)
5. Luo, Y.M.: Current status and trends in research on polluted soil remediation technologies. Prog. Chem. **21**(Z1), 558–565 (2009)
6. Chen, W.H., Ren, Y.P., Zhang, Y.Y., Xu, G.H., Xu, X.B.: Research on containment and isolation technologies for polluted sites. Environ. Sci. Technol. **42**(S2), 114–124 (2019)
7. Liu, W.: Analysis of Service Life and Thickness Calculation Method for Impermeable Curtain in Landfills. Dissertation, Zhejiang University (2011)
8. Wu, H.L.: Research and Development of Magnesium Oxide Activated Slag-Bentonite and High-Performance ECC Vertical Barrier Materials and Their Barrier Performance. Dissertation, Southeast University (2019)
9. Nguyen Ngoc, M., Dultz, S., Kasbohm, J.: Simulation of retention and transport of Copper, Lead, and Zinc in a paddy soil of the Red River Delta, Vietnam. Agric. Ecosyst. Environ. **129**(1), 8–16 (2008)
10. Šimůnek, J., van Genuchten, MTh., Šejna, M.: Development and applications of the HYDRUS and STANMOD software packages and related codes. Vadose Zone J. **7**(2), 587–600 (2008)
11. Wu, B., Chen, J.D.: Study on the migration patterns of contaminants in the unsaturated zone based on Hydrus-1D. J. North China Inst. Sci. Technol. **14**(06), 35–39 (2017)
12. Guo, H., Liu, W.X., Wang, C.R., Huang, Z.G.: Comparison of Matlab and neural network methods for fitting the soil water retention curve VG model. Jiangsu Agric. Sci. **47**(03), 232–235 (2019)
13. Zhang, H.K., Jiang, C.L., Li, L., Zhu, L.Q.: Sensitivity analysis of VG model parameters under variable head infiltration conditions simulated by HYDRUS-1D. J. Hohai Univ. (Nat. Sci.) **47**(01), 32–40 (2019)
14. van Genuchten, MTh.: A closed-form equation for predicting the hydraulic conductivity of unsaturated soils. Soil Sci. Soc. Am. J. **44**(5), 892–898 (1980)
15. Zhao, D.: Groundwater pollution prediction in chemical industrial parks based on hydrus and MODFLOW. Groundwater **39**(02), 56–57+124 (2017)

16. Zhang, W.J., Jia, W.Q., Zhang, G.F., Huang, Y.Y.: Study on the convection and diffusion of chloride ions in clay-bentonite barriers. Chin. J. Geotech. Eng. **35**(11), 2076–2081 (2013)
17. Cheng, Z.K., Cai, Q., Li, J.X.: Experimental study on the reinforcement of coastal soft soil layers using the GS new-type stabilizer. Eng. Qual. **38**(02), 47–51 (2020)
18. Ren, J., Zhang, W.P., Wang, G.F., Huang, M.J., Qiao, L.: Soil water retention curve under fixed bulk density obtained using centrifuge method. J. Irrig. Drainage **39**(01), 84–90 (2020)
19. Shang, M.T., Feng, J., Liu, P.G., Shi, Y.: Discussion on suction calculation formulas and optimal centrifuge time for SWCC determination. J. Hohai Univ. (Nat. Sci.) **37**(01), 12–15 (2009)
20. Pan, D.L., Ni, W.K., Yuan, K.Z., Zhang, Z.F., Wang, X.J.: Determination of basic parameters for soil-water characteristic curve based on the VG Model. J. Eng. Geol. **28**(01), 69–76 (2020)
21. Qian, X.D., Zhu, W., Shu, S., Xu, H.Q.: Design and Construction of Pollution Barriers for Landfills and Contaminated Sites, vol. 1, pp. 99–100, 138–144. Science Press, Beijing (2017)
22. Feng, Y.S.: Migration Patterns of Landfill Leachate and Its Impact on Surrounding Rock Stability Under Multi-Field Coupling Effects. Dissertation, Chongqing University (2019)
23. Wei, W.W.: Experimental Study on Heavy Metal Adsorption Properties of Gravel-Bentonite Pollution Barriers. Dissertation, Zhejiang University of Technology (2020)
24. Xu, Y.Y., Ai, Y.B., Fan, Z.H., Sha, X.J.: Study on the adsorption characteristics of Cu^{2+} during transport in Kaolinite columns. Henan Sci. **37**(11), 1790–1795 (2019)
25. Zhu, W., Shu, S., Wang, S.W., Xu, H.Q., Fan, X.H.: Study on indicator pollutants of leachate permeation failure in landfill barrier systems. Chin. J. Geotech. Eng. **38**(04), 619–626 (2016)

Methodological Analysis of Studying the Impact of Urban Geometry on Outdoor Thermal Comfort

Reem Abd Elraouf, Ashraf Elmokadem, Sara Eltarabily, Osama Abo Eleinen, and Naglaa Megahed

Abstract Urban design and planning studies face an increasing challenge in improving outdoor thermal comfort (OTC) and microclimate conditions of urban environments. Analyzing OTC is more challenging outdoors since there is a wider range of interrelated parameters. Additionally, each of these problems is covered by the Sustainable Development Goals (SDGs). Therefore, this paper aims to clarify parameters addressing the impact of urban geometry on OTC at pedestrian level as urban geometry is thought to be the most important and fundamental aspect of urban design that affects outdoor microclimate. The literature on numerous urban design studies in urban environment is first introduced and an overall analyzation of case studies was presented. Then, the paper identified a series of important urban geometries parameters, methods, and techniques which impact thermal comfort outdoor spaces. The study's findings boost the decision-making process and enable architects and urban planners to more effectively contribute to OTC and climate-sensitive projects.

Keywords Outdoor thermal comfort (OTC) · Urban geometry · Microclimate parameters · Methods and techniques · Urban design

1 Introduction

Urban environments have grown rapidly during the last few decades. The health, well-being, and thermal comfort of urban inhabitants in outdoor settings are at risk due to the negative effects of urbanization—such as rising temperatures, pollution, and climate change—which are felt more strongly in urban areas [1, 2].

Thermal comfort is a major factor in attracting people to use outdoor spaces in numerous regions [2]. Better outdoor living conditions, improved city perception, increased usage of outdoor areas for social events, leisure, and recreation, and lower

R. Abd Elraouf · A. Elmokadem · S. Eltarabily · O. Abo Eleinen · N. Megahed (✉)
Architecture and Urban Planning Department, Faculty of Engineering, Port-Said University, Port Said, Egypt
e-mail: naglaaali257@eng.psu.edu.eg

Y. Zhu (ed.), *Proceedings of the 4th International Symposium in Environmental Science and Industrial Ecology*, https://doi.org/10.1007/978-981-96-1578-0_2

energy use related to interior space use are only a few benefits of thermal comfort in outdoor spaces. Outdoor thermal comfort is a complex function of atmospheric conditions and physical, physiological, psychological, and behavioral elements [3]. The OTC is challenging to assess than studies comparable to those carried out for indoor conditions because of the lack of thermal barriers, strong direct radiation, and significant wind speed variations. The vast variety of outdoor activities and the physiological and psychological characteristics of each individual further contribute to the OTC evaluation's complexity. Several indicators have been created to quantify OTC level, including the Predicted Mean Vote [PMV], Physiological Equivalent Temperature [PET], Standard Effective Temperature [SET], Universal Thermal Climate Index [UTCI], and Thermal Sensation Vote [TSV]. These indexes all seek to describe the OTC, but they do so in different ways and with different levels of detail [4].

It is possible to improve OTC level and minimize negative microclimate effects on urban environment considering the urban design parameters. Urban planning and design principles can lower the energy requirements of nearby buildings and establish a comfortable and healthful microclimate [1]. Thus, prior evaluations have addressed the following mitigating techniques to improve OTC: the influence of urban geometry [2, 5], evaporative paving as a means of mitigation [6, 7], the thermal comfort impacts of green areas [8, 9], and the control of the built environment's components [10]. The OTC concept has been examined in other research under various geographical and meteorological conditions. Johansson, for example, examined the equipment and procedures utilized in field research conducted outside and looked at what was needed to standardize the instruments employed for subjective evaluation [11]. Chen et al. [12] examined OTC research about pedestrian experiences and perceptions. Jamei et al. [13] developed mitigating techniques for enhancing OTC at the pedestrian level by analyzing the effects of building geometry and urban greening. Lai et al. [14] looked at four mitigation measures and their cooling effects: altering built-up areas, green spaces, water bodies, cool surfaces, and roofing. Taleghani reviewed how human thermal comfort in urban areas is affected mitigating urban heat islands [15]. Aghamolaei et al. [16] examined how senior persons and other age groups differed in their OTC preferences and experiences. Azevedo assessed the comfort and climate policies in place and noted the challenges that arose while evaluating regional initiatives and guidelines [17]. Zhao et al. [18] reviewed data-driven methods for both indoor and outdoor comfort modelling while reviewing the adaptive thermal comfort models. In order to create outdoor spaces that are comfortable in regarding temperature, Lin et al. [19] examined techniques and recommendations in the field of landscape architecture. The evaluation of the previously described OTC review studies indicates that the majority of their attention was directed towards the study of OTC indices and metrics [12, 18], or mitigation techniques such material qualities and vegetation types [13, 15, 19]. Taking into account the vast range of intricate and confusing factors influencing thermal comfort in outdoor environments is a challenge in OTC studies.

When calculating outdoor comfort, it is crucial to take into account the impact of urban geometry on the microclimate and human comfort in urban areas. The processes that give rise to these climates are complicated. A few studies have examined and determined these factors in order to create a comprehensive OTC assessment framework that is specific to outdoor environments. In this regard, this paper addresses this gap [1] by classifying urban planning and design studies that investigated OTC to identify the most effective urban geometry parameters on OTC and environmental parameters, [2] by revealing suitable methodologies, integrated tools, and techniques. This study offers a comprehensive perspective on several aspects of OTC and related criteria, which can assist architects, urban designers, and urban planners in improving their climate-sensitive designs and strategies.

2 Methodology

The research methodology is divided into five components as seen in Fig. 1.

In order to locate, examine, and evaluate the relevant literature on the effects of urban geometry on OTC and microclimatic parameters at the pedestrian level, this paper employed an organized search technique. PRISMA technique was employed in the study to investigate the relationships between various dimensions and approaches and how these will contribute to the formation of the research framework. Four processes make up the suggested search approach based on the PRISMA method: identification, screening, eligibility, and ultimately included studies. These steps are addressed in more detail in Fig. 2.

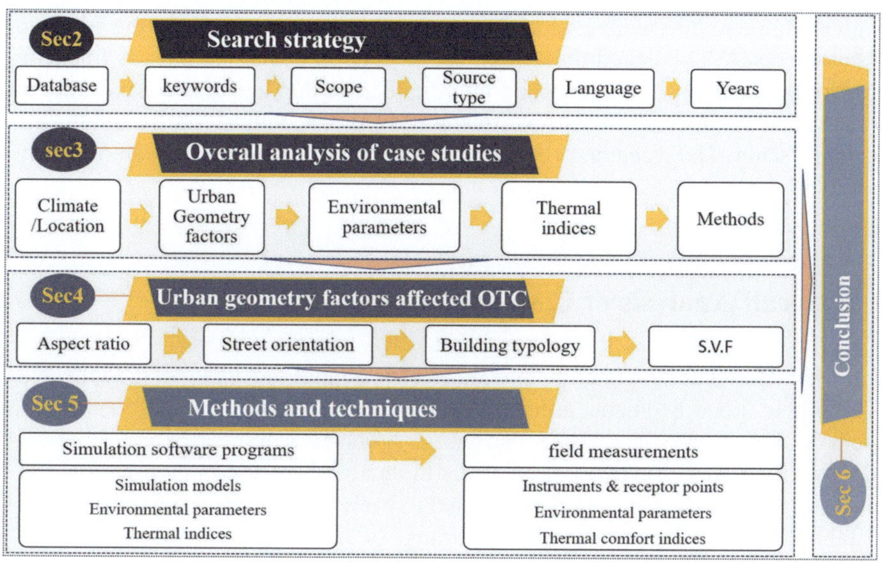

Fig. 1 Methodology flowchart (authors)

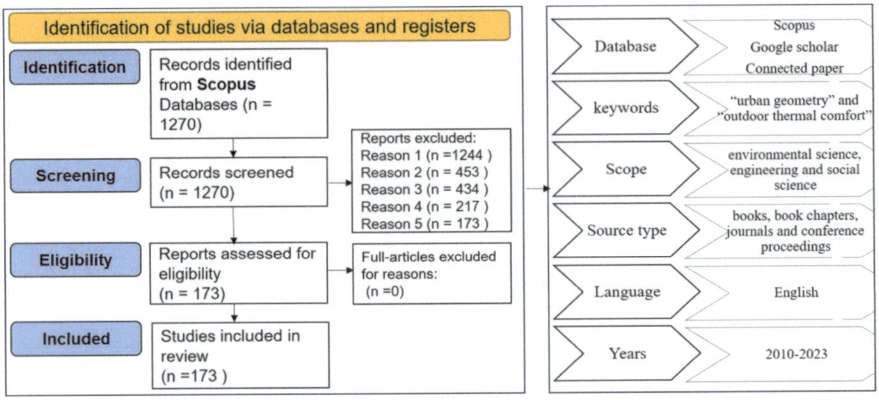

Fig. 2 Workflow of selection procedure based on the Prisma method (authors)

In the identification step, a keyword based on "urban geometry" and "outdoor thermal comfort" was first conducted in Scopus to obtain document relevant for this study. Database fast filters were used in the screening and eligibility stages to group the findings into general categories as seen in Fig. 2. In the first step, the initial results were narrowed down to 1270 publications, and duplicate results were discarded from additional analysis. The publications' titles, abstracts, and moreover the introduction of each paper were then carefully reviewed and evaluated for eligibility to preserve papers pertinent to the research topic. After elimination, the final dataset for this investigation contained 173 publications in total, as shown in Fig. 2. After a thorough reading, studies were evaluated to determine whether they met the following requirements: (1) evaluated the overall set of environmental parameters impacting outdoor thermal comfort; (2) looked into the relationship between outdoor thermal comfort and the geometry of urban streets; and (3) concentrated on the urban street canyon's scale. The research produced 29 document findings, which are displayed in Table 1.

3 Overall Analysis of Case Studies

The selected studies have been deeply reviewed to analyze the methodology for determining how urban geometry influences environmental factors and outdoor thermal comfort. Table 1 investigates the urban geometry and environmental parmeters which affected OTC. Also, the table summarizes thermal indices used in measuring OTC. Methods and instruments which were used to evaluate OTC were reviewed as seen in Table 1. These studies covered a wide range of climate zones as shown in Fig. 3.

Table 1 A survey of the literature on how urban geometry affects environmental factors and outdoor thermal comfort in the urban street canyon (authors)

References	Climate	Location	Urban Geometry factors	Environmental parameters	Thermal indices	Methods	
						Simulation	Field measurement
[20]	Hot-humid tropical	Chennai, India	Aspect Ratio—percentage of built up area	Ta–RH	PET	RayMan	HOBO dataloggers [HOBO U10 Temp/RH]
[21]	Temperate oceanic	Curitiba, Brazil	S.V.F	Ta–MRT–Va	PET	ENVI-met	Two HOBO weather stations, equipped with anemometer-humidity probes
[22]	Warm-humid	Dhaka	Aspect ratio- orientation—S.V.F	Ta–MRT–Ts	PET	ENVI-met RayMan	
[23]	Mediterranean subtropical	Gaza Strip	Aspect ratio- orientation		Total loads	ECOTECT and IES	
[24]	Warm-humid	Colombo	H/W-orientation—S.V.F	MRT	PMV	ENVI-met	
[25]	Temperate	Netherlands	Aspect ratio- orientation—typology	Ta–MRT–Va–RH	PET	ENVI-met RayMan	A wireless Vantage Pro2 weather station
[26]	Hot arid	Syria	Aspect ratio- S.V.F—density	Ta–MRT–Va–RH	PET- PMV	ENVI-met	Kestrel 4500 weather meter and 2 data logger TGU 4500
[27]	Hot-humid	Chennai, India	Aspect ratio—S.V.F—density—materials	Ta–MRT	PET	RayMan	HOBO dataloggers [HOBO U20 Temp/RH]—Anemometer
[28]	Humid continental	Massachusetts, USA	Aspect ratio- orientation—S.V.F	MRT	PET	ENVI-met RayMan	A wireless Vantage Pro2 weather station

(continued)

Table 1 (continued)

References	Climate	Location	Urban Geometry factors	Environmental parameters	Thermal indices	Methods	
						Simulation	Field measurement
[29]	Mediterranean subtropical	Tunis, Tunisia	Aspect ratio- orientation—S.V.F	Ta–MRT–Va–RH	UTCI	ENVI-met	
[30]	Warm-humid continental	Montreal, Canada	Aspect ratio- S.V.F—typology	Ta–Va–RH		ENVI-met	
[31]	Temperate oceanic climate	Melbourne, Australia	Aspect ratio- S.V.F—surface materials	Ta–MRT	PET	ENVI-met	A portable weather station (3 m above the ground), four HOBO data loggers (2 m)
[32]	Hot arid	New Cairo, Egypt	H/W-orientation—S.V.F		PMV	ENVI-met	
[33]	Hot and humid	South China	S.V.F—Typology	Ta–MRT	PET	ENVI-met RayMan	A small weather station
[34]	Moderately warm-wet	Pécs, Hungary	Aspect ratio—orientation	Ta–MRT–Va–RH	PMV	ENVI-met	
[35]	Moderate	Baltim, Egypt	Aspect ratio—orientation	Ta–MRT–Va–RH	PMV	ENVI-met	
[36]	Hot arid	New Aswan, Egypt	Orientation—S.V.F	Ta	PET	ENVI-met	Three HOBO data loggers [at approximately 1.5 m height]

(continued)

Table 1 (continued)

References	Climate	Location	Urban Geometry factors	Environmental parameters	Thermal indices	Methods	
						Simulation	Field measurement
[37]	Tropical savanna	Bangkok, Thailand	Aspect ratio—orientation	Ta–MRT–RH	MRT	ENVI-met RayMan	Hot-wire anemometer- Lutron WBGT-2010SD -Black-globe thermometer
[38]	Hot arid	New Aswan, Egypt	Aspect ratio—orientation—S.V.F—typology	Ta–Va–RH	PET	ENVI-met	The sensors were placed in solar radiation shields at 2 m height
[39]	Hot-humid	Jeddah, Saudi Arabia	Orientation	Ta–Va–RH	PMV	ENVI-met	
[40]	Humid subtropical	Sydney	Aspect ratio—orientation—typology	Ta–Va–RH-Tmrt	PET-Solar radiation	Honeybee, Eplus-ENVI	
[41]	Monsoon with a dry summer	Chennai, India	Aspect ratio—S.V.F	Ta–MRT–RH	PET	ENVI-met	HOBO data loggers [HOBO U23 Temp/ RH]
[42]	Mediterranean	Barcelona	Orientation		PET	RayMan	
[43]	Hot-humid	Bandung, Indonesia	Typology—density	Ta–Va–RH	Solar radiation	ENVI-met	
[44]	Temperate climate	London	Surface albedo—surface materials	Ta–MRT–Ts	PET-Solar radiation	ENVI-met—Eplus	

(continued)

Table 1 (continued)

References	Climate	Location	Urban Geometry factors	Environmental parameters	Thermal indices	Methods	
						Simulation	Field measurement
[45]	Hot and dry	Kashan, Iran	Aspect ratio—Surface albedo—surface materials	Ta–MRT–Va–RH	PMV-Solar radiation	ENVI-met	
[46]	Hot arid	Cairo, Egypt	Aspect ratio—orientation—typology—density		UTCI—Energy	Ladybug	
[47]	Hot and humid	Nanjing, China	H/W–typology—density	Ta–MRT–Va–RH	UTCI–UHI	ENVI-met	
[2]	Hot and humid	PortSaid, northeastern Egypt	Aspect ratio- orientation—typology	Ta–MRT–Va–RH	PET	ENVI-met	BAPPU EVO device

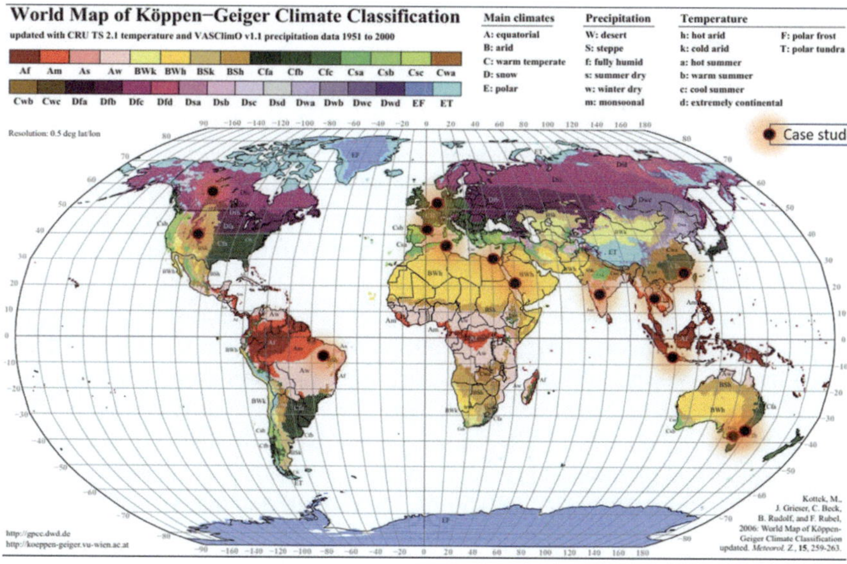

Fig. 3 Different climate zones of the reviewed case studies (authors)

4 Results of the Impact of Urban Geometry Factors on Outdoor Thermal Comfort

The microclimate and thermal comfort of urban areas are influenced by a number of urban design elements, including form and geometry, terrain, vegetation, the availability of water, and surface characteristics [48–51]. The majority of related studies conclude that the two most significant urban factors affecting microclimate at the scale of urban canyon are aspect ratio and street orientation. They directly affect how solar radiation is absorbed and emitted, how well metropolitan areas are ventilated, and ultimately, how comfortable it is for pedestrians to walk in the different thermal conditions. [5, 40, 52, 53]. Furthermore, the detrimental effects of the urban climate can be lessened with an appropriate arrangement of urban buildings [38, 43, 46].

Researchers have long agreed that building height [H], street width [W], and aspect ratio [H/W], which is the ratio of a building's height to the width of the street, indicate urban geometry on the urban canyon scale [20, 23, 24, 28, 41, 54]. Street orientation is a significant aspect that influences canyon performance and comfort levels [35, 36, 42, 46]. Street orientation has been discovered to have an effect on the microclimate conditions due to solar radiation, passive solar heat, absorption, reflection against wind movement, and potential urban cooling [39]. Moreover, altering the street layout of a building typology—a group of buildings that have a common form—has a complex impact on thermal comfort and environmental factors [25, 30, 33, 43, 46].

Numerous literature studies have discovered that **aspect ratio and street orientation** are crucial for establishing a comfortable microclimate and thermal comfort in urban canyons. According to Rose et al. (2011), Zakhour (2015), and Jamei and Rajagopalan (2017), increasing the H/W ratio decreased the PET values, which raised the comfort levels during the day [26, 31, 55]. According to Sharmin and Steemers [22] PET value is mostly influenced by the presence of shade, with N-S streets having lower Tmrt than E-W streets. According to Muhaisen and Abed [23] a H/W of 5 is the ideal choice for E-W orientation, however a H/W of 4 is preferable for N-S streets. Furthermore, Perera [24] discovered that heat stress is decreased by more shade, which is produced by N-S oriented street' greater H/W ratio potential compared to E-W streets. Additionally, Achour-Younsi and Kharrat [29] discovered that the most comfortable streets are those with the largest H/W ratio, whereas the least pleasant streets are those that are oriented E–W. A N-S oriented street has the smallest Tmrt, followed by NE-SW and NW-SE streets, according to Srivanit and Jareemit [37]. The E–W canyon has the worst condition. With the exception of E-W, raising H/W also significantly lowers Tmrt [37]. Additionally, Abdollahzadeh and Biloria [40] discovered that street orientation [46.42%] has the greatest influence among the four design elements, followed by H/W [30.59%]. They also found that Va had the largest influence on OTC [40]. On the other hand, while street canyon shape primarily influences airflow and speed in fairly warm, rainy regions, Albdour and Baranyai [34] found that it has little effect on PMV, Tmrt, and CO_2. The best results were found in street canyons oriented parallel to the direction of the prevailing wind [34].

Few research has looked at the relationship between **building typology** and thermal comfort. It was shown by Wang et al. [30] that courtyard models are more affected by the H/W ratio than tower and flat forms. This also relates to Taleghani et al. [25] that the best microclimate is achieved in courtyards. In addition, arcades serve as efficient buffer spaces because of the street profile's shading strategy design [33], and different urban geometry can enhance the thermal environment more than a homogenous street canyon, according to Yin and Xiao [33]. Furthermore, the thermal performance of a hybrid block combined with a closed courtyard was consistently superior to alternative configurations, according to Dwiputra et al. [43]. Compact high-density distributed forms are preferred when taking thermal comfort into account, according to Ibrahim et al. [46], while courtyards perform better overall and in terms of energy efficiency than other typologies. Subsequent research by Zhang et al. [47] revealed that differences in building typology, building arrangement, and enclosure clearly affected the microclimate.

5 Methods and Techniques

5.1 Simulation Software Programs

Simulation Models. Numerical simulation models are now used to measure and forecast environmental factors, climate change, and thermal comfort rather than just to compute thermal comfort indices. These models give accurate results and are easy

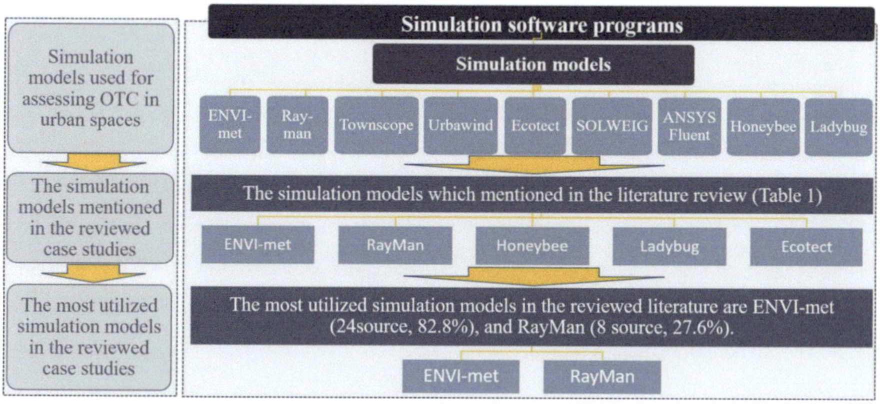

Fig. 4 Simulation software programs (authors)

to use. They are gaining popularity because to their adaptability in handling various assets and weather situations [40, 56, 57]. However, very few simulation models take into account both microclimate changes and the ensuing thermal comfort. This is largely because it is challenging to measure radiation fluxes from the environment around a human body in complex metropolitan settings [58]. Each of the simulation models shown in Fig. 4 has been developed, enhanced, and is currently in use [35, 59].

The simulation models ENVI-met, RayMan, Honeybee, Ladybug, and E-Plus were listed in the literature review (Table 1). In the studied literature, ENVI-met [24 source, 82.8%] and RayMan [8 source, 27.6%] are the most often used simulation models.

Environmental Parameters. Results from previous studies showed that simulation modelling can predict environmental parameters such as air temperature, wind velocity, relative humidity and mean radiant temperature. ENVI-Met among several microclimate simulation models is capable of predicting previous environmental parameters [39, 40, 43].

In reference to the connection between environmental parameters and thermal comfort, [62] talked about how different factors interact to determine thermal comfort outdoors. Ta was found to have the most effect on outdoor thermal comfort in several investigations [60, 61]. It was found in earlier research [62, 63] that radiation affected thermal comfort more than wind. According to contract, [21] discovered that wind had a bigger impact than radiation. Previously, the least significant environmental factor affecting outdoor thermal comfort has been humidity [64].

Thermal Comfort Indices. Several thermal comfort indices can be used by simulation models to measure outdoor thermal comfort. The indices which used in the literature studies reviewed are PET, UTCI, PMV/PPD and MRT as seen in Fig. 5 [22, 26, 29, 42, 46, 47]. PET is the most often used thermal comfort index, with 15 sources and 51.7%.

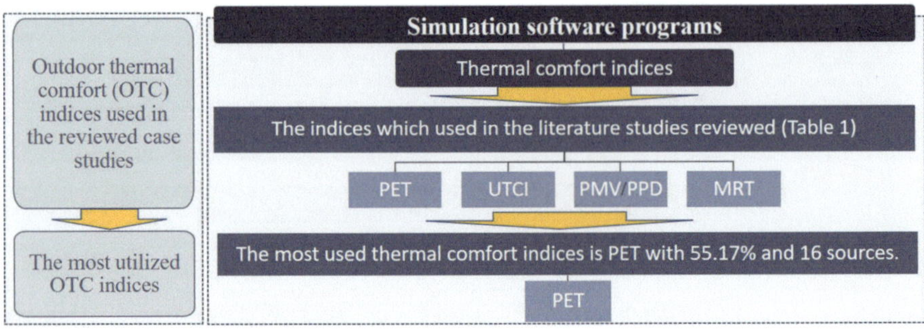

Fig. 5 Thermal comfort indices (authors)

5.2 *Field Measurements*

The benefit of field measurements is that they enable the investigation of real-world scenarios, guaranteeing that the complexity of a problem is taken into account from every perspective. Although the procedures can be expensive and time-consuming, this strategy is rather easy to apply [12–15].

A number of standards and guidelines address measures and meteorological variables related to human biometeorological research in general, but there aren't many that address field surveys of outdoor thermal comfort [65]. These guidelines addressed a wide range of subjects, including kinds, equipment accuracy and measuring range, MRT calculation, and thermal comfort evaluation [66]. However, there is no guidance available on how to plan the field survey in terms of choosing the proper sites, having the right number of sites, needing a proper number of subjects, describing and classifying the site characteristics, choosing the right times of day, ensuring that the survey lasts the minimum amount of time, and taking seasonal climate variations into account [66].

Instruments Measuring Range and Accuracy. As shown in Table 1, a wide range of instruments were employed to measure the environmental parameters [Ta, Va, RH and global temperature] such as HOBO weather stations, wireless Vantage Pro2 weather station, Kestrel 4500 weather meter, and Bappu-Evo [2, 21, 25–28, 31, 55]. High-quality and high-resolution instruments are required for obtaining accurate measurements. For both moderate and thermally difficult conditions, ISO 7726 [67] specifies standards for measuring range and accuracy of instruments; see Table 2.

Receptor Points. Sensors typically measure heights between 1.1 and 1.5 m, which is the same as the human body's center of gravity [33, 36].

Environmental Parameters. Many studies investigated the fundamentals of thermal comfort, such as air temperature, relative humidity, wind speed, and mean radiant temperature or solar radiation, even if they employed different models of instruments and meteorological stations [25, 26, 28, 31, 37]. A set of recommendations for measuring environmental parameters was created by Johansson et al. [65].

Table 2 Specifications for measuring devices' accuracy and range in both comfortable and challenging environments according to ISO 7726 [65, 67]

Environmental parameters	Thermal conditions	Accuracy		
		Measuring range	required	Desirable
Air temperature (Ta)	Comfort	+ 10 to + 40 °C	± 0.5 °C −40 °C: ± 0.9 °C	± 0.2 °C ± 0.45 °C
	Stress	−40 to + 120 °C	0 to + 50 °C: ± 0.5 °C + 120 °C: ± 3.3 °C	± 0.25 °C ± 1.65 °C
Mean radiant temperature (Tmrt)	Comfort	+ 10 to + 40 °C	± 2 °C −40 °C: ± 5.8 °C	± 0.2 °C ± 0.9 °C
	Stress	−40 to + 150 °C	0 to + 50 °C: ± 5 °C + 150 °C: ± 13 °C	± 0.5 °C ± 4.5 °C
Air velocity (Va)	Comfort	0.05 to 1 m/s	± [0.05 + 0.05 Va] m/s	± [0.02 + 0.07 Va] m/s
	Stress	0.2 to 20 m/s	± [0.1 + 0.05 Va] m/s	± [0.05 + 0.05 Va] m/s
Absolute humidity	Comfort	0.5 to 3.0 kPa	± 0.15 kPa	
	Stress	0.5 to 6.0 kPa	± 0.15 kPa	

Thermal Comfort Indices. As previously demonstrated, the majority of research employed PMV, UTCI, or PET in their analysis. The thermal comfort indices were typically calculated using RayMan Pro [25, 27, 33, 37].

6 Conclusion

Designing outdoor spaces is a vital issue in creating thermal comfort for urban environmental design. Changing urban geometry factors can improve and achieve thermal comfort. The geometry of the street aspect ratio and street orientation, which directly affect solar radiation absorption and emission, urban ventilation, and thermal comfort, are the primary urban characteristics that influence the microclimate at the street canyon scale. Additionally, a suitable building typology assists in reducing the adverse urban microclimate.

The utilization of simulation programs in urban planning, urban design, and architecture has provided precise results and simplified the management of microclimate and thermal comfort issues. Among several microclimate simulation models, ENVI-met is the most complete model in terms of the calculation of outdoor thermal comfort and the most utilized model. Together with typical thermal comfort indices (PMV and PET), it generates the four primary environmental factors (Ta, Va, Tmrt, and RH). On the other side, the thermal comfort indices were often determined using RayMan Pro. The outdoor thermal comfort was measured using a variety of thermal indices, including PMV, UTCI, MRT and PET. Because the results are still expressed in degrees, the comfort temperature is clearly indicated by the PET and UTCI, which are reliable indices.

The benefit of field measurements is that they enable the investigation of real-world scenarios, guaranteeing that the complexity of a problem is taken into account from every perspective. A variety of filed measurement instruments were used with different accuracies and measuring ranges to measure environmental parameters such as Ta, Va, RH and Tmrt.

The majority of examined research neglected thermal comfort in outdoor environments because the thermal indices and environmental characteristics fell short of the comfort level. Therefore, by considering thermal comfort considerations early in the urban design process, urban planners and designers can help close the gap that exists between thermal comfort, environmental parameters, and urban geometry.

References

1. Aghamolaei, R., Azizi, M.M., Aminzadeh, B., O'Donnell, J.: A comprehensive review of outdoor thermal comfort in urban areas: effective parameters and approaches. Energy Environ. **34**(6), 2204–2227 (2023)
2. Abd Elraouf, R., Elmokadem, A., Megahed, N., Abo Eleinen, O., Eltarabily, S.: The impact of urban geometry on outdoor thermal comfort in a hot-humid climate. Build. Environ. **225**, 109632 (2022). https://doi.org/10.1016/j.buildenv.2022.109632
3. Zhu S, Chen M, Lu S, Mai X (2022) Influence of urban geometry on thermal environment of urban street Canyons in Hong Kong. Buildings **12**(11) (2022)
4. Ruefenacht, L.A., Adelia, A.S., Acero, J.A., Nevat, I.: Urban design guidelines for improvement of outdoor thermal comfort in tropical cities. CyberGeo **2022** (2022)
5. Ali-Toudert, F., Mayer, H.: Numerical study on the effects of aspect ratio and orientation of an urban street canyon on outdoor thermal comfort in hot and dry climate. Build. Environ. **41**(2), 94–108 (2006)
6. Angeliki Chatzidimitriou, S.Y.: Microclimate design for open spaces: ranking urban design effects on pedestrian thermal comfort in summer. Sustain. Cities Soc. **26**, 2016
7. Donny Koerniawan, M., Gao, W.: Thermal comfort investigation in three hot-humid climate theme parks in Jakarta. Am. J. Environ. Sci. **11**(3), 133–144 (2015)
8. Aljawabra, F., Nikolopoulou, M.: Influence of hot arid climate on the use of outdoor urban spaces and thermal comfort do cultural and social backgrounds matter? Intell. Build. Int. **2** (2010)
9. Potchter, O., Cohen, P., Lin, T.P., Matzarakis, A.: Outdoor human thermal perception in various climates: a comprehensive review of approaches, methods and quantification. Sci. Total Environ. **631–632**, 390–406 (2018). https://doi.org/10.1016/j.scitotenv.2018.02.276
10. Aghamolaei, R., Azizi, M.M., Aminzadeh, B., Mirzaei, P.A.: A tempo-spatial modelling framework to assess outdoor thermal comfort of complex urban neighbourhoods. Urban Clim. **33** (2020)
11. Johansson, E., Emmanuel, R.: The influence of urban design on outdoor thermal comfort in the hot, humid city of Colombo, Sri Lanka. Int. J. Biometeorol. **51**(2), 119–133 (2006)
12. Nouri, A.S., Matzarakis, A., Wetterdienst, D.: Approaches to Outdoor Thermal Comfort Thresholds through Public Space Design: A Review (2018)
13. Jamei, E.J.J.R.R.S.S.: Review on the impact of urban geometry and pedestrian level greening on outdoor thermal comfort. Renew Sustain Energy **54**, 2016

14. Lai, D., Liu, W., Gan, T., Liu, K., Chen, Q.: A review of mitigating strategies to improve the thermal environment and thermal comfort in urban outdoor spaces. Sci. Total. Environ. **661**, 337–353 (2019)
15. Taleghani, M.: Outdoor thermal comfort by different heat mitigation strategies—a review. Renew. Sustain. Energy Rev. **81**(June), 2011–2018 (2018)
16. Aghamolaei Azadeh, R., Lak, L.: Outdoor thermal comfort for active ageing in urban open spaces: reviewing the concepts and parameters. Ageing Int. **48** (2022)
17. Azevedo, I., Leal, V.M.S.: Methodologies for the evaluation of local climate change mitigation actions: a review. Renew. Sustain. Energy Rev. **79** (2017)
18. Zhao, Q., Lian, Z., Lai, D.: Thermal comfort models and their developments: a review. Energy Built. Environ. **2**(1), 21–33 (2021). https://doi.org/10.1016/j.enbenv.2020.05.007
19. Lin, J., Brown, R.D.: Integrating microclimate into landscape architecture for outdoor thermal comfort: a systematic review. Land **10**(2), 1–12 (2021)
20. Rose, L., Horrison, E., Jothi, L., Venkatachalam: Influence of built form on the thermal comfort of outdoor urban spaces. In: 5th International Conference on International Forum Urban [Internet] (2011). Available from: http://www.ifou.org/globalvisions2011/Index/Group2/FOUA00137-00244P2.pdf
21. Krüger, E.L., Minella, F.O., Rasia, F.: Impact of urban geometry on outdoor thermal comfort and air quality from field measurements in Curitiba, Brazil. Build. Environ. **46**(3), 621–634 (2011)
22. Sharmin, T., Steemers, K.: Effect of Canyon geometry on outdoor thermal comfort: a case-study of high-density, warm-humid climate. In: PLEA2013—29th International Conference Proceedings: Sustainable Architecture for a Renewable Future, Munich, Germany [Internet]. Available from: https://orca.cardiff.ac.uk/126461/
23. Muhaisen, A.S., Abed, H.M.: Effect of urban geometry and spacing on the thermal performance in the Mediterranean climate of the Gaza strip. J. Archit. Plann. **26**(1), 1–14 (2014). Riyadh [Internet]. Available from: http://hdl.handle.net/20.500.12358/24568
24. Perera, N.: The effect of street canyon geometry on outdoor thermal comfort in Colombo. In: Proceedings of the International Conference on 'Cities, People and Places' (ICCPP-2014), October 31th–November 02nd, 2014, Colombo, Sri Lanka [Internet] (2014). Available from: http://dl.lib.uom.lk/handle/123/11430
25. Taleghani, M., Eerekoper, L., Tenpierik, M., Den, A.: Outdoor thermal comfort within five different urban forms in the Netherlands. Build. Environ. **83** (2015). https://doi.org/10.1016/j.buildenv.2014.03.004
26. Zakhour, S.: The impact of urban geometry on outdoor thermal comfort conditions in hot-arid region. J. Civ. Eng. Archit. Res. **2**, 862–875 (2015). Available from: https://www.researchgate.net/publication/322330055_The_impact_of_urban_geometry_on_outdoor_thermal_comfort_conditions_in_hot-arid_region
27. Amirtham, L.R., Horrison, E., Rajkumar, S.: Impact of urban morphology on microclimatic conditions and outdoor thermal comfort—a study in mixed residential neighbourhood of Chennai, India. ICUC9—9th International Conference on Urban Climate jointly with 12th Symposium on Urban Environment [Internet] (2015). Available from: https://jesaun.journals.ekb.eg/article_115472.html
28. Targhia, Z.M., Van, D.S.: Potential contribution of urban developments to outdoor thermal comfort conditions: the influence of urban geometry and form in Worcester, Massachusetts, USA. Procedia Eng. 118, 1153–1161 (2015). https://doi.org/10.1016/j.proeng.2015.08.457
29. Achour-Younsi, S., Kharrat, F.: Outdoor thermal comfort: impact of the geometry of an urban street Canyon in a Mediterranean subtropical climate—case study Tunis, Tunisia. Procedia Soc. Behav. Sci. **216**, 689–700 (2016)
30. Wang, Y., Akbari, H., Chen, B.: Urban geometry and environmental urban policy development. Procedia Eng. **1**(169), 308–315 (2016)
31. Jamei, E., Rajagopalan, P.: Urban development and pedestrian thermal comfort in Melbourne. Sol. Energy **144**, 681–698 (2017). https://doi.org/10.1016/j.solener.2017.01.023

32. Saad, S.A., Shafey, A.M.: Optimizing the thermal performance of street canyons in New Cairo, Egypt. Int. J. Adv. Res. Sci. Eng. Technol. **5** (2018) [Internet]. Available from: https://www.researchgate.net/publication/326973081_Optimizing_the_Thermal_Performance_of_Street_Canyons_in_New_Cairo_Egypt
33. Yin, S., Xiao, Y.: Research on the impact of traditional urban geometry on outdoor thermal environment, case study of neighbourhoods with arcade street in South China. In: Proceedings of the 23rd International Conference of the Association for Computer-Aided Architectural Design Research in Asia [CAADRIA] 2018, vol. 2, pp. 503–512 (2018) [Internet]. Available from: https://doi.org/10.52842/conf.caadria.2018.2.503
34. Albdour, M.S., Baranyai, B.: Impact of street canyon geometry on outdoor thermal comfort and weather parameters in Pécs. Pollack Period 14(3), 177–187 (2019) [Internet]. Cited 28 Feb 2022. Available from: https://akjournals.com/doi/10.1556/606.2019.14.3.17
35. Limona, S.S., Al-Hagla, K.S., El-Sayad, Z.T.: Using simulation methods to investigate the impact of urban form on human comfort. Case study: Coast of Baltim, North Coast, Egypt. Alexandria Eng. J. 58(1), 273–282 (2019)
36. Mahmoud, H.: Effect of urban form on outdoor thermal comfort of governmental residential buildings: new Aswan as a case study, Egypt. J. Eng. Sci. Assist. Univ. Fac. Eng. **47** (2019) [Internet]. Available from: https://doi.org/10.21608/JESAUN.2019.115472
37. Srivanit, M., Jareemit, D.: Modelling the urban microclimate effects of street configurations on thermal environment in the residential townhouse of Bangkok. In: Conference: International Congress on Recent Advances in Sciences and Technology at Kuala Lumpur, Malaysia (2019) [Internet]. https://doi.org/10.13140/RG.2.2.17499.21283
38. Galal, O.M., Sailor, D.J., Mahmoud, H.: The impact of urban form on outdoor thermal comfort in hot arid environments during daylight hours, case study: new Aswan. Build. Environ. 184, 107222 (2020). https://doi.org/10.1016/j.buildenv.2020.107222
39. Hegazy, I.R., Qurnfulah, E.M.: Thermal comfort of urban spaces using simulation tools exploring street orientation influence of on the outdoor thermal comfort: a case study of Jeddah, Saudi Arabia. Int. J. Low-Carbon Technol. **15**(4), 594–606 (2020)
40. Abdollahzadeh, N., Biloria, N.: Outdoor thermal comfort: analyzing the impact of urban configurations on the thermal performance of street canyons in the humid subtropical climate of Sydney. Front. Archit. Res. **10**(2), 394–409 (2021). https://doi.org/10.1016/j.foar.2020.11.006
41. Salal Rajan, E.H., Amirtham, L.R.: Impact of building regulations on the perceived outdoor thermal comfort in the mixed-use neighbourhood of Chennai. Front. Archit. Res. **10**(1), 148–163 (2021)
42. Rodríguez-Algeciras, J., Tablada, A., Nouri, A.S., Matzarakis, A.: Assessing the influence of street configurations on human thermal conditions in open balconies in the Mediterranean climate. Urban Clim. **1**(40), 100975 (2021)
43. Dwiputra, I.D., Winarso, H.: Impact of urban block typology on microclimate performance in a hot-humid high-density city. IOP Conf. Ser. Earth Environ. Sci. **738**(1), 012067 (2021). https://doi.org/10.1088/1755-1315/738/1/012067
44. Salvati, A., Kolokotroni, M., Kotopouleas, A., Watkins, R., Giridharan, R., Nikolopoulou, M.: Impact of reflective materials on urban canyon albedo, outdoor and indoor microclimates. Build. Environ. **1**(207), 108459 (2022)
45. Ahmadpour, N., Yousefian, S., Moshfeghi, V., Alizadeh, H.: Analysing the impact of urban morphology on the thermal comfort of the street canyon in a hot and dry climate [Case Study of Kashan, Iran]. Ital. J. Plan. Pract. **12** (2022) [Internet]. Available from: http://www.ijpp.it/index.php/it/article/view/99
46. Ibrahim, Y., Kershaw, T., Shepherd, P., Coley, D.: On the optimisation of urban form design, energy consumption and outdoor thermal comfort using a parametric workflow in a hot arid zone. Energies **14**(13), 40 (2021). https://doi.org/10.3390/en14134026
47. Zhang, M., You, W., Qin, Q., Peng, D., Hu, Y., Gao, Z., et al.: Investigation of typical residential block typologies and their impact on pedestrian-level microclimate in summers in Nanjing, China. Front. Archit. Res. **11**(2), 278–296 (2022)

48. Yahia, M.W., Johansson, E., Thorsson, S., Lindberg, F., Rasmussen, M.I.: Effect of urban design on microclimate and thermal comfort outdoors in warm-humid Dar es Salaam, Tanzania. Int. J. Biometeorol. **62**(3), 373–385 (2018)
49. Erell, E.: The application of urban climate research in the design of cities. Adv. Build. Energy Res. 2(1), 95–112 (2008) [Internet]. Available from: https://www.researchgate.net/publication/233143951
50. Dissanayake, C., Weerasinghe, U.G.D.: Urban microclimate and outdoor thermal comfort of public spaces in warm-humid cities: a comparative bibliometric mapping of the literature. Am. J. Clim. Chang. **10**(04), 433–466 (2021)
51. Tapias, E., Schmitt, G.: Climate-sensitive urban growth: outdoor thermal comfort as an indicator for the design of urban spaces. WIT Trans. Ecol. Environ. **191**, 623–634 (2014)
52. Shishegar, N.: Street design and urban microclimate: analyzing the effects of street geometry and orientation on airflow and solar access in urban canyons. J. Clean Energy Technol. **1**(1), 52 (2013). https://doi.org/10.7763/JOCET.2013.V1.13
53. Peng, M., Huang, H.: The synergistic effect of urban canyon geometries and greenery on outdoor thermal comfort in humid subtropical climates. Front. Environ. Sci. **10**, 1–22 (2022)
54. Ridha, S.: Effect of aspect ratio and symmetrical distribution on urban design in Baghdad City, and the impact of greenery strategies on improving outdoor thermal comfort. IOP Conf. Ser. Earth Environ. Sci. **151** (2018). https://doi.org/10.1088/1755-1315/151/1/012035
55. Rose, L., Horrison, E., Venkatachalam, L.J.: Influence of built form on the thermal comfort of outdoor urban spaces. In: The 5th International Conference of the International Forum on Urbanism [IFoU] 2011 National University of Singapore, Department of Architecture Global Visions: Risks and Opportunities for the Urban Planet (2011)
56. Setaih, K., Hamza, N., Mohammed, M.A., Townshend, T.G.: CFD Modeling As a Tool for Assessing Outdoor Thermal Comfort Conditions in Urban Settings in Hot Arid Climates (2014)
57. El Bat, A.M., Romani, Z., Bozonnet, E., Draoui, A.: Integration of a practical model to assess the local urban interactions in building energy simulation with a street canyon. J. Build. Perform. Simul. 13(6), 720–739 (2020). https://doi.org/10.1080/19401493.2020.1818829
58. Salata, F., Golasi, I., de Lieto Vollaro, R., de Lieto Vollaro, A.: Urban microclimate and outdoor thermal comfort. A proper procedure to fit ENVI-met simulation outputs to experimental data. Sustain. Cities Soc. **26**, 318–343 (2016). https://doi.org/10.1016/j.scs.2016.07.005
59. Albdour, M., Baranyai, B.: An overview of microclimate tools for predicting the thermal comfort. Meteorolog. Parameters Des. Strateg. Outdoor Spaces **14**(2), 109–118 (2019)
60. Lai, D., Liu, W., Chen, Q., Calculation, T.: A review of mitigating strategies to improve the thermal environment and thermal comfort in urban outdoor spaces. Sci. Total Environ. (2019)
61. Detommaso, M., Costanzo, V., Nocera, F.: Application of weather data morphing for calibration of urban ENVI-met microclimate models. Results and critical issues. Urban Clim. **38**, 100895 (2021). https://doi.org/10.1016/j.uclim.2021.100895
62. Acero, J.A., Herranz-Pascual, K.: A comparison of thermal comfort conditions in four urban spaces by means of measurements and modelling techniques. Build. Environ. **93**(P2), 245–257 (2015). https://doi.org/10.1016/j.buildenv.2015.06.028
63. Yang, X., Zhao, L., Michael Bruse, Meng, Q.: Evaluation of a microclimate model for predicting the thermal behavior of different ground surfaces. Build. Environ. **60** (2013)
64. Ayyad, Y.N., Sharples, S.: Envi-MET validation and sensitivity analysis using field measurements in a hot arid climate. IOP Conf. Ser. Earth Environ. Sci. **329**(1) (2019)
65. Johansson, E., Thorsson, S., Emmanuel, R., Krüger, E.: Instruments and methods in outdoor thermal comfort studies—the need for standardization. Urban Clim. **10**(P2), 346–366 (2014). https://doi.org/10.1016/j.uclim.2013.12.002
66. Turner, S.C., Paliaga, G., Lynch, B.M., Arens, E.A., Aynsley, R.M., Brager, G.S., et al.: American society of heating, refrigerating and air-conditioning engineers. Int. J. Refrig. **2**(1), 56–57 (2010)
67. Standardization EC for. SLOVENSKI STANDARD iTeh STANDARD PREVIEW iTeh STANDARD PREVIEW. 2005;

Study on Efficient Removal Technology of Calcium Carbonate in Oilfield

Chunlong Du, Jianzhong Li, Jing Rui, Xuhui Wang, Zhenzhen Li, and Meiyu Dong

Abstract The deposition of calcium carbonate scale ($CaCO_3$) is a serious problem during oilfield extraction, which not only increases the maintenance cost of the oilfield, but also may lead to the instability of oilfield production or even shutdown. The development of an efficient and environmentally friendly calcium carbonate scale remover is of great significance to ensure the normal operation of oilfields and improve the oil recovery efficiency. In this study, an efficient calcium carbonate scale remover formulation for oilfield was successfully screened by compounding chelating agent and other additives. The formulation of $CaCO_3$ removal descaler was determined: TG, EDTA-2Na: citric acid was 20:7, and the descaling rate reached 98.50% at the dosage of 3%, 40 °C, pH 5, and reaction time of 12 h. This achievement not only effectively solves the problem of removing calcium carbonate scale from oilfields, but also demonstrates great potential in guaranteeing the efficiency and environmental friendliness of oilfield extraction.

Keywords Calcium carbonate · Descalers · Chelating agents

1 Introduction

Scale deposits are a common problem in the development and production of oil fields. These scale deposits not only affect the normal operation of the oilfield and reduce the oil recovery efficiency, but also may cause serious corrosion to the equipment, thus increasing the cost of maintenance and replacement of equipment. Among

C. Du · J. Li · J. Rui · X. Wang
No. 11 Oil Production Plant, Changqing Oilfield Company, Qinyang, Gansu, China

Z. Li
Engineering Research Center of Oil and Gas Field Chemistry, Universities of Shaanxi Provence, Xi'an Shiyou University, Xi'an, China

M. Dong (✉)
Dongying Huatai Chemical Group Co. Ltd., Dongying, China
e-mail: 1443449071@qq.com

Y. Zhu (ed.), *Proceedings of the 4th International Symposium in Environmental Science and Industrial Ecology*, https://doi.org/10.1007/978-981-96-1578-0_3

them, calcium carbonate scale is one of the most common scale deposits in oilfields. Calcium carbonate scale forms on the surfaces of pipes and equipment primarily due to changes in pressure and temperature in the wellbore, or due to obstructions to downhole flow. The deposition of this scale not only reduces the flow capacity of the pipeline, but also may trigger pipeline blockage, which greatly affects the normal production of the oilfield [1, 2]. Therefore, the development of an efficient and environmentally friendly calcium carbonate scale remover is of great significance to ensure the normal operation of oilfields and improve oil recovery efficiency. By removal calcium carbonate scale, it can not only restore the flow capacity of pipelines and reduce the risk of pipeline blockage, but also reduce the corrosion of equipment, prolong the service life of equipment, and reduce the operating costs of oilfields [3].

In the process of removal of calcium carbonate scale in oilfield, the commonly using descaling agents mainly include acid cleaning descaling agent, physical descaling agent and chemical descaling agent. These descaling agents can effectively remove calcium carbonate scale to a certain extent, but each has different limitations. For example, although acid cleaning descaling agent has a good removal effect, but the equipment corrosion is serious, and easy to produce secondary pollution; physical descaling agent although no corrosion of the equipment, but the removal effect is limited; chemical descaling agent needs to be used under specific conditions, and may have a certain impact on the environment. And chelating agent as a new type of descaling agent, has unique advantages [4, 5]. The chelating agent is able to complex with metal ions to form a stable compound, thus effectively removing calcium carbonate scale from the surface of pipes and equipment. In addition, chelating agent also has the advantages of environmental protection, non-toxic, non-corrosive, etc., no damage to the equipment, and do not cause pollution to the environment. Despite the obvious advantages of chelating agents in removing calcium carbonate scale, the single use of chelating agents may not be able to meet the descaling needs in complex oilfield environments [6, 7]. Therefore, by compounding other additives, the descaling effect of descaling agents can be further improved and adapted to the descaling needs in different oilfield environments.

The aim of this paper is to screen out an efficient and environmentally friendly calcium carbonate scale remover through the compounding of chelating agent and other auxiliaries. In the study, we first analyze the formation mechanism and characteristics of calcium carbonate scale in oilfield, and then combine the advantages and disadvantages of the commonly used descaling agents, and select suitable chelating agents and additives for compounding. Through experiments to verify the descaling effect of the compound descaling agent, the environmental protection performance and the impact on the equipment, etc., and finally determine a kind of high-efficiency and environmental protection descaling agent suitable for the removal of calcium carbonate scale in the oilfield.

2 Experimental Section

2.1 *Materials and Instruments*

HPMA, PASN, Sodium Polyphosphate, analytically pure, Tianjin Kemei Chemical Reagent Factory; Citric Acid, Sodium Pyrophosphate, PESA, Ethylene diamine tetra acetic acid (EDTA), disodium ethylene diamine tetra acetic acid (EDTA-2Na), Oxalic Acid, Lactic Acid, Acrylic Acid, analytically pure, Tianjin Damao Chemical Reagent Factory.

2.2 *Scale Dissolution Experiment*

A small amount of the drug was prepared as an aqueous solution and $CaCO_3$ scale was added to observe its scale solubility. The scale dissolution rate is calculated by the following formula:

$$w = 1 - \frac{m_3 - m_2}{m_1} \times 100\% \tag{1}$$

where: w—scale dissolution rate, %;

m_1—total mass of scale sample, g;
m_2—mass of filter paper before filtration, g;
m_3—mass of filter paper after extraction and drying, g.

3 Results and Discussion

3.1 *Influence of Chelating Agent Concentration on Descaling Effect*

EDTA, EDTA-2Na, sodium polyphosphate and citric acid, which are commonly used as descaling agents in oil fields, were used as research objects to investigate the descaling effect of different types of descaling agents on $CaCO_3$ scale. As required, 100 mL of descaler solution with mass fraction of 1%, 2%, 3%, 4%, 5%, 6% were prepared respectively, and NaOH was used to adjust the pH of the solution to about 5. 1 ± 0.01 g of $CaCO_3$ powder was added, and placed in a constant-temperature water bath at 40 °C to stand and react for 12 h, and the descaling rate was calculated. The results are shown in Fig. 1. From the experimental data, it can be seen that under the same experimental conditions, there are differences in the effect of the four descaling agents on the removal of $CaCO_3$, of which EDTA, EDTA-2Na and citric acid show excellent descaling performance on $CaCO_3$, and sodium polyphosphate is not effective in removing $CaCO_3$. When the mass fraction of EDTA was increased from 1 to 3.5%, the descaling rate increased from 32.60 to 99.60%. When the mass fraction was increased from 3.5 to 4%, the descaling rate was 98.80%. The descaling rate of EDTA-2Na increased and then stabilized with the increase of the mass fraction, and the descaling rate of $CaCO_3$ was 43.20% when the mass fraction was 2%. When the mass fraction was 5%, the descaling rate increased greatly to 95.70%, and continued to increase to 6%, the descaling rate changed slightly. When the concentration of the descaler reached a certain value, the descaling effect did not change significantly when the concentration continued to increase. This may be due to the spatial site-blocking effect of excess chelator molecules [8, 9]. The variation of the descaling rate of citric acid was similar to that of the previous two, and the descaling rate increased from 36.50 to 96.20% when the mass fraction was increased from 0.5 to 1.5%. Compared with organic acids such as citric acid, the chelating agents EDTA and EDTA-2Na have better stability, high efficiency, environmental protection and non-corrosiveness, and thus are more widely used in practical applications. EDTA-2Na contains two more sodium ions than EDTA, and the presence of sodium ions increases solubility [10]. This also means that EDTA-2Na is able to react more quickly with metal ions in water in a coordination reaction, which promotes the dissolution and removal of scaling materials and improves the removal effect [11, 12].

To examine the three descaling agents mentioned above in terms of cost-effectiveness, EDTA-2Na > EDTA > citric acid, the unit price of EDTA-2Na is relatively high, but considering its excellent descaling effect, stability, and sustainability, it can reduce the time of oilfield treatment and maintenance cost in long-term use. The efficient removal of $CaCO_3$ was achieved by utilizing the advantages of EDTA-2Na and compounding it with other additives under low concentration conditions to effectively reduce the cost. After comprehensive analysis, the mass fraction of 2% EDTA-2Na was selected as the main agent for $CaCO_3$ removal.

3.2 *Influence of Efficient Descaling Additives on Descaling Effect*

In order to obtain a descaler system with good solubilization ability for $CaCO_3$, the single agent of descaler was considered for compounding experiment. On the one hand, the descaling effect of the main agent in high concentration can be achieved while reducing the dosage of the main agent, thus reducing the cost. On the other hand, different single agents can play a synergistic effect when cooperating with the main agent for descaling, which can further improve the descaling effect. The effects of different concentrations of citric acid, sodium polyphosphate, hydrolyzed poly maleic anhydride (HPMA) and sodium pyrophosphate on the descaling effect of

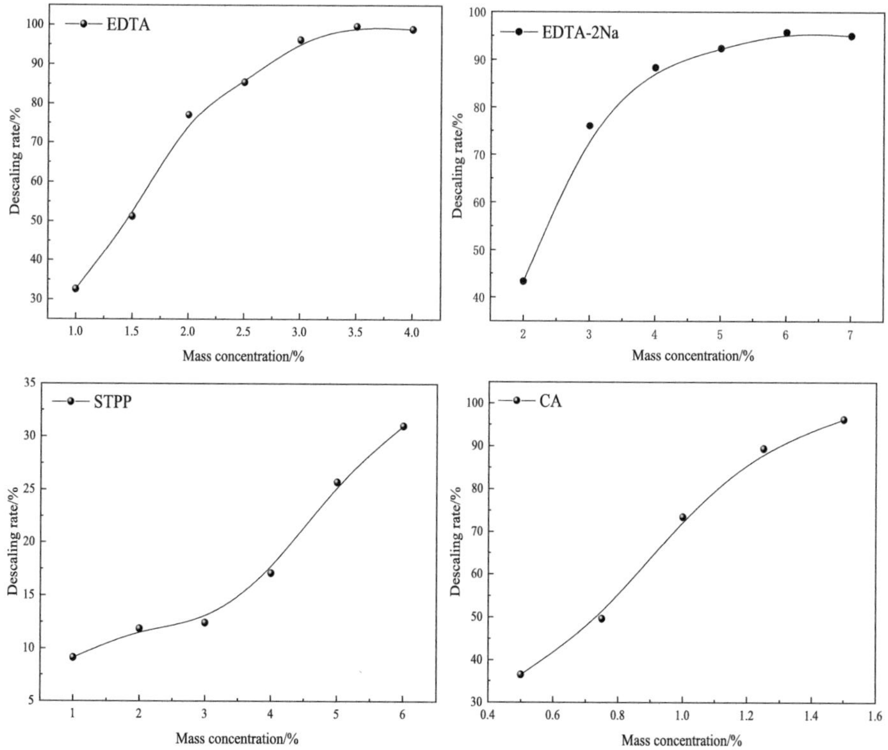

Fig. 1 Evaluation of $CaCO_3$ scale removal performance of different descaling agents

EDTA-2Na were investigated under the condition of 2% EDTA-2Na concentration, and the experimental results are shown in Fig. 2.

According to Fig. 2, it can be seen that all four additives increased the ability of EDTA-2Na to dissolve $CaCO_3$, in which sodium citrate shows a good synergistic effect and sodium pyrophosphate has the worst synergistic effect. When the citric acid concentration was increased from 0.5 to 0.7%, the descaling rate increased from 91.50 to 98.50%. After continuing to increase the concentration, the descaling effect did not change significantly, which may be due to the fact that when the citric acid concentration is too high, the concentration of H^+ in the solution also increases, which can lead to a decrease in the pH of the solution [13, 14]. Under acidic conditions, the carboxyl group in EDTA-2Na tends to lose its negative charge, which reduces the ability to form complexes with metal ions, thus affecting the descaling efficiency. When the concentration of sodium polyphosphate was increased from 0.2 to 1%, the scale removal rate increased from 59.15 to 89.59%, and when the concentration continued to be increased, instead, there was a slightly decreasing trend, which may be due to the fact that sodium polyphosphate is a macromolecular compound with good complexing ability, however, when the concentration of sodium polyphosphate is too high, it can lead to its excessive complexing with EDTA-2Na and the formation

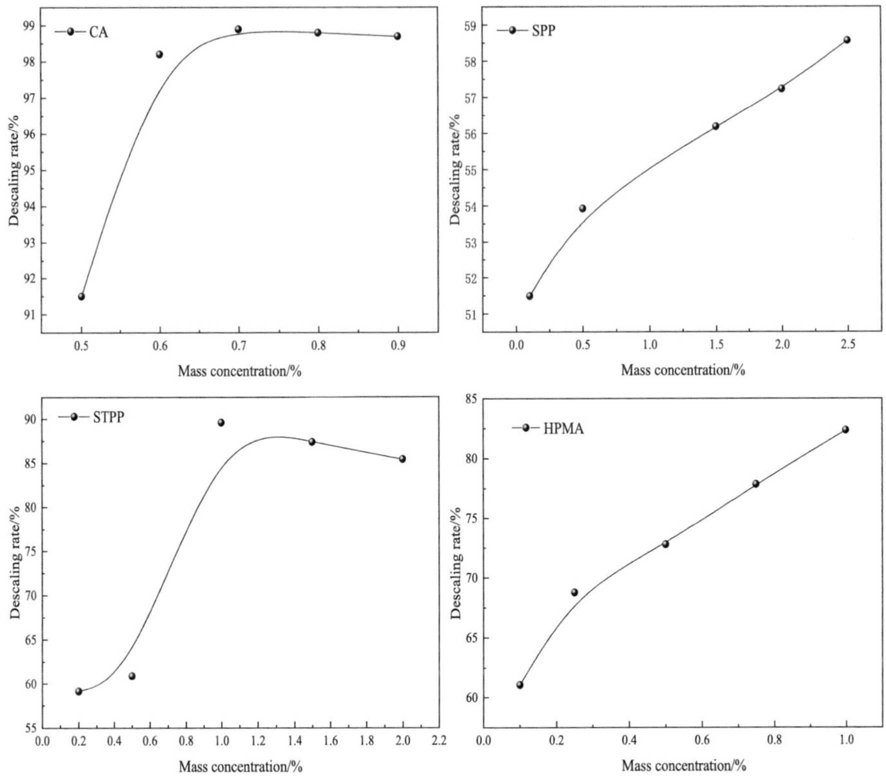

Fig. 2 Effect of different additives on the effect of EDTA-2Na on $CaCO_3$ removal

of a large number of complexes, These complexes may form insoluble precipitates in the solution, thus affecting the effectiveness of scale removal. HPMA is a polymeric dispersant that effectively prevents the reaggregation of $CaCO_3$ particles, increases the contact area between EDTA-2Na and $CaCO_3$ particles, and promotes descaling efficiency.

3.3 Experimental Study of the Effect of Concentration on the Descaling Effect

The concentration of the descaling agent is closely related to the descaling effect. If the concentration of descaling agent is too low, it cannot dissolve the scale effectively. On the contrary, if the concentration of descaling agent is too high, it can cause unnecessary economic losses. In order to consider the influence of descaling agent concentration on the descaling effect, the effect of different concentrations of descaling agents on the descaling effect of $CaCO_3$ was thoroughly investigated under

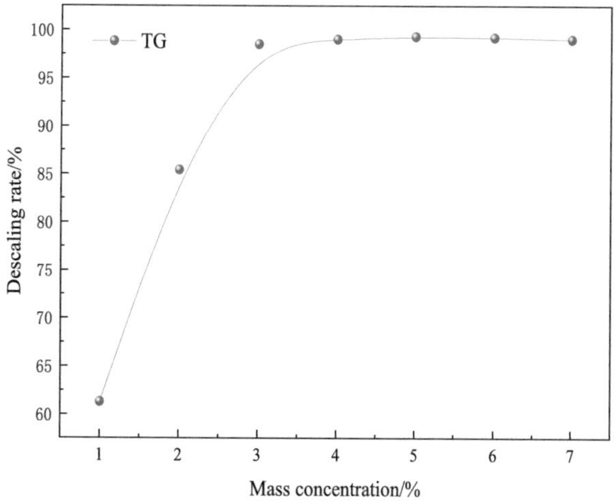

Fig. 3 Effect of different concentrations on descaling effect

the conditions of 50 °C, pH 5, and descaling time of 12 h. The results are shown in Fig. 3.

Figure 3 shows the effect of descaler concentration on the descaling rate of $CaCO_3$ respectively. It can be seen that the descaling rate increases rapidly with large variations when the concentration of descaling agent TG increases from 1 to 3, 4, 5 and 6%. After the descaler concentration exceeds this value, the descaling rate shows a stable trend. It was analyzed that TG, the main descaling mechanism is chelation. As the concentration of the scale dissolver increases, the concentration of ligands that react with the metal ions increases accordingly. This means that there is an increased likelihood of the ligand binding to the metal ions, which enhances the descaling effect. However, when the concentration of descaler reaches a specific value, if its concentration continues to increase, it leads to the formation of a protective film on the surface of the scale, which reduces the effective contact area between the descaler and the scale, which in turn affects the rate of descaler removal. When the concentration of TG was 3%, the descaling rate was 98.50% respectively as the concentration continued to increase, the descaling rate changed slowly.

3.4 Experimental Study on the Effect of Temperature on Descaling Effect

Changes in temperature have a significant impact on the performance and efficiency of descalers, and in descaler systems temperature determines the rate at which chelating agents form complexes with metal ions. The effect of the temperature

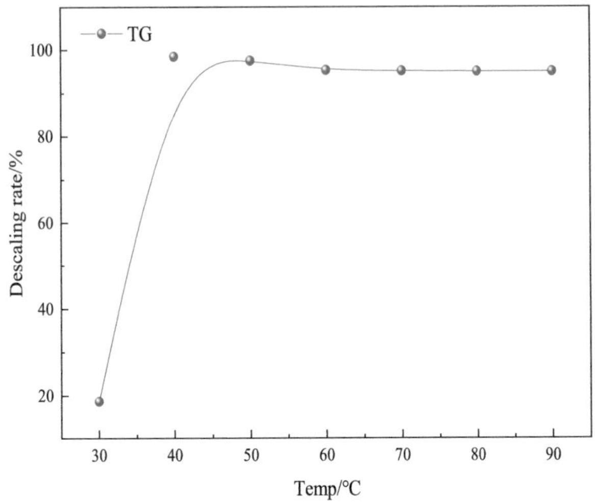

Fig. 4 Effect of different temperatures on descaling effect

at 30, 40, 50, 60, 70, 80 and 90 °C on the descaling effect of $CaCO_3$ was investigated by choosing the LB concentration at 5%, pH 5 and the descaling reaction time at 12 h. The experimental results are shown in Fig. 4.

Figure 4 shows the effect of temperature on the descaling rate, and it can be seen that the temperature has a significant effect on the descaling rate of the reaction system. With the gradual increase in temperature, the descaling rate shows a trend of increasing and then stabilizing. When the temperature is lower than 40 °C, the descaling rate changes more with the increase of temperature. At 40 °C, the descaling rate reached 98.50% and 96.00%, after which the descaling rate stabilized with the increase of temperature.

3.5 Experimental Study on the Effect of pH on Descaling Effect

pH is one of the key factors affecting the descaling effect, in order to investigate the effect of pH on the scale effect, the concentration of the dissolvent TG was selected at 3% and the temperature was 40 °C, and the time of dissolution was 12 h. The effect of pH on the removal effect of $CaCO_3$ scale was investigated when the pH was at 4–9. The experimental results are shown in Fig. 5.

As can be seen from Fig. 5, pH has a significant effect on the descaling rate. In each descaling system, with the increase of pH, the descaling rate shows a trend of first increasing and then decreasing. In the removal of $CaCO_3$ by TG descaler, the peak value of 98.50% was reached at pH 5. The removal of $CaCO_3$ is better under acidic conditions. This is because $CaCO_3$ is more soluble under acidic conditions and the optimal pH range of EDTA-2Na is 8–9, whereas citric acid may decompose,

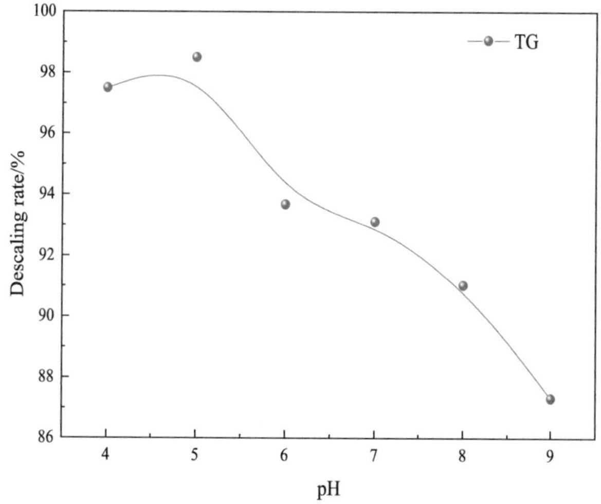

Fig. 5 Effect of different pH on descaling effect

inactivate, or be oxidized when the pH is too high, which reduces the descaling ability of the descaling system. Therefore, controlling the pH around 5 is an optimal reaction condition when removing $CaCO_3$.

3.6 Experimental Study on the Effect of Time on Descaling Effect

Under the optimal reaction conditions examined above, the effect of different reaction times on the descaling effect was examined at 4 h intervals. Since the descaling process usually involves chemical reactions or physical changes, and the rate of these reactions and changes is closely related to time. By determining the optimal reaction time, in order to avoid the waste of resources. The experimental results are shown in Fig. 6. It is obvious from Fig. 6 that the descaling rates of all descaling systems increase with time, but after reaching a certain value, the descaling rate remains basically unchanged. In the reaction time at 8 h descaling rate of 91.30% at 12 h, the descaling effect are improved, but the growth rate gradually slows down, and finally 98.50% thereafter, the descaling rate is almost no longer occurring significant changes, which is mainly due to the relatively high concentration of descaling agent at the beginning of the reaction, the reaction rate is faster, and the effect of descaling is also more pronounced, and thus the 12 h is the optimum reaction time.

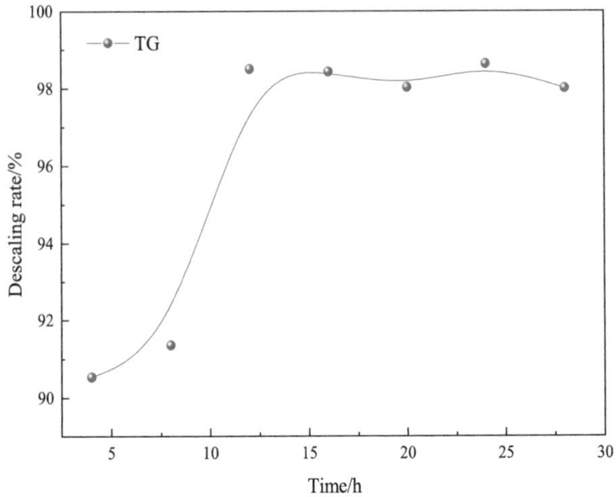

Fig. 6 Effect of different reaction time on descaling effect

4 Conclusions

After a series of experimental studies, we successfully determined the formulation of an efficient descaler for removing $CaCO_3$ scale layer, it is compounded from TG, EDTA-2Na and citric acid in the ratio of 20:7. Under the conditions of 3% dosage, 40 °C, pH 5 and 12 h reaction time, the descaler shows excellent descaling performance with a high descaling rate of 98.50%. This conclusion not only verifies the effectiveness of the formulation, but also provides strong technical support for the removal of calcium carbonate scale in oilfield extraction and other industrial fields. In the future, we will further optimize the formulation and explore its performance in practical applications, with a view to playing a greater role in safeguarding industrial production safety and improving equipment operation efficiency.

Acknowledgements This work was financially supported by grants from Youth Innovation Team of Shaanxi University, the Key Research and Development Program of Shaanxi (No. 2021KW-10) and the Innovation Capability Support Program of Shaanxi (No. 2022PT-08).

References

1. Li, H.: Research on Comprehensive Anti-scaling Technology of Scaling in Gathering and Transportation System of Hua Qing Oilfield. Xi'an Shiyou University (2021)
2. Kumar, S., Naiya, T.K., Kumar, T.: Developments in oilfield scale handling towards green technology-a review. J. Petrol. Sci. Eng. **169**, 428–444 (2018)
3. Xiang, G.H.: Research on Anti-scaling Technology of Oil and Water Wells in Yushulin Oilfield. Northeast Petroleum University (2011)
4. Olajire, A.A.: A review of oilfield scale management technology for oil and gas production. J. Petrol. Sci. Eng. **135**, 723–737 (2015)

5. Zhai, J., Ren, D., Xiao, Y., et al.: Synthesis of phosphorus-free terpolymer and its inhibitory performance on scale formation. J. Polym. Res. **30**(6), 232 (2023)
6. Yan, M., Tan, Q., Liu, Z., et al.: Synthesis and application of a phosphorous-free and non-nitrogen polymer as an environmentally friendly scale inhibition and dispersion agent in simulated cooling water systems. ACS Omega **5**(25), 15487–15494 (2020)
7. Zhou, Z., Yong Gang, Y., Hang, T., et al.: Optimization of corrosion and scale inhibitors for CO_2-driven surface production and transportation system. Nat. Gas Petrol. **39**(3), 88–94 (2021)
8. Wu, N.: Corrosion Law and Corrosion Inhibitor Screening Evaluation of Fractured Pipe Column in Daqing Oilfield. Northeast Petroleum University, Daqing (2017)
9. Alahmad, M.: Factors affecting scale formation in sea water environments–an experimental approach. Chem. Eng. Technol. Ind. Chem.-Plant Equip.-Process Eng.-Biotechnol. **31**(1), 149–156 (2008)
10. Bernstein, M.: Particle adhesion theory and experiment. Adv. Coll. Interface. Sci. **1**, 111–239 (1967)
11. Bohnet, M.: Influence of the transport properties of the crystal/heat transfer surface interfacial on fouling behavior. Chem. Eng. Technol. **26**(10), 1055–1060 (2003)
12. Li, Z., Zhang, D., Qin, G., et al.: Studies on the scaling of high pressure and low permeability oil reservoir water injection well. Adv. Petrol. Explor. Dev. **8**(1), 1–8 (2014)
13. Zhang, N., Wang, X., Zhang, J., et al.: Experimental and statistical study on wellbore scaling mechanisms and characteristics for Huan Jiang Oilfield. Geofluids **2022** (2022)
14. Liu, G., Li, S., Cheng, C.: Study on scaling mechanism of injected water in Yu Jiaping Oilfield. In: Journal of Physics: Conference Series, vol. 2594, no. 1, p. 012056. IOP Publishing (2023)

Research on the Formula for Neutral Removal of Barium Sulfate Scale in Oil Fields

Xiaolong Wan, Xuhui Wang, Yarui Zhao, Peifu Su, Le Qu, Meiyu Dong, and Zhenzhen Li

Abstract In the process of oilfield exploitation, barium sulfate scaling seriously affects production efficiency, increases maintenance costs, and has potential safety hazards. It is of great significance to study effective barium sulfate removal technology to ensure stable production and reduce operating costs in oilfields. The purpose of this paper is to explore the background of barium sulfate removal in oilfields and its practical application value. This study focused on the development of high-efficiency $BaSO_4$ descaler, and successfully developed LB descaler by combining DTPA as the main agent and combining oxalic acid and other synergists, which had a significant descaling effect on $BaSO_4$. After experimental verification, the optimal formulation of LB descaler was determined, in which the ratio of DTPA to oxalic acid was 1:1. The descaling rate was as high as 82.00% after 12 h of reaction under the conditions of 5% addition, 80 °C reaction temperature, and pH 5. This result provides a new and effective method in the field of $BaSO_4$ descaling.

Keywords Barium Sulfate · Descalers · Chelating agents

1 Introduction

Oilfield scaling is one of the most serious problems encountered in oilfield water quality control. After the oilfield enters the water-bearing period of development, scaling occurs in the water-injection stratum, oil casing, downhole, surface equipment, and gathering pipelines, which results in a decrease in the amount of liquid

X. Wan · X. Wang · Y. Zhao · P. Su
No. 11 Oil Production Plant, Changqing Oilfield Company, Qinyang, Gansu, China

L. Qu · M. Dong
Dongying Huatai Chemical Group Co. Ltd., Dongying, China

Z. Li (✉)
Engineering Research Center of Oil and Gas Field Chemistry, Universities of Shaanxi Provence, Xi'an Shiyou University, Xi'an, China
e-mail: 2839410462@qq.com

Y. Zhu (ed.), *Proceedings of the 4th International Symposium in Environmental Science and Industrial Ecology*, https://doi.org/10.1007/978-981-96-1578-0_4

produced from the wells, an increase in the pressure of water injection, and clogging of the pipelines and heating furnaces, and at the same time, scaling also affects the downhole operation of the wells, and in some cases, it causes the shutdown of the wells, which results in a great economic loss. As the offshore oil field environment is completely different from the land. Oil field scaling has seriously affected the normal exploitation of offshore oil wells [1–4]. Chemical descaling is a widely used descaling method in oilfield descaling measures, in which organic complexing agent descaling is more efficient than acid and alkali descaling. The most commonly used organic complexing agents are ammonia-carboxylic complexing agents, a class of carboxylic acid derivatives dominated by the $-N-[(CH_2)nCOOH]_2$ group, which can form stable, water-soluble chelates with a wide range of metal ions [5–7]. The simplest structure of this type of complexing agent is ethylenediamine tetra acetic acid (EDTA). In order to enhance the chelating ability of EDTA, it has also been modified, but the process is more complicated and the cost of the product is higher [8–10]. Diethylenetriaminepentaacetic acid (DTPA) is also a good chelating agent for metal ions, which has gained extensive research in oilfield descaling, especially in barium sulfate scale removal applications due to its high descaling efficiency, low toxicity, and good thermal stability [11]. However, when chelating agents are used alone, the following problems may exist: chelating agents may not be able to cover all metal ions to the extent that some metal ions are missed and complete descaling cannot be achieved; chelating agents are slow in descaling and it takes a longer time to remove the inorganic salt scale completely, which increases the time and cost of cleaning [12, 13]. Therefore, this chapter utilizes the synergistic effect of additives and chelating agents by adding other additives in conjunction with chelating agents, which helps to achieve a more efficient, stable and comprehensive descaling effect while protecting oilfield equipment and pipelines [14, 15].

2 Experimental

2.1 Materials and Instruments

HPMA, PASN, Sodium Polyphosphate, analytically pure, Tianjin Kemei Chemical Reagent Factory; Citric Acid, Sodium Pyrophosphate, PESA, Ethylene Diamine Tetra acetic Acid (EDTA), Disodium Ethylene diamine Tetra acetic Acid (EDTA-2Na), Oxalic Acid, Lactic Acid, Acrylic Acid, analytically pure, Tianjin Damao Chemical Reagent Factory.

2.2 Scale Dissolution Experiment

A small amount of the drug was prepared as an aqueous solution and $BaSO_4$ scale was added to observe its scale solubility. The scale dissolution rate is calculated by the following formula:

$$w = 1 - \frac{m_3 - m_2}{m_1} \times 100\% \quad (1)$$

where: w—scale dissolution rate, %;

m_1—total mass of scale sample, g;
m_2—mass of filter paper before filtration, g;
m_3—mass of filter paper after extraction and drying, g.

3 Results and Discussion

3.1 Effect of Chelating Agent Concentration on Descaling Effect

1 ± 0.01 g of $BaSO_4$ powder was added to 100 mL of solutions prepared with mass fractions of 1%, 2%, 3%, 4% and 5%, respectively, and NaOH was used to adjust the pH of the solutions to about 5. The reaction was placed in a 90 °C thermostatic water bath for 12 h. After the reaction was completed, the reaction was cooled to room temperature, and the descaling rate was calculated according to the method of Eq. (1) after filtration and drying, and the experimental results are shown in Fig. 1.

Figure 1 shows that under the same experimental conditions, the descaling rates all increased slowly with the increase of descaling agent concentration, and the descaling effect of DTPA was significantly better than that of the other three descaling agents, followed by EDTA, EDTA-2Na, and sodium pyrophosphate, respectively [16, 17]. When the concentration of descaling agent was increased from 1 to 5%, the descaling rate increased from 48.60%, 17.00%, 2.40% and 0.90% to 95.20%, 49.00%, 14.00% and 11.00%, respectively. In conclusion, 2% DTPA was selected as the main agent for subsequent experimental studies.

3.2 Influence of Efficient Descaling Additives on Descaling Effect

DTPA is a commonly used chelating agent with excellent scavenging effect, but at the same time its price is relatively high, which limits its wide application in industry to some extent. Therefore, in order to further optimize the scavenging effect and

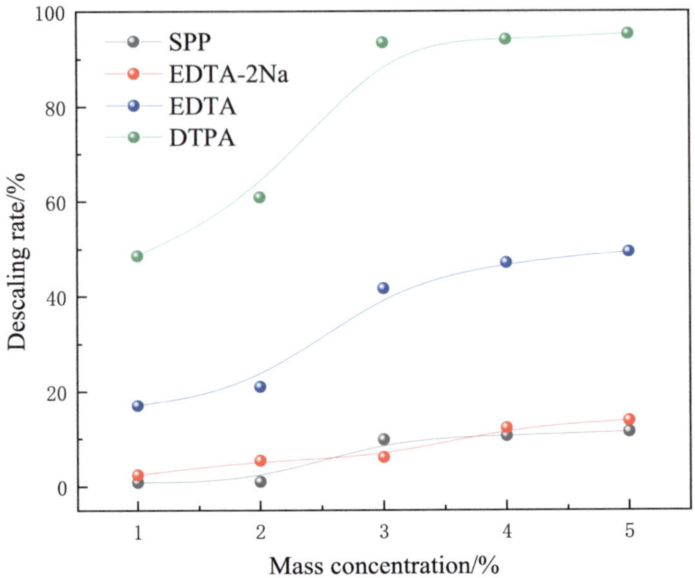

Fig. 1 Evaluation of $BaSO_4$ removal effect by different descaling agents

cost-effectiveness, we considered compounding low concentrations of DTPA with other additives. The effect of oxalic acid, lactic acid, citric acid and acrylic acid on $BaSO_4$ descaling was investigated at 2% DTPA concentration at 0.5%, 1%, 1.5% and 2% mass concentrations, respectively. NaOH was used to adjust the solution pH to about 5. The solution was placed in a 90 °C thermostatic water bath for 12 h. At the end of the reaction, the solution was cooled down to room temperature, and the descaling rate was calculated after filtration and drying, and the experimental results are shown in Fig. 2.

As can be seen from the comparison of the experimental results in Fig. 2, several synergists have obvious synergistic effects on DTPA. The addition of oxalic acid, lactic acid, citric acid and acrylic acid agents significantly improved the descaling effect of DTPA by 20.43%, 19.71%, 20.12% and 14.57%, respectively. Among them, oxalic acid and citric acid had better solubilizing effect than lactic acid and acrylic acid. The relative increase in the descaling ability of DTPA was significant when the concentration of oxalic acid was low, and the best descaling effect of the formulation was achieved when the concentration was increased to 2%, with a descaling rate of 82.00%. The analysis shows that the functional groups (e.g., carboxylic acid group COOH) in organic acids can form complexes with sulphate ions during the $BaSO_4$ removal reaction, which helps to improve the solubility of $BaSO_4$ [18–20]. When the DTPA host agent was introduced into the reaction system, it formed stable complexes with barium ions, which further promoted the dissolution of $BaSO_4$. The $C_2O2\text{-}4$ contained in oxalic acid is able to form a relatively stable complex when reacting with $BaSO_4$, thus showing a better descaling effect. Comprehensively analyzing the

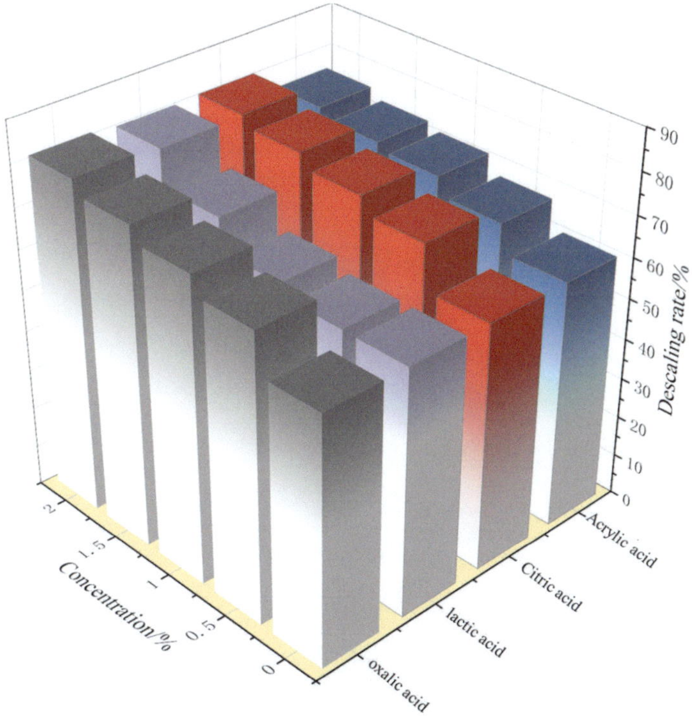

Fig. 2 Effect of different potentiation on the effectiveness of DTPA in scavenging $BaSO_4$

synergistic effect and cost-effectiveness, DTPA: oxalic acid of 1:1 was selected as the formula for removing $BaSO_4$ (named LB) in this thesis.

3.3 *Experimental Study of the Effect of Concentration on the Descaling Effect*

The concentration of the descaling agent is closely related to the descaling effect. If the concentration of descaling agent is too low, it will not be able to dissolve the scale effectively. On the contrary, if the concentration of descaling agent is too high, it will cause unnecessary economic losses. In order to consider the influence of descaling agent concentration on the descaling effect, the influence of different concentrations of descaling agent on the descaling effect of $BaSO_4$ was studied in depth under the conditions of 50 °C, pH 5, and descaling time of 12 h. The results are shown in Fig. 3.

Figure 3 shows the effect of descaler concentration on the $BaSO_4$ descaling rate, and it can be seen that the descaler LB concentration increases from 1 to 3%, 4%, 5%, and 6%, respectively, the descaling rate increases rapidly with a large variation.

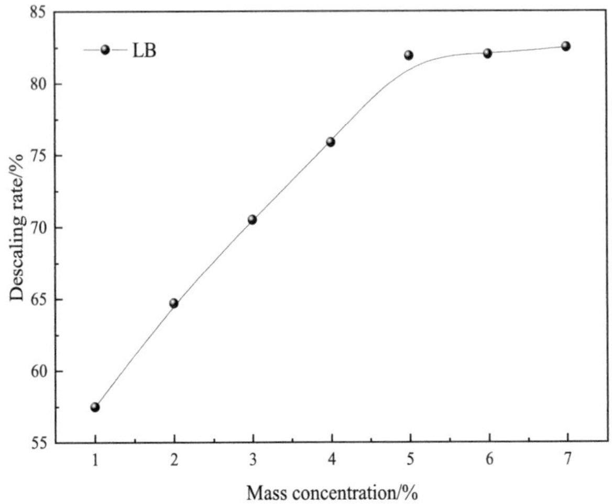

Fig. 3 Effect of different concentrations on descaling effect

The main descaling mechanism was chelation. With the increase of scale dissolver concentration, the concentration of ligands reacting with metal ions increased accordingly. This means that the possibility of ligand binding to metal ions increases, thus enhancing the descaling effect. When the LB concentration was 5%, the descaling rate was 82.00%, and as the concentration continued to increase, the descaling rate changed slowly.

3.4 *Experimental Study on the Effect of Temperature on Descaling Effect*

Changes in temperature have a significant impact on the performance and efficiency of descalers, and in descaler systems temperature determines the rate of complex formation between chelating agents and metal ions. The influence of the temperature on the descaling effect of $BaSO_4$ at 30, 40, 50, 60, 70, 80 and 90 °C was investigated by selecting the LB concentration at 5%, pH 5 and the descaling reaction time at 12 h. The results are shown in Fig. 4. Figure 4 shows the effect of temperature on the descaling rate, and it can be seen that the temperature has a significant effect on the descaling rate of the reaction system. With the gradual increase of temperature, the descaling rate shows a trend of increasing and then stabilizing. At the reaction temperature of 80 °C, the descaling rate was 82%, and the variation of descaling rate became smaller when exceeding 80 °C. The increase in temperature helps to increase the molecular activity and subsequently accelerate the chelation [13].

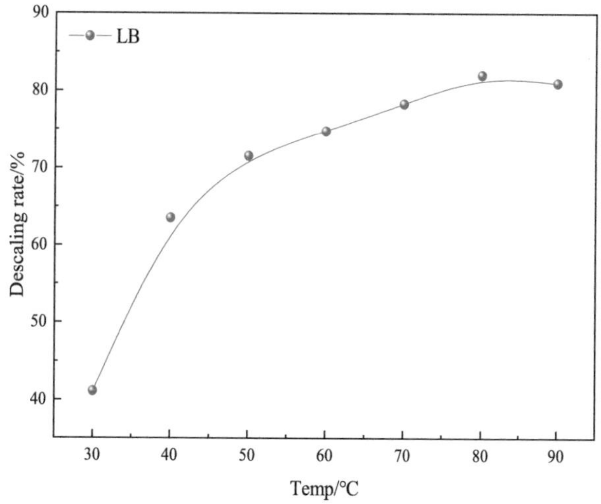

Fig. 4 Effect of different temperatures on descaling effect

3.5 *Experimental Study on the Effect of pH on Descaling Effect*

pH is one of the key factors affecting the descaling effect, in order to investigate the effect of pH on the scale effect, the effect of pH at 4, 5, 6, 7, 8 and 9 on the scale removal effect was investigated under the conditions of concentration of 5%, 80 °C, and the time of scale dissolution of 12 h, respectively. The experimental results are shown in Fig. 5. As can be seen from Fig. 5, pH has a significant effect on the descaling rate. In each descaling system, the descaling rate shows a trend of increasing and then decreasing with the increase of pH. The dissolution of $BaSO_4$ was strongly dependent on the pH of the chelating agent, and DTPA and EDTA showed the maximum solubility of $BaSO_4$ at pH 5.

3.6 *Experimental Study on the Effect of Time on Descaling Effect*

Under the optimal reaction conditions examined above, the effect of different reaction times on the descaling effect was examined at 4 h intervals. Since the descaling process usually involves chemical reactions or physical changes, the rates of these reactions and changes are closely related to time. By determining the optimal reaction time, in order to avoid the waste of resources. The experimental results are shown in Fig. 6. From Fig. 6, it can be clearly seen that the descaling rate increases with time, but after reaching a certain value, the descaling rate remains basically unchanged. At the reaction time at 8 h, the descaling rate was 81.20%. At 12 h, the descaling effect of each system increased, but the growth rate gradually slows down to 82.00%.

Fig. 5 Effect of different pH on descaling effect

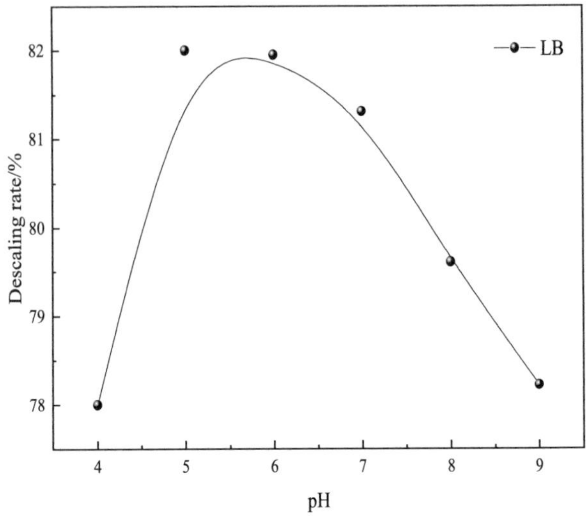

After that, the descaling rate almost no longer changes significantly, which is mainly because at the beginning of the reaction, the concentration of the descaling agent is relatively high, the reaction rate is faster, and the descaling effect is more obvious. Thus, 12 h is the optimal reaction time.

Fig. 6 Effect of different reaction time on descaling effect

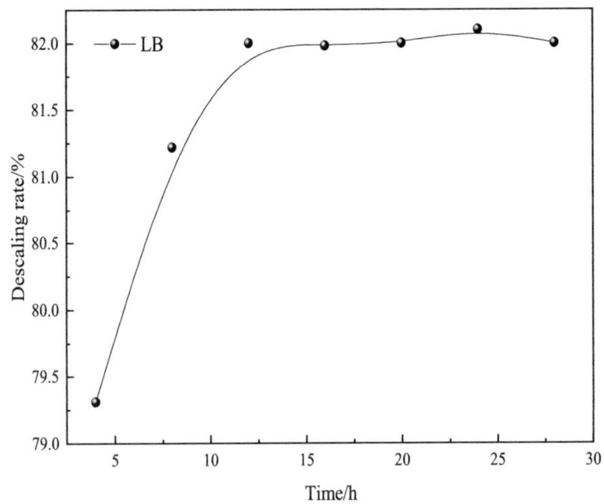

4 Conclusions

In this chapter, a LT descaling agent with good descaling effect on $BaSO_4$ was optimized through a series of experimental compounding by using DTPA as the main agent, combined with synergists such as oxalic acid and sodium pyrophosphate. After systematic experimental verification, we successfully determined the formulation of LB descaler, in which DTPA: oxalic acid is 1:1, and under the optimized experimental conditions, the descaler dosage of 5% (relative to the volume of the solution to be treated), the reaction temperature of 80 °C, the pH value of 5, and the reaction time of 12 h, the descaling rate of LB descaler on $BaSO_4$ reached 82.00%, which shows good descaling effect and practical application potential. This research result provides a new efficient solution for $BaSO_4$ descaling and a valuable reference for the development of descaling agents in related fields.

Acknowledgements This work was financially supported by grants from Youth Innovation Team of Shaanxi University, the Key Research and Development Program of Shaanxi (No. 2021KW-10) and the Innovation Capability Support Program of Shaanxi (No. 2022PT-08).

References

1. Huang, J.M., Kehua, L., Lei, Z.: Progress of research on oilfield scaling and prevention and its mechanism. Chem. Technol. Dev. **4**, 18–21 (2013)
2. Peng, Y., Yue, Z., Ozu Rui, G., Bo, C., et al.: Carbonate scale control under high level of dissolved iron and calcium in the Bak ken formation. In: SPE International Conference on Oilfield Chemistry. One Petro (2015)
3. Reyes-Garcia, E.A., Holan, K.H.: Are scale-dissolver enhancers necessary for calcium sulfates. In: SPE International Oilfield Scale Conference and Exhibition. One Petro (2020)
4. Putnis, A., Putnis, C.V., Paul, J.M.: The efficiency of a DTPA-based solvent in the dissolution of barium sulfate scale deposits. In: SPE International Conference on Oilfield Chemistry. One Petro (1995)
5. Weng, Z.W., Xie, G.S., Li, Y.: Application of oilfield scaling and chemical descaling agents. Chem. Eng. **34**(3), 61–65 (2020)
6. Ona Wole, A.T., Hussein, I.A., Saad, M.A., et al.: Effect of pH on acidic and basic chelating agents used in the removal of iron sulfide scales: a computational study. J. Petrol. Sci. Eng. **178**, 649–654 (2019)
7. Zhou, W., Tong, D.H., Li, L.P.: Research progress on oilfield produced water treatment methods. Ind. Water Treat. **28**(12), 5–8 (2008)
8. Qin, Y.: Formation of sulphate scale in oil and gas field development and progress of research and application of anti-scaling and descaling agents. Eng. Technol. Res. **5**(10), 51–53 (2023)
9. He, S., Ying, C., Juan, Q.H.: Research progress of oilfield descaling technology. Chem. Reagents **34**(11), 991–994 (2012)
10. Zhao, Z.H.: Research progress of oilfield fouling prediction and scale prevention technology. J. Liaoning Shi Hua Univ. **38**(05), 19 (2018)
11. Gamal, H., Elk Atat Ny, S., Al-Afnan, S., et al.: Development of a unique organic acid solution for removing composite field scales. ACS Omega **6**(2), 1205–1215 (2021)
12. Zhang, N., Wang, X., Zhang, J., et al.: Experimental and statistical study on wellbore scaling mechanisms and characteristics for Huan Jiang Oilfield. Geofluids **2022** (2022)

13. Yan, S.M., Zhang, C.S., Wang, B.Y.: Preparation and performance evaluation of barium sulfate scale inhibitor in oilfield. Prog. Fine Petrochem. Ind. **13**(9), 29-32 (2012)
14. Ghalib, H.B., Al Mallah, I.A.R.: Scaling simulation resulting from mixing predicted model between Mish Rif formation water and different waters injection in Basrah oil field, southern Iraq. Model. Earth Syst. Environ. **3**, 1557–1569 (2017)
15. Cui, C., Zhang, S.: Preparation, characterization and performance evaluation of a novel scale inhibiting and dispersing copolymer containing natural tannin. J. Polym. Environ. **28**(7), 1869–1879 (2020)
16. Zhang, Z., Lu, M., Liu, J., et al.: Fluorescent-tagged hyper-branched polyester for inhibition of $CaSO_4$ scale and the scale inhibition mechanism. Mater. Today Commun. **25**, 101359 (2020)
17. Yang, J.H., Song, H.F., Yang, S.N.: Chemical anti-scaling and descaling technology in Pu Cheng oilfield. Clean. World **33**(1), 13–18 (2017)
18. Krupp, H.: Particle adhesion theory and experiment. Adv. Colloid Interface Sci. **1**, 111–239 (1967)
19. Zheng, H.A., Xu, B., Ma, L., Wei, Z., et al.: Development and evaluation of barium-strontium scale descaler in oilfield. Appl. Chem. Ind. **41**(12), 2211–2213 (2012)
20. Wang, S., Wei, L., Li, C., et al.: Performance and mechanism study of IA-AMPS-HPA as an oilfield scale inhibitor: experimental and molecular dynamics simulation. J. Appl. Polym. Sci.Polym. Sci. **140**(41), e54516 (2023)

Energy-Saving Strategies for Buildings Based on Optimization of Enclosure Structures

Yaping Chen, Peng Li, and Yinze Hu

Abstract The construction sector in China is a significant contributor to the nation's energy consumption, representing approximately 30% of the total energy usage, and this proportion is increasing year by year, making the construction sector gradually become the primary area of energy consumption. In regions with hot summers and cold winters, passive residential buildings face a difficult dilemma between energy conservation and providing a comfortable living environment. This study is dedicated to optimizing the design of the enclosure structure of residential buildings in hot-summer and cold-winter regions. By using SketchUp and Energy Plus software for modelling and analysis, the study deeply explores the impact of the heat transfer coefficients of external walls and roofs on the heating, cooling load, and the total annual load of the building, aiming to find the best design scheme for the enclosure structure. In terms of material selection, this study chooses polyurethane insulation panels as the insulation material for external walls, with an insulation layer thickness set at 130 mm; for the roof insulation, extruded polystyrene boards are used, with an insulation layer thickness of 160 mm, which can achieve the lowest total annual load. The results of this study provide a theoretical foundation and practical guidance for the design of the enclosure structure of low-energy buildings in regions with hot summers and cold winters.

Keywords Energy-saving · Enclosure structures · Buildings

Y. Chen · P. Li (✉)
School of Design and Art, Hunan University of Technology and Business, Changsha, China
e-mail: Leepeng0258@163.com

Y. Hu
Hunan Lugu Architectural Technology Co. Ltd., Changsha, China

Y. Zhu (ed.), *Proceedings of the 4th International Symposium in Environmental Science and Industrial Ecology*, https://doi.org/10.1007/978-981-96-1578-0_5

1 Introduction

The peripheral protective structure of a building, composed of elements such as external walls, roofs, floors, windows, doors, and glass, serves to demarcate the indoor space from the outdoor environment. The external thermal and humidity conditions, adhering to the fundamental principle of heat flow from high to low temperatures, exert certain influences on the internal environment of the building through three primary modes of heat transfer: conduction, radiation, and convection. Consequently, the thermal performance of the peripheral protective structure directly impacts the energy consumption for heating and cooling, playing a significant role in the building's overall energy usage.

To conduct a more comprehensive study on the impact of the peripheral protective structure's thermal performance on the energy consumption of passive residential buildings, this research takes a typical residential building in Changsha City as a case study. By varying parameters such as the thickness of the external wall insulation, the type of wall material, the thickness of the roof insulation, and the type of roof material, the study simulates the influence of the roof's thermal performance on the building's energy consumption.

In the theoretical realm of energy-saving retrofitting for protective structures, Chvatal [1] have conducted research indicating that during the summer, the energy consumption for indoor air conditioning increases with the thickness of the insulation layer. They suggest that effective methods to reduce air conditioning energy consumption involve controlling the intensity of heat dissipation from indoor heat sources and reducing the amount of solar heat gain by the building. In the theoretical research on external wall insulation technology, Naouel [2] have monitored the energy consumption of buildings over a 30-year period. Their findings reveal that the orientation of the building has the least impact on determining the thickness of the insulation layer. They have also identified the optimal thickness for insulation layers and the payback period for investment in protective structure insulation. Bojoc [3] have explored the energy consumption of sandwich insulation, external insulation, and internal insulation, all utilizing polystyrene insulation boards, when applied to high-rise buildings. Their research demonstrates that the external insulation method results in the lowest energy consumption.

In the theoretical research on window and shading transformation, Czeslaw [4] have investigated the impact of shading louvers on heat loss from windows at night in Central Europe. Their findings indicate that shading louvers can effectively prevent heat loss, with external louvers offering superior energy-saving performance. Andria [5] has examined the effects of four types of glazing on the heating and cooling loads in cities such as Rome, Paris, and Nice, under varying window-to-wall ratios. This study provides insights into how different glazing options can influence energy demands in both summer and winter. Elisabeth [6] has explored the impact of shading systems on office buildings that utilize double-glazed curtain walls as part of their protective structure. By adjusting factors such as the position, color, and material of the louver shading system, her research assesses the influence on air conditioning energy consumption during the summer months. These studies collectively contribute to a deeper understanding of how architectural features like windows and shading devices can be optimized to enhance energy efficiency and comfort in various climatic conditions.

In recent years, domestic universities have made significant strides in the theoretical research of energy-saving renovations. Zhang [7] conducted a simulation analysis of energy consumption before and after the renovation of office building exterior walls in the Changsha area, along with an economic analysis post-insulation renovation. Their study determined the optimal thickness for exterior insulation layers. Long and Fu [8] utilized the characteristic temperature method to explore the impact of the same insulation system on energy consumption across different climate zones. Their research indicated that the same insulation technology yields varying energy-saving effects under different climatic conditions. Hao and Gao [9] employed the DEST simulation to investigate the influence of different exterior wall insulation forms and various insulation layer thicknesses on air conditioning loads in public buildings across various climate zones. They concluded that there is a relationship between the thickness of the exterior insulation layer and the energy-saving rate, and that the energy-saving effect of the same thickness of insulation layer varies under different climatic conditions. Liu et al. [10] took office buildings as the basis for his research model and used the DEST energy consumption simulation software to analyze cities such as Taiyuan, Jinan, and Beijing. He determined the impact of different insulation layer thicknesses on both winter and summer loads. Pan [11] conducted research on the impact of building insulation renovations in different climate zones on the load, proposing metrics such as load gain ratio, load-weighted ratio, and the change rate of total annual load to assess the energy-saving effects. These studies collectively enrich the body of knowledge on energy-saving renovations, offering valuable insights into optimizing building insulation to achieve greater energy efficiency across diverse climatic conditions.

2 Method

This simulation utilizes the typical meteorological year hourly data for Changsha, which was sourced from the CSWD format typical meteorological year data available for download on the EnergyPlus software's meteorological data website, produced by the U.S. Department of Energy. The data, known as "China Standard Meteorological Data," is a collaborative publication by Tsinghua University and the China Meteorological Administration. In accordance with the "Technical Guidelines for Passive Ultra-Low Energy Consumption Green Buildings (Trial) (Residential Buildings)," ultra-low energy buildings should opt for highly energy-efficient lighting sources with a lighting power density of 3 W/m^2. The internal heat gain from the building, excluding lighting, is set at 2 W/m^2. The indoor design parameters for air-conditioned rooms in the building model are 26 °C for both winter and summer. The forthcoming analysis will reference the exterior wall insulation material, extruded polystyrene

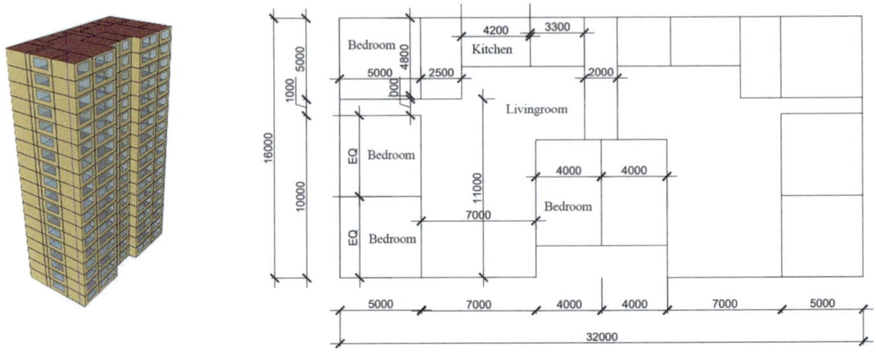

Fig. 1 Model parameters

(XPS) board, used in the model building (Fig. 1), while maintaining all other parameters constant. To investigate the influence of exterior wall insulation layer thickness on a building's energy consumption, seven datasets were meticulously chosen. These datasets correspond to insulation thicknesses varying at 20 mm increments, spanning the range from 30 to 150 mm.

3 Results

For regions with hot summers and cold winters, to achieve the energy-saving level of passive residential buildings, the improvement of the thermal performance of the peripheral protective structure should be combined with the climatic characteristics and economic principles of the region to optimize the selection and combination of insulation materials and thicknesses, thereby enhancing the thermal performance of the protective structure.

3.1 The Impact of the Thermal Performance of Exterior Wall on Energy Consumption

The exterior walls constitute a significant portion of a building's peripheral protective structure. Consequently, the energy-saving efforts focused on these walls are of paramount importance. The thermal transmittance coefficient serves as a critical metric for evaluating the insulation and thermal barrier performance of exterior walls. This coefficient directly influences the indoor thermal environment, which in turn affects the load on the air conditioning system and the overall energy consumption of the building. Adjusting the thickness of the insulation material and selecting insulation materials with varying thermal conductivities can significantly alter the wall's thermal transmittance coefficient.

The impact of the thickness of the exterior wall insulation layer on building energy consumption. Each insulation material has a specific thermal resistance. The greater the thickness of the material, the greater its thermal resistance. There is a definite relationship between thermal resistance and the material's thermal transmittance coefficient: the higher the thermal resistance, the lower the thermal transmittance coefficient, which means better insulation and thermal barrier effect. Taking the extruded polystyrene board used for the exterior wall insulation of the model building as a reference, we selected insulation layer thicknesses ranging from 30 to 150 mm at 20 mm intervals to explore the relationship between the thermal transmittance coefficient and the insulation layer thickness, as shown in Table 1. Based on the energy consumption simulation results for varying thicknesses of the exterior wall insulation, the following conclusions can be drawn:

(1) As the insulation thickness of the building's exterior walls increases from 30 to 150 mm, the reduction in the specific cooling energy consumption during the cooling season is as follows: 7.98%, 5.84%, 4.6%, 3.15%, 1.73%, and 1.28%, respectively. Similarly, during the heating season, the reduction in specific heating energy consumption per unit area is more pronounced, with rates of 15.48, 13.54, 10.66, 9.97, 8.05, and 3.10%. Collectively, the total energy consumption per unit area of the building decreases from 54.53 to 45.21 $KW \cdot h/m^2$, achieving a total energy consumption reduction rate of 16.84%.
(2) Within the 30–150 mm insulation thickness range, a significant downward trend in heating energy consumption per unit area is observed during the building's heating season, accentuating the overall energy savings. Conversely, cooling energy consumption during the cooling season exhibits a gradual decline, with the energy-saving rate approaching a plateau, yet still indicating a reduction. This suggests that enhancing the exterior wall's thermal performance by adjusting insulation thickness is advantageous for winter insulation, as the substantial indoor-outdoor temperature differential is mitigated by insulation,

Table 1 Energy consumption per unit area with different exterior wall insulation layer thickness

Insulation layer thickness (mm)	Total energy consumption per unit area ($KW \cdot h/m^2$)	Heating energy consumption per unit area ($KW \cdot h/m^2$)	Cooling energy consumption per unit area ($KW \cdot h/m^2$)
30	54.53	10.54	19.9
50	50.69	9.57	18.5
70	49.14	8.43	17.29
90	46.34	7.67	16.58
110	46.39	6.68	15.06
130	45.63	5.47	14.79
150	45.21	5.36	14.67

minimizing indoor heat loss. Moreover, when insulation thickness extends from 130 to 150 mm, the annual heating energy consumption trend plateaus, implying that beyond a certain thickness, further increases yield minimal gains in energy efficiency.

The impact of exterior wall insulation material types on building energy consumption. Different choices of exterior wall insulation materials lead to varying degrees of energy-saving effects, as each insulation material has a distinct thermal conductivity coefficient. Thermal conductivity, also known as the coefficient of thermal conduction, is an inherent characteristic of the material. The lower the thermal conductivity, the greater the thermal resistance, and there is a certain relationship between thermal resistance and the material's thermal transmittance coefficient: the greater the thermal resistance, the lower the thermal transmittance coefficient. This means that the thermal transmittance coefficients of insulation materials with different thermal conductivities also vary. Materials with a thermal conductivity less than 0.23 W/m^2K are generally referred to as insulating materials. Currently, commonly used insulation materials on the market can be categorized based on material type into organic types, such as phenolic boards, expanded polystyrene boards, polyurethane insulation boards, etc.; and inorganic types, such as rock wool boards, foam glass boards, traditional insulating mortars, etc. The thermal conductivity of the materials and their corresponding thermal transmittance coefficients when applied to exterior walls are shown in Table 2.

Taking the thickness of the insulation material on the exterior walls of the model building as a reference, while keeping other parameters consistent with the original model, different insulation materials are selected to alter the thermal conductivity coefficient. Simulation studies were conducted on the insulation and thermal barrier performance of the seven materials, and the results are shown in Table 3.

Based on the energy consumption simulation results for different exterior wall insulation materials, the following conclusions can be drawn:

(1) The highest energy consumption is observed with the hydro composite foamed cement board, and the lowest with the polyurethane (PU) insulation board. The difference in total energy consumption per unit area of the building between

Table 2 Thermal conductivity and heat transfer coefficient of roof insulation layer materials

Insulation layer materials	Thermal conductivity coefficient ($W/m \cdot K$)	Heat transfer coefficient ($W/m^2 \cdot K$)
PU	0.02	0.25
XPS	0.03	0.29
SEPS	0.03	0.32
Rock wool board	0.04	0.36
EPS	0.04	0.37
Foam glass	0.05	0.43
Composite foam cement board	0.06	0.49

Table 3 Energy consumption per unit area with different exterior wall insulation layer materials

Insulation layer thickness ($W/m^2 \cdot K$)	Total energy consumption per unit area ($KW \cdot h/m^2$)	Heating energy consumption per unit area ($KW \cdot h/m^2$)	Cooling energy consumption per unit area ($KW \cdot h/m^2$)
0.25	46.33	5.43	15.05
0.29	46.54	6.43	15.76
0.32	48.21	6.50	15.65
0.36	49.45	7.37	16.09
0.37	49.46	7.84	16.27
0.43	51.57	8.53	16.73
0.49	54.30	9.60	17.78

these two materials is 7.03 $KW \cdot h/m^2$, with a reduction rate of 14.26%. This indicates that using insulation materials with a lower thermal conductivity coefficient can significantly enhance the thermal insulation performance of the exterior walls, thereby reducing the building's energy consumption.

(2) Employing polyurethane (PU) insulation boards for wall insulation, the building's total annual energy consumption per unit area drops from 47.56 $KW \cdot h/m^2$ in the reference model to 46.34 $KW \cdot h/m^2$, reflecting a 2.57% reduction. The annual cooling energy consumption per unit area of the building decreases from 15.54 $KW \cdot h/m^2$ in the reference model to 15.03 $KW \cdot h/m^2$, a reduction of 0.51 $KW \cdot h/m^2$. The annual heating energy consumption per unit area of the building decreases from 6.62 $KW \cdot h/m^2$ in the reference model to 5.91 $KW \cdot h/m^2$, a reduction of 0.71 $KW \cdot h/m^2$. These findings indicate that insulation materials with lower thermal conductivity, such as PU, may not substantially impact energy consumption reduction.

3.2 *The Impact of Roof Thermal Performance on Energy Consumption*

The roof, as the building's uppermost peripheral component, significantly receives solar radiation and, despite its modest proportion, its thermal performance notably influences the building's energy use. The thermal transmittance coefficient, a key metric for evaluating the roof's insulation and thermal barrier efficacy, directly impacts indoor thermal conditions, air conditioning load, and overall energy demand. Selecting insulation and thermal barrier materials with varying thermal conductivities or adjusting the thickness of identical materials can alter the roof's overall thermal transmittance. Consequently, judicious material and thickness selection for the roof is pivotal for energy conservation in passive housing, especially in regions with extreme seasonal temperatures.

The impact of roof insulation layer thickness on building energy consumption. Taking the extruded polystyrene (XPS) board used for the roof insulation of the model building as a reference and keeping all other parameters consistent with the original model, seven different insulation layer thicknesses were selected at 20 mm intervals ranging from 60 to 180 mm. The relationship between the thermal transmittance coefficient and the insulation layer thickness is shown in Table 4.

Based on the energy consumption simulation results for different roof insulation layer thicknesses, the following conclusions can be drawn:

(1) As the insulation layer thickness on the building's roof increases from 60 to 180 mm, the reduction rate in the cooling season's energy consumption per unit area for cooling is 1.72%, 1.0%, 1.65%, 0.45%, 0.39%, and 0.19%, respectively. The reduction rate in the heating season's energy consumption per unit area for heating is 3.72%, 2.72%, 2.5%, 1.21%, 0.61%, and 0.31%, respectively. The total energy consumption per unit area of the building decreases from 48.98 to 47.34 $KW \cdot h/m^2$, with a total energy consumption reduction rate of 3.23%.

(2) When the insulation layer thickness is between 60 and 180 mm, the reduction trends for both the heating season's heating energy consumption per unit area and the cooling season's cooling energy consumption per unit area are slow, resulting in an inconspicuous decrease in the total building energy consumption. This indicates that the roof has a minor impact on the building's cooling and heating demands. When the insulation layer thickness is between 160 and 180 mm, the total annual energy consumption of the building only decreases by 0.06%, with the annual heating and cooling energy consumptions stabilizing at approximately 6.48 $KW \cdot h/m^2$ and 15.38 $KW \cdot h/m^2$, respectively.

(3) As the insulation layer gradually thickens, the trend of heat dissipation from the roof decreases and slows down, similar to the exterior walls. Considering cost and energy-saving effects, it is not significantly meaningful to simply increase the thickness of the insulation layer.

The impact of roof insulation material types on building energy consumption. Similar to exterior walls, different choices of roof insulation and thermal barrier

Table 4 Energy consumption per unit area with different roof insulation layer thickness

Insulation layer thickness	Total energy consumption per unit area ($KW \cdot h/m^2$)	Heating energy consumption per unit area ($KW \cdot h/m^2$)	Cooling energy consumption per unit area ($KW \cdot h/m^2$)
60	48.98	7.32	16.56
80	48.45	6.97	15.97
100	47.96	6.78	15.76
120	47.56	6.64	15.58
140	47.43	6.53	15.45
160	47.38	6.48	15.36
180	47.34	6.45	15.32

materials, that is, insulation materials with different thermal conductivity coefficients, will also have different thermal transmittance coefficients. Therefore, the energy-saving effects of the roof will also vary. Common insulation materials are the same as those for exterior walls, such as organic types including phenolic boards, expanded polystyrene boards, polyurethane (PU) insulation boards, etc.; and inorganic types including rock wool boards, foam glass boards, traditional insulating mortars, etc. Taking the thickness of the insulation material on the roof of the model building as a reference, different insulation layer materials are selected to change the thermal conductivity coefficient. The inherent thermal conductivity of the materials and their thermal transmittance coefficients when applied to the roof are shown in Fig. 2.

Keeping all other parameters consistent with the original model, a simulation study was conducted on the insulation and thermal barrier performance as well as the energy-saving level of the roof for the aforementioned seven materials. The results are shown in Table 5.

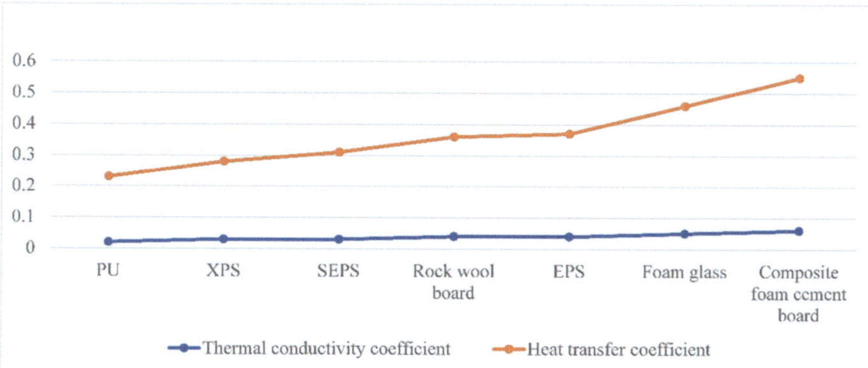

Fig. 2 Thermal conductivity and heat transfer coefficient of roof insulation layer materials

Table 5 Energy consumption per unit area with different heat transfer coefficient

Heat transfer coefficient (W · h/m^2)	Total energy consumption per unit area (KW · h/m^2)	Heating energy consumption per unit area (KW · h/m^2)	Cooling energy consumption per unit area (KW · h/m^2)
0.23	47.35	6.46	14.50
0.28	47.45	6.41	14.12
0.31	47.54	6.45	14.25
0.36	47.62	6.51	14.35
0.37	47.67	6.55	14.49
0.46	48.03	6.81	14.70
0.55	48.28	7.11	14.85

Based on the energy consumption simulation results for different roof insulation materials, the following conclusions can be drawn:

(1) The highest energy consumption is observed with cement expanded perlite, and the lowest with polyurethane (PU) insulation board. The difference in total energy consumption per unit area of the building between these two materials is 0.93 kWh/m^2, with a reduction rate of 2.17%. This indicates that using insulation materials with a lower thermal conductivity coefficient can to some extent enhance the insulation and thermal barrier performance of the roof. However, compared to exterior walls, the effect on reducing the total building energy consumption is not very significant.
(2) When PU is used as the roof insulation material, the total annual energy consumption per unit area of the building decreases from 48.28 kWh/m^2 in the reference model to 47.35 kWh/m^2. The annual cooling energy consumption per unit area of the building decreases from 14.85 kWh/m^2 in the reference model to 14.50 kWh/m^2. The annual heating energy consumption per unit area of the building decreases from 7.11 kWh/m^2 in the reference model to 6.46 kWh/m^2. This suggests that different types of insulation materials with lower thermal conductivity coefficients do not have a significant effect on energy consumption reduction.

Energy consumption simulation analysis reveals that the exterior wall's thermal transmittance coefficient significantly influences building energy use. The primary energy-saving strategy for exterior walls is to minimize this coefficient. Given its close relationship with the wall's insulation layer, the design of this layer is crucial for enhancing the wall's thermal performance (Table 6).

Based on the energy consumption simulation analysis of different types of exterior wall insulation materials, it can be seen that compared with using extruded polystyrene (EPS) as the wall insulation material, using polyurethane (PU) insulation board, extruded polystyrene (XPS), and graphite polystyrene (SEPS), as well as rock wool board as wall insulation materials can all achieve certain energy-saving effects. Among them, the energy-saving effect of polyurethane (PU) insulation board is more significant, which is related to its lower thermal conductivity coefficient that can enhance the insulation and thermal barrier performance of the exterior wall. The

Table 6 Enclosure structure parameters

Enclosure structure	Materials	Heat transfer coefficient (W/m^2 · K)
Exterior wall	5 mm cement mortar + 200 mm glass wool board + 20 mm cement mortar + 200 mm porous brick + I5mm mixed sand fan	0.21
Roof	40 mm fine stone mixed soil + 80 mm aerated concrete + 60 mm hard foam polyurethane + 20 mm cement mortar + 150 mm steel reinforcement + 10 mm mixed mortar	0.20

choice of other insulation materials does not have a significant impact on energy consumption. In addition, according to the characteristics of insulation materials, extruded polystyrene (EPS) has characteristics such as light weight, heat insulation, sound insulation, low temperature resistance, and low cost, making it also suitable for exterior wall insulation systems. Extruded polystyrene (XPS) has poor breathability, poor dimensional stability, and poor adhesion to inorganic bonding mortar, making it unsuitable for exterior wall insulation systems. Compared with rock wool board, graphite polystyrene (SEPS) has better waterproof performance and is more suitable for hot summer and cold winter areas. In summary, it is advisable to choose polyurethane (PU) insulation board, graphite polystyrene (SEPS), and extruded polystyrene (EPS) as exterior wall insulation materials.

Additionally, based on the energy consumption simulation analysis for different types of roof insulation materials, it can be concluded that using insulation materials with low thermal conductivity coefficients to improve the insulation and thermal barrier performance of the exterior walls does not result in very significant energy-saving effects. Among them, the polyurethane (PU) insulation board and extruded polystyrene (XPS) have relatively good energy-saving effects. In addition, according to the characteristics of insulation materials described in Table 5.1, foam glass boards have good insulation and thermal barrier performance, high tensile strength, and good durability. Compared with rock wool, foam glass does not absorb water and is more suitable for use on roofs. Therefore, for roof insulation materials, it is advisable to choose polyurethane (PU) insulation boards, extruded polystyrene (XPS), and foam glass boards.

4 Conclusions

In regions with hot summers and cold winters, it is advisable to optimize the thermal performance of exterior walls by choosing an external wall insulation method for residential buildings. The preferred materials for the exterior wall insulation layer are polyurethane (PU) insulation boards, graphite polystyrene boards, and extruded polystyrene (EPS) boards. When the exterior wall insulation layer material is an EPS board, the thickness of the roof insulation layer using EPS boards should be close to 130 mm, and exceeding 130 mm should be avoided. For optimizing roof thermal performance: it is advisable to choose an inverted insulation method for the roof insulation. The preferred materials for the roof insulation layer are polyurethane (PU) insulation boards, extruded polystyrene (XPS) boards, and foam glass boards. When the roof insulation layer uses XPS boards, the thickness should be close to 160 mm, and exceeding 160 mm should be avoided.

Acknowledgements The study was supported by the Natural Science Foundation of Hunan Province [No. 2023JJ40228], the Natural Science Foundation of Changsha [No. kq2208058], 2023 Hunan University of Technology and Business "Digital Intelligence +" Interdisciplinary Research

Project [No. 2023SZJ25], 2024 Teaching Research Project of Hunan University of Technology and Business [No.6104411CYP01].

References

1. Chvatal, K. M. S., Corvacho, H.: The impact of increasing the building envelop insulation upon the risk of overheating in summer and an increased energy consumption. J. Build. Perform. Simul. **2**(4), 267–282 (2019)
2. Naouel, D.: A study on optimum insulation thickness in walls and energy savings in Tunisian buildings based on analytical calculation of cooling and heating transmission loads. Appl. Energy **88**(1), 156–164 (2020)
3. Bojie, M., Yik, F., Leung, W.: Thermal insulation of cooled spaces in high rise residential buildings in Hong Kong. Energy Convers. Manage. **43**, 165–183 (2020)
4. Czeslaw, O.-P., Marta, S.: Effect of the roller blinds on heat losses through a double-glazing window during heating season in Central Europe. Energy Build. **73**, 48–58 (2021)
5. Andrea, G., Giovanni, P., Francesca, C., et al.: Analysis and modeling of window and glazing systems energy performance for a well-insulated residential buildings. Energy Build. **43**(4), 1030–1037 (2020)
6. Elisabeth, G., Andre, D.H.: The most efficient position of shading devices in a double-skin facade. Energy Build. **39**, 364–373 (2019)
7. Zhang, N.: A Study on the Impact of Exterior Wall Insulation on the Energy Consumption of Multi-story Office Buildings in Changsha. Master's Thesis, Hunan University, pp. 53–55 (2020)
8. Long, E.S., Fu, X.Z., Ma, X.F.: Comparison of energy-saving rates of the same external insulation system under different climatic conditions. Insul. Mater. Build. Energy Effic. **38**(8), 37–40 (2019)
9. Hao, M.H.: The impact of exterior wall insulation on the air conditioning load of public buildings in various building climate zones in China. China Hous. Facil. **10**(9), 51–54 (2019)
10. Liu, L., Jing, Y.Y., Du, H.: Optimization of thermal performance of building envelope for office buildings in cold regions. Refrig. Air Cond. **25**(4), 411–414 (2019)
11. Pan, D.M.: A study on the impact of building insulation under different climatic conditions on cooling and heating loads. HVAC **38**(8), 11–16 (2018)

Enhanced Blast Resistance of Slabs Using Integrated Steel Wire Mesh and Conventional Reinforcement in Response to Touch-Off Explosion

S. M. Anas, Rayeh Nasr Al-Dala'ien, Mohd. Shariq, and Mehtab Alam

Abstract Explosions have evolved beyond military applications, now impacting various contexts such as industrial accidents and terrorism, each with unique challenges. In particular, urban infrastructure and critical facilities are increasingly at risk, necessitating advanced protective measures. Touch-off explosions can cause widespread destruction when an explosive directly strikes a target, significantly increasing its destructive potential. Concrete slabs in buildings are especially vulnerable to these forces, as they bear structural loads and maintain overall integrity. Their propensity to fracture under the pressure of such explosions highlights the need for improved structural design and safety protocols. Enhancing their blast resistance is crucial for ensuring occupant safety and preventing progressive collapse. This research evaluates and enhances the blast resilience of concrete slabs by integrating steel wire mesh reinforcement. A baseline analysis of a reference slab reinforced solely with tension steel is validated against experimental data from Zhao et al., establishing a benchmark for further enhancements. The study examines two configurations of steel wire mesh, welded into the upper and lower layers of the slabs with nominal pitches of 15 mm and 10 mm, aiming to optimize blast performance. Using an Eulerian–Lagrangian method combined with finite element analysis in Abaqus, the blast loading is simulated to accurately capture the interaction between the blast wave and the slabs' structural response. Results show that both mesh configurations reduce deformation, peak stresses, and perforation sizes compared to the standard slab. Additionally, analysis of plastic damage metrics indicates that finer mesh spacing enhances energy absorption and structural integrity under blast conditions. The study also underscores the importance of reinforcement placement, as it

S. M. Anas (✉) · Mohd. Shariq
Department of Civil Engineering, Jamia Millia Islamia (A Central University), New Delhi, India
e-mail: s1910521@st.jmi.ac.in

R. N. Al-Dala'ien
Civil Engineering Department, College of Engineering, Al-Balqa Applied University (BAU), Salt, Jordan

M. Alam
Department of Civil Engineering, Netaji Subhas University of Technology, New Delhi, India

Y. Zhu (ed.), *Proceedings of the 4th International Symposium in Environmental Science and Industrial Ecology*, https://doi.org/10.1007/978-981-96-1578-0_6

directly influences failure patterns and overall slab stability. These findings demonstrate the effectiveness of steel wire mesh reinforcement in mitigating blast damage and strengthening concrete structures in hazardous environments. Future research can explore hybrid reinforcement strategies and material innovations to further improve blast resistance in critical infrastructure.

Keywords Blast resilience · Concrete slabs · Reinforcement · Steel wire mesh · Structural integrity · Touch-off explosion

1 Introduction

In the realm of building construction, slabs denote horizontal, structural components composed of reinforced concrete or steel [1–6]. Their essential function lies in bearing downward loads (gravity and live loads), such as floors or ceilings, and in effectively distributing these loads to the underlying beams and columns [7]. This distribution ensures the integrity and steadfastness of the structure. Slabs play a pivotal role in endowing a building with substantial stiffness and stability, qualities indispensable for enduring diverse forces, notably lateral ones such as wind or seismic activities. Nevertheless, their significance is particularly tested in situations where lateral stiffness is deficient, emphasizing the imperative of their capacity to withstand dynamic forces, including explosions [1, 2, 8–13].

Explosions, stemming predominantly from military origins, have undergone a multifaceted evolution in recent times, encompassing a spectrum of scenarios: local skirmishes, clandestine acts, cutting-edge warfare, unintended mishaps, and global confrontations [1, 11]. Events like the tragic Beirut port blast of 2020, triggered by stored ammonium nitrate, and the harrowing assault on Kabul airport in 2021, starkly underscore the profound devastation that explosions can inflict upon urban landscapes [14]. Furthermore, ongoing geopolitical struggles for supremacy among nations have increasingly targeted critical infrastructure, employing explosions as strategic implements to sow chaos and assert authority. These conflicts span from localized clashes to expansive international rivalries, illustrating the diverse contexts where explosives are wielded to achieve pivotal aims. Structural elements, notably slabs due to their expansive surfaces, are particularly susceptible to the force of explosions unleashed by devices like suitcase bombs, parcel bombs, and vehicle-borne improvised explosive devices [1, 11]. These high-energy occurrences impose sudden, intense strain rates upon slabs, presenting formidable structural challenges and potentially catastrophic outcomes for occupants and adjacent infrastructure alike.

Strengthening the resilience of slabs to withstand high-strain-rate loads from explosions is crucial for urban safety and the durability of infrastructure [1, 2, 11]. Progress in materials science and structural engineering offers avenues to mitigate blast impacts by refining concrete compositions, optimizing reinforcement methods, and pioneering construction techniques that resist blasts [11]. Implementing design principles that guard against progressive collapse and enhance energy dissipation can markedly fortify slab performance in explosive scenarios. Committing to research and development focused on boosting slab resilience not only safeguards buildings but also fortifies community resilience against the dynamic challenges posed by contemporary explosives and their deployment strategies.

Li et al. [15] studied slabs under simultaneous fire and blast loads, focusing on thermally induced stress, deformation, and material degradation. They used a calibrated numerical model to assess the impact of fire duration, blast scenarios, and slab parameters on blast performance, comparing concurrent and sequential load responses. Zhou et al. [16] developed deep learning models—a multi-layer perceptron (MLP) for maximum displacement and a one-dimensional convolutional neural network (1D-CNN) for failure modes—to predict explosion-induced responses in RC slabs. Their 1D-CNN outperformed traditional machine learning models, aided by permutation feature importance analysis for better prediction interpretability. Mohotti et al. [17] investigated the effects of close-range detonations on band-beam slabs in modern high-rise buildings. Using 2.5 kg to 4.5 kg of Composition-4 explosives, they found that varying reinforcement ratios and shear link configurations significantly influenced damage severity, aligning with UFC-3-340-02 guidelines. Design adjustments were proposed to enhance resilience against spalling and other forms of damage. López et al. [18] conducted an experimental investigation on a RC frame measuring 7×14 m^2, subjecting it to explosive charges equivalent to 10–20 kg of TNT at proximities ranging from 0.5 to 1.5 m. The suite of monitoring apparatus employed encompassed pressure transducers, accelerometers, a high-velocity camera, and a three-dimensional scanner. Minor cracking occurred in two tests; a 20 kg TNT equivalent detonated 50 cm from the slab caused localized damage, verified by LS-DYNA simulations. Xu et al. [19] studied one-way square ultra-high toughness cementitious composite (UHTCC) slabs under close-in blast loads, demonstrating superior performance over ordinary concrete. UHTCC endured 200 g TNT blasts with breaching and cracking but maintained load-bearing capacity, unlike ordinary concrete slabs that fractured severely. Empirical formulas and LS-DYNA models analyzed failure modes, highlighting UHTCC's potential for protective structures. Wang et al. [20] investigated Polyisocyanate-oxazodone coatings on concrete slabs under varying contact explosions. Thicker coatings improved blast resistance, requiring a higher TNT equivalent for critical failure. They developed empirical formulas for crater and spall diameters and assessed blast resilience. Anas et al. [1] reviewed numerous studies on slab responses to blasts, analyzing spall damage and parameters, discussing close-in detonation effects, and identifying damage modes like spalling, crater formation, and various cracks. They also examined numerical modeling for blast resistance enhancement.

Draganić et al. [21] reinforced RC slabs with commercial glass fibers, textiles, and epoxy resin, conducting lab and field tests that successfully reduced debris and deflection under blast conditions. Zhao et al. [22] analyzed precast concrete slab behavior under blasts using finite element models validated by experiments, focusing on factors affecting slab dynamics and anti-blast performance. Rasouli et al. [23] investigated the impact of confined water above RC slabs during near-field air explosions, showing significant increases in explosion energy and damage severity, suggesting potential for enhancing controlled demolitions with less explosives in urban areas. Meng et al. [24] researched alkali-activated geopolymer concrete as a green construction material, comparing its blast resistance with steel-reinforced conventional concrete. The investigation revealed that geopolymer concrete, when fortified with a steel wire mesh, exhibited diminished damage and fragmentation under the impact of a 50 kg TNT detonation at distances of 3 m, 5 m, and 7 m. Yang et al. [25] tested steel-reinforced rubber concrete slabs with varying rubber content (0, 10, 30, and 50% as NC, RC10, RC30, and RC50), validating empirical models with overpressure measurements. Their findings demonstrated less tensile damage compared to standard concrete under large explosives, confirmed by computational simulations highlighting the slabs' effectiveness against high-energy detonations.

In our current study, we have implemented a pioneering methodology to fortify slabs against explosive detonations. By incorporating a steel wire mesh alongside traditional reinforcements within the slab, our objective is to markedly enhance its resistance to explosive impacts. The steel wire mesh assumes a pivotal role by proficiently disseminating and absorbing the formidable kinetic energy from explosions across the entire slab surface. This mechanism effectively diffuses the blast forces, thereby mitigating localized harm and augmenting the overall structural robustness. Furthermore, the integration of steel wire mesh diminishes the direct transfer of energy to the rebars, resulting in reduced deformation and mitigating potential structural impairments. Moreover, this endeavor remains unprecedented within existing literature and holds profound implications for the progression of blast-resistant construction materials. Additionally, the accessibility and cost-effectiveness of steel wire mesh in the market render it a pragmatic alternative compared to more intricate and costly options. The amalgamation of rebars and steel wire mesh in slabs thus emerges as a pragmatic choice for fortifying against sudden and severe loading conditions induced by explosive occurrences.

2 Computational Modeling and Validation

The research methodology employed in this study utilizes an Eulerian–Lagrangian approach synergized with the finite element method using Abaqus [26] software to simulate the impact of touch-off explosions on a slab. This innovative approach integrates the Eulerian framework for modeling fluid dynamics during explosions with the Lagrangian method for tracking the movement of solid structures like the slab [1, 11, 26]. It accurately captures complex explosive physics like shock wave propagation, material deformation, and structural response, crucial for safety improvements and resilient structure design against blast loads. Figures 1 and 2 illustrate the detailed numerical model used for validation and analysis, showing mesh details, boundary conditions, and visualizations of pressure boundaries, displacement fields, and damage models. Moreover, the authors have adhered to a prior experimental investigation by Zhao et al. [13] in the numerical simulation of the slab, concurrently substantiating their simulations with the extant experimental data from Zhao et al. [13]. This practice not only guarantees coherence and dependability in the modeling methodology but also enhances the relevance of the discoveries to practical, real-world contexts.

The authors have created a total of three slab configurations to investigate their structural responses. The first slab embodies a conventional design, incorporating solely tension steel re-bars, and adheres to the experimental framework established by Zhao et al. [13]. The second slab integrates a square steel wire mesh, characterized by a 3 mm diameter and a nominal pitch of 15 mm, welded to both the upper and lower layers of rebars. The third slab similarly incorporates a square steel wire mesh, albeit with a denser nominal pitch of 10 mm, affixed to the upper and lower rebar layers. For both the second and third slabs, the steel wire mesh is confined to the central 450 mm × 450 mm area, predicated on the assumption that the touch-off explosion effect on the slab is localized. The mechanical properties of the steel are maintained

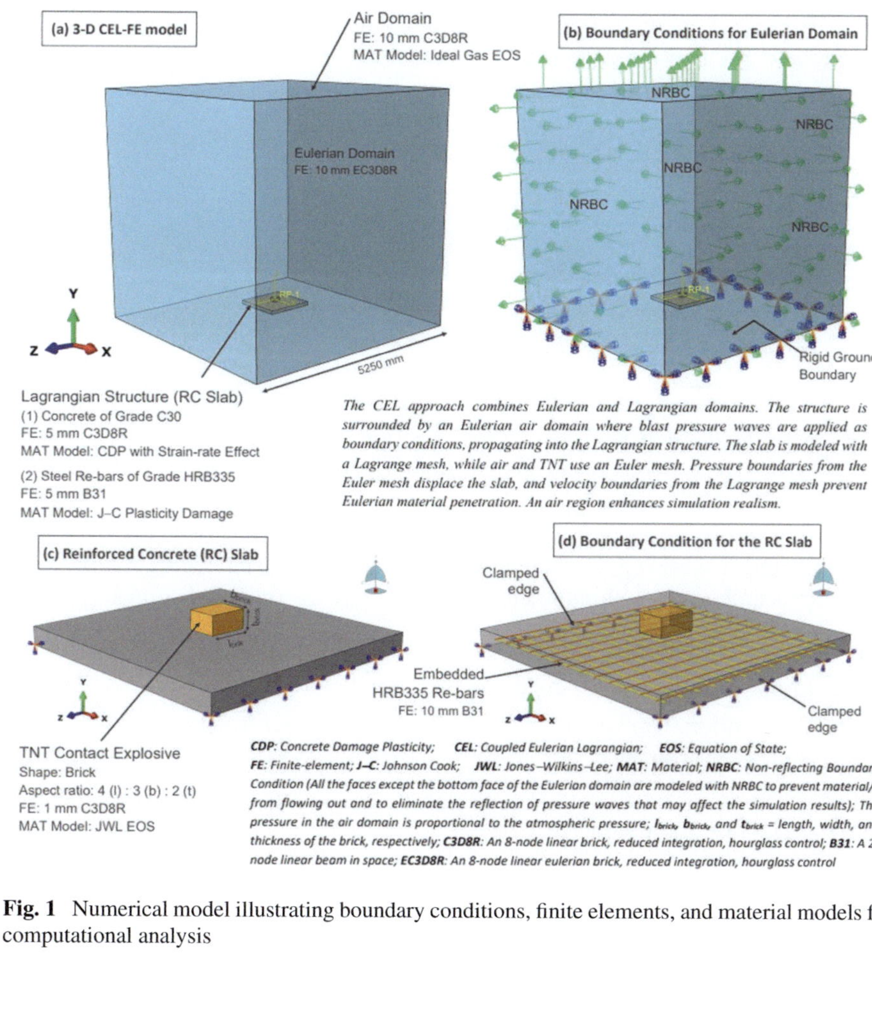

Fig. 1 Numerical model illustrating boundary conditions, finite elements, and material models for computational analysis

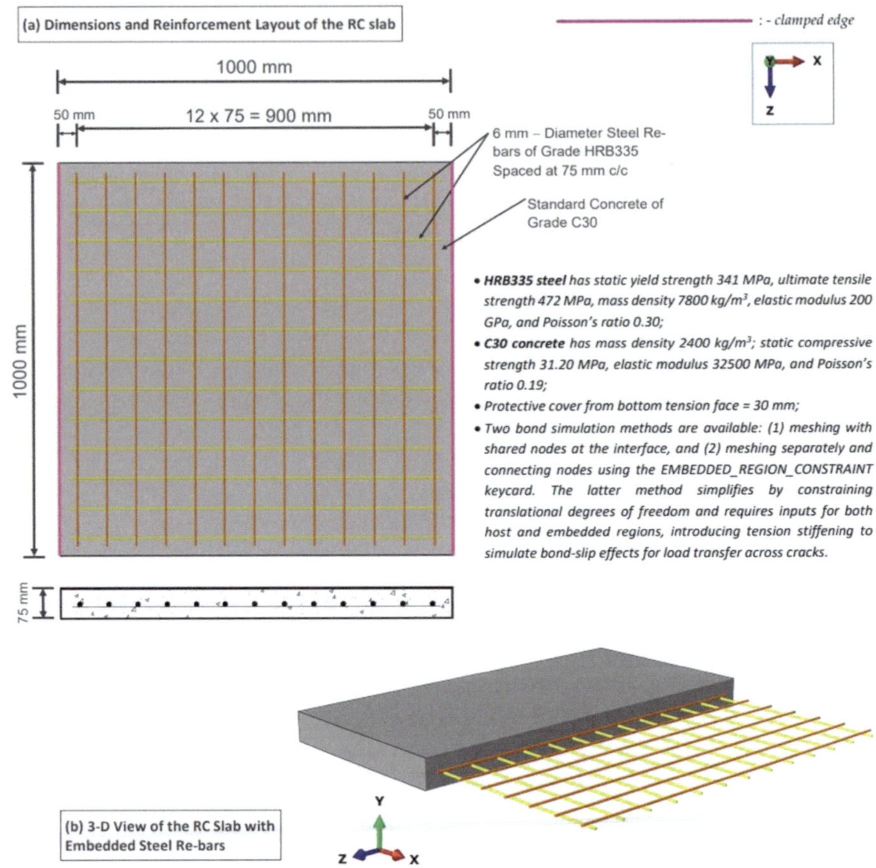

Fig. 2 Detailing of the reinforcement of the reference slab "CS-TSR-Zhao"

consistently across all slab configurations to ensure a fair comparison of their structural behaviors. Following are the model numbers for each slab configuration, with descriptions embedded within the numbers to convey the specific features:

(1) Model 1: Conventional Tension Steel Rebars (Zhao et al. [13] Framework)

- **Model Number**: CS-TSR-Zhao
- **Description**: Conventional Slab with Tension Steel Rebars following Zhao et al.'s [13] framework.

(2) Model 2: Conventional Tension Steel Rebars + Steel Wire Mesh (3 mm Diameter, 15 mm Pitch)

- **Model Number**: CS-TSR-SWM3-15

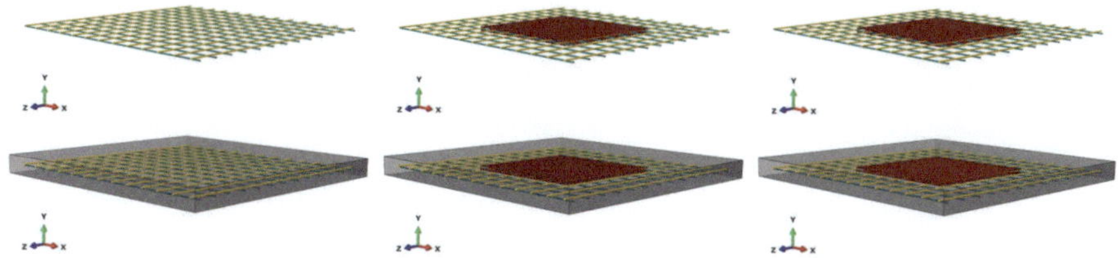

Fig. 3 Slab models used in the study

- **Description**: Conventional Slab with Tension Steel Rebars and Steel Wire Mesh of 3 mm diameter and 15 mm pitch on both upper and lower layers of rebars.

(3) Model 3: Conventional Tension Steel Rebars + Steel Wire Mesh (3 mm Diameter, 10 mm Pitch)

- **Model Number**: CS-TSR-SWM3-10
- **Description**: Conventional Slab with Tension Steel Rebars and Steel Wire Mesh of 3 mm diameter and 10 mm pitch on both upper and lower layers of rebars.

The discretization choices for different segments of the numerical model are delineated in Fig. 1. Additionally, Fig. 1 elucidates the selection of material models employed to characterize these segments, with the pertinent material properties derived from reference [1, 7, 13]. In the realm of blast loading, accounting for strain-rate effects is pivotal due to their profound influence on material behavior at elevated strain rates. In this study, strain rate effects are addressed through stress augmentation factors detailed in reference [7]. To adhere to the conference's prescribed page constraints, the authors have presented a succinct account of their numerical modeling efforts, ensuring avoidance of redundant material property data by citing appropriate sources (Figs. 3 and 4).

An essential aspect of finite element modeling and simulation lies in conducting meticulous mesh refinement analysis to ensure the precision and dependability of computational findings [11]. By methodically adjusting the dimensions of finite elements and monitoring resulting changes in outcomes such as slab deformation and damage, researchers can verify the efficacy of simulation software. In this investigation, the authors undertook a comprehensive analysis of mesh refinement, scrutinizing various sizes of finite elements and cross-referencing simulated outcomes with experimental data sourced from Zhao et al. [13]. Figures 5 and 6, through these comparisons, they identified the optimal element size of 5 mm as crucial for accurately forecasting slab behavior under blast loading conditions.

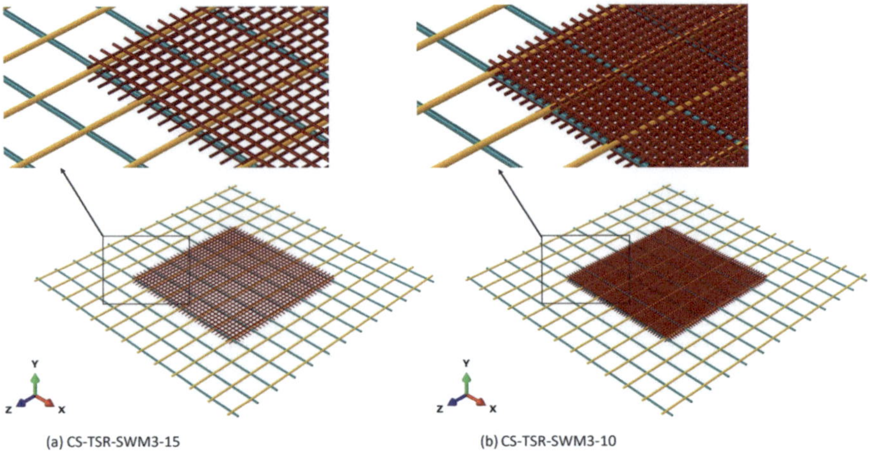

Fig. 4 Steel wire mesh welded with the upper and lower rebar layers of the slabs

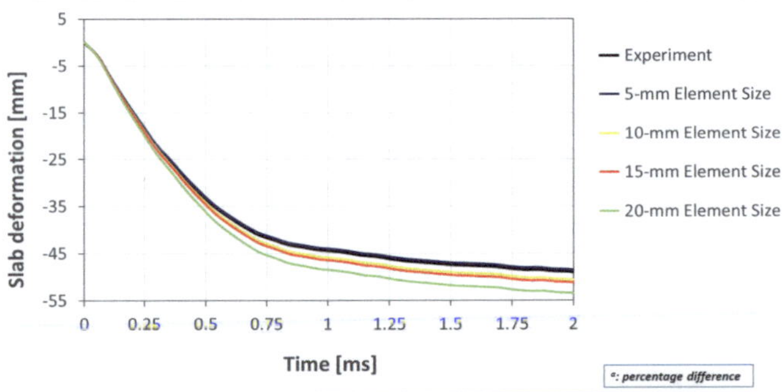

Finite Element Size [mm]	Total Elements	Total Nodes	Maximum Displacement [mm] of the Slab	Computational Time [x 10^4 ms]	Disk Space [GB] used
5	290,178,029	317,687,244	-48.55 ([a]0.92%)	2640	11.84
10	289,658,029	315,055,537	-50.76 (3.52%)	1440	8.68
15	289,600,474	312,423,075	-51.34 (4.66%)	840	7.98
20	289,588,029	310,918,447	-53.57 (8.90%)	720	7.68
Experiment	---	---	-49.0	---	---

Fig. 5 Deformation histories for varying element sizes compared with experimental data

3 Results

When an explosive detonates upon contact with a slab that is supported on two edges, it sets off an intricate chain of interactions involving stress waves within the materials [13]. Initially, the detonation generates compressive stress waves that travel through the slab, engaging with both the concrete matrix and the reinforcement bars embedded within it. The concrete is subjected to intense compressive forces, resulting in puncturing and crushing effects, while the reinforcement bars undergo

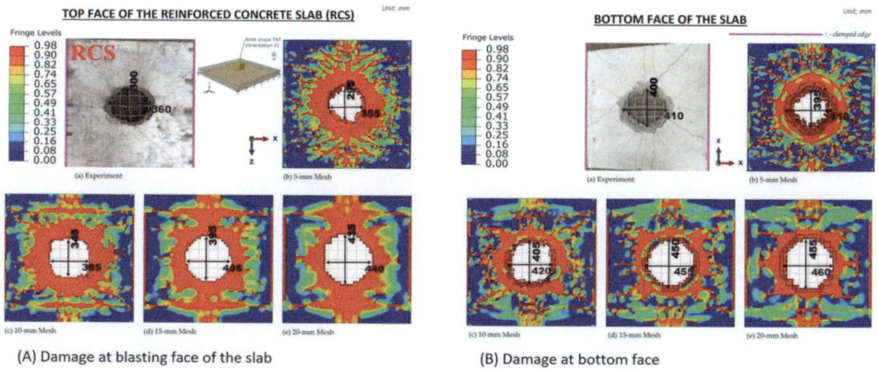

Fig. 6 Comparison of slab damage at various element sizes with experimental observations

substantial deformation [13]. As these compressive waves rebound from the opposite side (tension face) of the slab, they metamorphose into high-stress tensile waves due to concrete's intrinsic weakness in tension. This transformation induces the emergence of cracks, usually originating in the vicinity of the perforation, a phenomenon known as scabbing, and progressively extending toward the restrained peripheries of the slab. The cracks originate initially at the edge of the perforation and then propagate along trajectories determined by how the slab responds structurally to the dynamic loading induced by the explosive detonation. The nature of concrete, which is stronger in compression than in tension, leads to more extensive cracking on the tension side upon wave reflection. This occurrence engenders an elevated density of cracks on the slab's inferior surface, where tensile stress concentrations are exacerbated by the reflection and transformation of compressive stress waves into tensile forces. Consequently, this underscores the structural response to the explosive event.

Figure 7 illustrates visual depictions of slab damage. The reference slab denoted as "CS-TSR-Zhao," when subjected to a detonation initiated at its centroid, underwent substantial deformation with a deflection measuring 48.55 mm (Fig. 8). Figure 9 illustrates the distribution of stress in the reinforcement of this slab. The upper concrete surface experienced peak compressive stress reaching 22.71 MPa, reflecting the concentrated forces exerted during the explosion (Fig. 10). In the vicinity of the blast, the concrete endured maximum shear stress of 24.08 MPa, underscoring its pronounced susceptibility to shear forces near the epicenter of the blast. The resulting perforation dimensions on the upper surface measured 355 mm in the X-direction and 295 mm in the Z-direction, vividly demonstrating the extent of material displacement and fragmentation caused by the explosive impact, Fig. 7. Conversely, at the lower surface, the perforation dimensions were measured at 410 mm (X-direction) and 395 mm (Z-direction), highlighting the asymmetric propagation of blast effects through the structure of the slab.

The second slab, designated as "CS-TSR-SWM3-15," incorporated a 15 mm-pitch steel wire mesh welded to both upper and lower rebar layers, resulting in substantial enhancements in blast resistance compared to the reference slab "CS-TSR-Zhao."

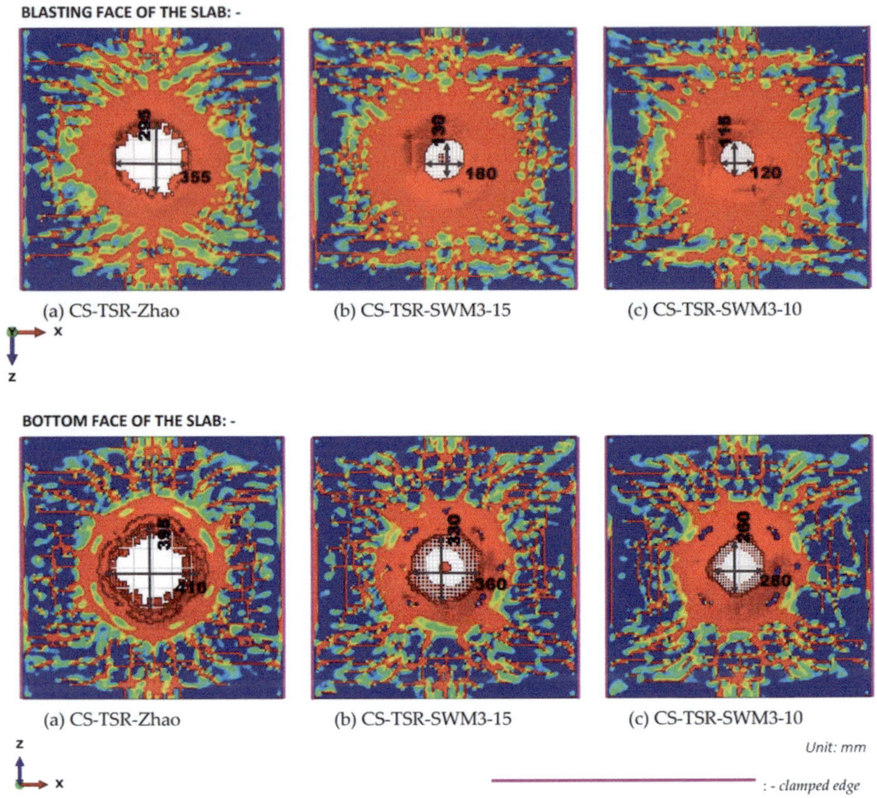

Fig. 7 Visual representation of slab damage assessed

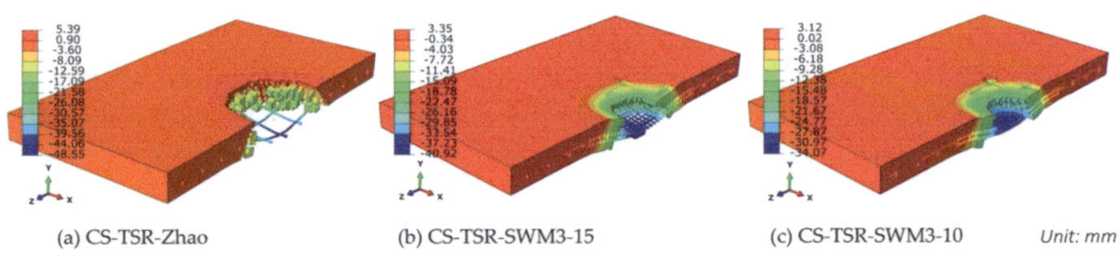

Fig. 8 Deformation patterns of slab structures under touch-off explosion

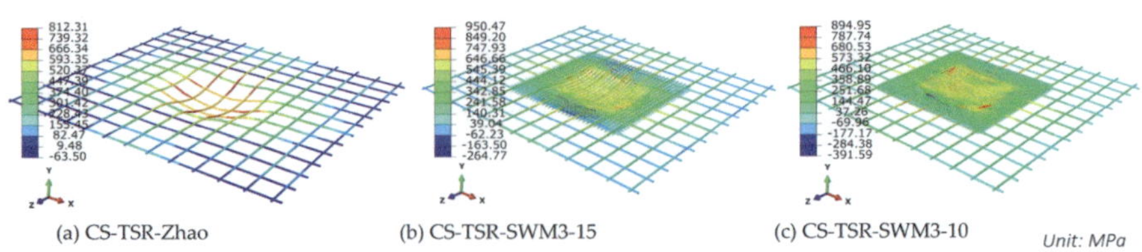

Fig. 9 Maximum principal stress distribution in slab reinforcements

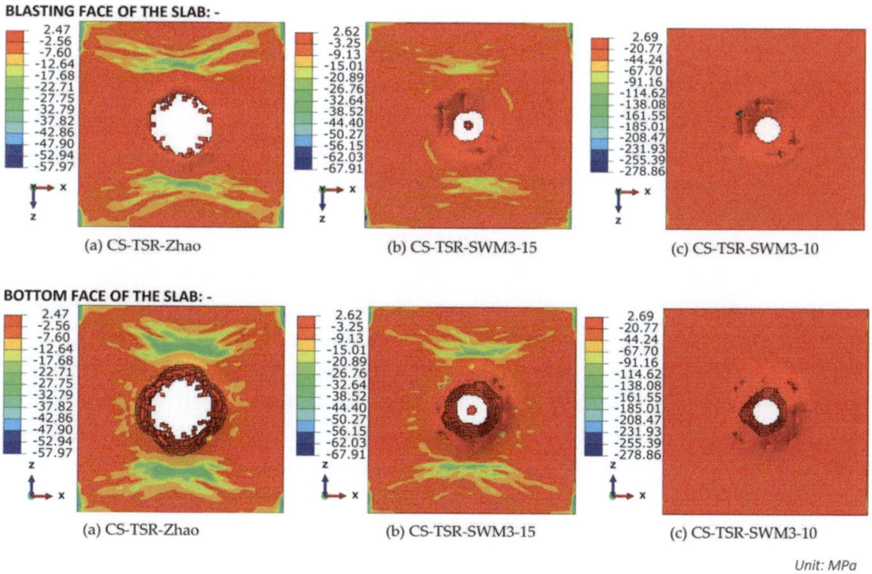

Fig. 10 Maximum principal stress distribution in concrete of the slabs

It displayed a deformation of 40.92 mm, significantly lower by 15.71% in comparison, underscoring its bolstered structural integrity. Peak tensile stress in the mesh at the blast epicenter reached 747.93 MPa, emphasizing its capability to redistribute and absorb energy. The upper concrete surface endured peak compressive stress of 20.89 MPa and shear stress of 22.11 MPa. Perforation dimensions on this surface measured 180 mm in the X-direction and 130 mm in the Z-direction, showcasing reduced fragmentation relative to the reference slab. Conversely, on the lower surface, perforation dimensions were 360 mm (X-direction) and 330 mm (Z-direction), indicating effective dissipation of blast energy through the structure. The 15 mm-pitch steel wire mesh effectively fortified the slab, augmenting its resilience against blast forces by managing stress distribution and controlling damage propagation through flexural damage mechanisms.

Similarly, the third slab, "CS-TSR-SWM3-10," integrated a 10 mm-pitch steel wire mesh welded to upper and lower rebar layers, further refining blast performance beyond the second slab. It exhibited a deformation of 34.07 mm, signifying heightened rigidity and decreased susceptibility to blast-induced deformations. Peak tensile stress in the mesh at the blast zone measured 680.53 MPa, marginally less by 9.01% than the second slab and markedly reduced by 16.22% compared to the reference slab. The upper concrete surface registered peak tensile stress of 2.69 MPa and shear stress of 18.51 MPa, indicating enhanced stress distribution and attenuation of peak forces. Perforation dimensions on this surface were 120 mm in the X-direction and 115 mm in the Z-direction, demonstrating further fragmentation reduction compared to both preceding slabs. On the lower surface, perforation dimensions measured 280 mm (X-direction) and 260 mm (Z-direction), illustrating effective dissipation of blast energy. The 10 mm-pitch steel wire mesh densified the slab structure, fortifying its resistance against blast effects through enhanced control of perforations and mechanisms for energy absorption, predominantly via flexural damage modes.

To further quantify the impact severity caused by the explosive force, the authors also evaluated the parameter termed Plastic Damage Energy (PDE). The benchmark slab "CS-TSR-Zhao" demonstrated a PDE of 43.55 Joules, signifying the complete energy dissipation due to plastic deformation during the explosion. In contrast, the subsequent slab "CS-TSR-SWM3-15" exhibited diminished damage at 36.91 Joules, while the third slab "CS-TSR-SWM3-10" displayed an even lower total of 32.45 Joules. This reduction in damage is directly linked to the inclusion of steel wire mesh reinforcement within the slabs. The steel wire mesh effectively absorbs and redistributes explosive energy, thereby mitigating stress concentration and reducing material penetration compared to the reference slab. By bolstering the slab's capacity to endure and dissipate energy through controlled damage formation and absorption mechanisms, the steel wire mesh diminishes plastic deformation and subsequently lowers the Plastic Damage Energy. This underscore enhanced structural resilience against blast-induced loads.

4 Conclusions

This research aimed to evaluate the blast resistance of concrete slabs reinforced with two different configurations of steel wire mesh, comparing them to a baseline slab lacking wire mesh reinforcement. The study underscores significant enhancements in structural resilience and blast resistance upon integrating steel wire mesh, evident in both mechanical responses and plastic damage energy metrics.

The reference slab, designated as "CS-TSR-Zhao," exhibited notable deformation and concentrated stress zones under blast loading, resulting in a Plastic Damage Energy (PDE) of 43.55 Joules. This indicates substantial energy dissipation through plastic deformation during the explosive event. In contrast, the second slab, "CS-TSR-SWM3-15," reinforced with a 15 mm pitch steel wire mesh, demonstrated improved blast performance with a reduced PDE of 36.91 Joules. This decrease in plastic damage energy correlates with diminished deformation, lower peak stresses in both concrete and reinforcement, and smaller perforation dimensions compared to the reference slab. The steel wire mesh effectively absorbed and redistributed energy, mitigating stress concentrations and controlling damage propagation via flexural mechanisms. Furthermore, the third slab, "CS-TSR-SWM3-10," reinforced with a denser 10 mm pitch steel wire mesh, exhibited further enhancements with a minimized PDE of 32.45 Joules. This slab experienced minimal deformation, reduced stress concentrations, and even smaller perforation dimensions compared to both the reference and second slabs. The finer mesh pitch facilitated more efficient energy absorption and stress redistribution, thereby improving overall structural integrity and mitigating damage from blast effects.

According to the results of this research, integrating steel wire mesh reinforcement proves advantageous not only for freshly engineered concrete slabs but also offers a feasible option for upgrading established structures. In the case of pre-existing slabs requiring heightened blast resilience, a pragmatic strategy entails the removal of the lower concrete layer followed by precise welding of the steel wire mesh onto the rebar layer positioned at the slab's tension side. This retrofitting technique significantly bolsters the slab's capacity to endure blast impacts by augmenting energy absorption and redistributing forces via the mesh's flexural attributes.

References

1. Anas, S.M., Alam, M., Umair, M.: Experimental and numerical investigations on performance of reinforced concrete slabs under explosive-induced air-blast loading: a state-of-the-art review. Structures **31**, 428–461 (2021). https://doi.org/10.1016/j.istruc.2021.01.102
2. Iqbal, M.A., Kumar, V., Mittal, A.K.: Experimental and numerical studies on the drop impact resistance of pre-stressed concrete plates. Int. J. Impact Eng **123**, 98–117 (2019)
3. Al-Dala'ien, R.N., Syamsir, A., Abu Bakar, M.S., et al.: Effect of the stirrup shear reinforcement on the dynamic behavior and failure modes of two-way reinforced concrete slab subjected to the low-velocity impact loading. Arab. J. Sci. Eng. **49**, 5599–5624 (2024)
4. Al-Dala'ien, R.N., Syamsir, A., Bakar, M.S.A., Usman, F., Abdullah, M.J.: Failure modes behavior of different strengthening types of RC slabs subjected to low-velocity impact loading: a review. J. Compos. Sci. **7**(6) (2023)
5. Abbas, H., Gupta, N.K., Alam, M.: Nonlinear response of concrete beams and plates under impact loading. Int. J. Impact Eng **30**(8–9), 1039–1053 (2004)
6. Al-Dala'ien, R.N., Anas, S.M., Kodrg, A.-F.J.: Anti-impact performance enhancement of two-way spanning slab through the implementation of steel trussed bars. J. Eng. **2024** (2024)
7. Anas, S.M., Al-Dala'ien, R.N., Shariq, M., Alam, M.: Fortifying slab resilience against touch-off explosions: integration of innovative stud reinforcements and computational analysis. Buildings **14**(5) (2024)
8. Tai, Y.S., Chu, T.L., Hu, H.T., Wu, J.Y.: Dynamic response of a reinforced concrete slab subjected to air blast load. Theoret. Appl. Fract. Mech. **56**(3), 140–147 (2011)
9. Wu, C., Hao, H.: Modelling of simultaneous ground shock and airblast pressure on nearby structures from surface explosions. Int. J. Impact Eng **31**(6), 699–717 (2005)
10. Yao, S., Zhang, D., Chen, X., Lu, F., Wang, W.: Experimental and numerical study on the dynamic response of RC slabs under blast loading. Eng. Fail. Anal. **66**, 120–129 (2016)
11. Hao, H., Hao, Y., Li, J., Chen, W.: Review of the current practices in blast-resistant analysis and design of concrete structures. Adv. Struct. Eng. **19**(8), 1193–1223 (2016)
12. Zhao, C., Wang, Q., Lu, X., Wang, J.: Numerical study on dynamic behaviors of NRC slabs in containment dome subjected to close-in blast loading. Thin-Walled Struct. **135**, 269–284 (2019)
13. Zhao, C., Lu, X., Wang, Q., Gautam, A., Wang, J., Mo, L.Y.: Experimental and numerical investigation of steel-concrete (SC) slabs under contact blast loading. Eng. Struct. **196** (2019)
14. Anas, S.M., Al-Dala'ien, R.N., Akram, S., Alam, M.: Accidental explosions, Beirut ANFO blast, Ghana truck explosion, global terrorism index 2021–22, and blast mitigation: a review. In: 5th International conference on energetics, civil and agricultural engineering (ICECAE 2024), vol. 497, Tashkent, Uzbekistan, May 13–14, 2024. E3S Web Conferences (2024)

15. Li, H., Chen, W., Hao, H.: Performance of reinforced concrete slabs subjected to simultaneous fire and blast loads. Eng. Struct. **311** (2024). https://doi.org/10.1016/j.engstruct.2024.118133
16. Zhou, X.-Q., Huang, B.-G., Wang, X.-Y., Xia, Y.: Deep learning-based prediction of structural responses of RC slabs subjected to blast loading. Eng. Struct. **311** (2024)
17. Mohotti, D., Wijesooriya, K., Fernando, P.L.N., Nishshanka, B., Lee, C.-K., Remmenikov, A.: Experimental investigation of band-beam slabs subjected to close-range blast loading. Eng. Struct. **309** (2024)
18. López, L.M., Caldentey, A.-P., Santos, A.P., Diego, Y.G., Castedo, R., Chiquito, M.: Experimental response and numerical modelling of a full-scale two-span concrete slab frame subjected to blast load. Eng. Struct. **296** (2023)
19. Xu, S., Chen, B., Li, Q., Zhou, F., Yin, X., Jiang, X., Wu, P.: Experimental and numerical investigations on ultra-high toughness cementitious composite slabs subjected to close-in blast loadings. Cem. Concr. Compos. **126** (2022)
20. Wang, W., Huo, Q., Yang, J.-C., Wang, J.-H., Wang, X., Gao, W.-L.: Damage analysis of POZD coated square reinforced concrete slab under contact blast. Defence Technol. **18**(9), 1715–1726 (2022)
21. Draganić, H., Gazic, G., Lukic, S., Jelec, M.: Experimental investigation on blast load resistance of reinforced concrete slabs retrofitted with epoxy resin impregnated glass fiber textiles. Compos. Struct. **274** (2021)
22. Zhao, C., Ye, X., He, K., Gautam, A.: Numerical study and theoretical analysis on blast resistance of fabricated concrete slab. J. Build. Eng. **32** (2020)
23. Rasouli, A., Toopchi-Nezhad, H.: The influence of confined water on blast response of reinforced concrete slabs: experimental investigation. J. Build. Eng. **30** (2020)
24. Meng, Q., Wu, C., Su, Y., Li, J., Liu, J., Pang, J.: A study of steel wire mesh reinforced high performance geopolymer concrete slabs under blast loading. J. Clean. Prod. **210**, 1150–1163 (2019)
25. Yang, F., Feng, W., Liu, F., Jing, L., Yuan, B., Chen, D.: Experimental and numerical study of rubber concrete slabs with steel reinforcement under close-in blast loading. Constr. Build. Mater. **198**, 423–436 (2019)
26. Abaqus, C.A.E.: Damage plasticity, explicit platform, material library, interactions, constraints, boundary conditions, loads, post-processing. Help & Learning Manual, ABAQUS DS-SIMULIA (2020)

Upcycling PET Plastic Bottle and Nylon 6,6 to Functional Metal Organic Frameworks for CO_2 Adsorption

Chawala Jermkuntod, Thilina Rajeendre Katugampalage, Preeti Waribam, Wanida Chooaksorn, Chalita Ratanatawanate, Pakorn Opaprakasit, and Paiboon Sreearunothai

Abstract This study explores the upcycling of polyethylene terephthalate (PET) plastic bottles into functional metal–organic frameworks (MOFs) for CO_2 adsorption application, aiming to address environmental problem of plastic waste accumulation. Specifically, it focuses on synthesizing MIL-88B(Fe) MOF derived from PET chemical depolymerization. MIL-88B(Fe) was further functionalized with hexamethylenediamine (HMD) obtained from nylon 6,6 digestion. Using microwave-assisted acidic hydrolysis, PET and nylon 6,6 were rapidly depolymerized into their constituent monomers, terephthalic acid (TPA) and HMD, respectively. MIL-88B(Fe) MOF was synthesized from reaction between TPA and ferric chloride and was then further functionalized with HMD. The obtained MOFs were characterized via XRD, FTIR, FE-SEM, and elemental analysis. The results reveal that the crystal structure of MIL-88B(Fe) undergoes significant changes post-functionalization with HMD. The CO_2 adsorption capacity of the synthesized MOFs was also assessed, showing a marked difference between MIL-88B(Fe) and MIL-88B(Fe)-HMD. This research demonstrates the potential of converting waste into high-value MOFs, addressing environmental concerns and contributing to sustainable waste management and the development of advanced materials for gas storage, particularly CO_2.

Keywords PET depolymerization · Metal–organic framework · MIL-88B(Fe) · CO_2 adsorption · Hexamethylenediamine · Nylon 6,6

C. Jermkuntod · T. R. Katugampalage · P. Waribam · P. Opaprakasit · P. Sreearunothai (✉)
Functional Advanced Materials Engineering Research Center (FAME), School of Integrated Science and Innovation (ISI), Sirindhorn International Institute of Technology (SIIT), Thammasat University, Bangkok, Thailand
e-mail: paiboon@siit.tu.ac.th

C. Ratanatawanate
National Nanotechnology Center (NANOTEC), National Science and Technology Development Agency (NSTDA), Khlong Nueng, Thailand

W. Chooaksorn
Department of Environmental Science, Faculty of Science and Technology, Thammasat University, Bangkok, Thailand

Y. Zhu (ed.), *Proceedings of the 4th International Symposium in Environmental Science and Industrial Ecology*, https://doi.org/10.1007/978-981-96-1578-0_7

1 Introduction

In recent years, the increasing demand for sustainable materials and the need to mitigate environmental pollution have spurred research into the chemical recycling of synthetic polymers and the development of advanced materials for pollutant adsorption applications [1]. Polyethylene terephthalate (PET) and nylon 6,6 widely used in various industries contribute significantly to plastic waste. Effective recycling methods are crucial for reducing this waste and recovering valuable chemical components. At the same time, metal–organic frameworks (MOFs) like MIL-88B(Fe) have gained attention for their potential in gas adsorption, catalysis, and separation processes due to their high surface areas and tunable pore structures [2].

Recent research has shown that functionalizing MOF-5 with ethylenediamine (EDA) [3] can lead to a significant increase in CO_2 adsorption. The MOF-5 modified with EDA demonstrated an adsorption capacity of 2.9 mmol/g at 25 °C and 0–15 bar, compared to 0.9 mmol/g for unfunctionalized MOF-5. This enhancement highlights the effectiveness of diamine functionalization in improving the performance of MOFs for gas adsorption applications.

Furthermore, this research aims to evaluate the potential of utilizing waste materials, PET and nylon 6,6, to enhance CO_2 adsorption in MOFs, aligning with zero waste management goals. The study delves into the chemical recycling of PET and nylon 6,6 plastics to recover valuable components like terephthalic acid (TPA) and hexamethylenediamine (HMD). Additionally, it explores the synthesis and functionalization of MIL-88B(Fe) with HMD to increase its CO_2 adsorption capacity. The recycled materials and synthesized MOFs are characterized using techniques such as FTIR, 1H-NMR, XRD, SEM, and CO_2 gas adsorption to gain insights into their structural and functional properties.

2 Experiment

2.1 Chemical and Materials

Nylon 6,6, Iron (III) chloride hexahydrate ($FeCl_3 \cdot 6H_2O$, 97%), terephthalic acid (TPA, 99%), hydrochloric acid (HCl, 37%), N, N-dimethylformamide (DMF, 98%) and ethanol (ETOH, 99.9%), were purchased from Merck and Sigma-Aldrich.

2.2 Chemical Recycling of PET

The recycling process started with washing PET bottles to remove oil, glue, and dirt, followed by machine cutting and grinding PET into particles smaller than 500 μm. A 25% v/v concentration of hydrochloric acid (HCl) was prepared. The recycling reaction was carried out in a microwave reactor (Milestone flexi WAVE) using 1 g of PET flakes and 50 mL of acid in a 100 mL Teflon-lined pressure vessel, heated to 200 °C for 30 min. The reactor operated at an input power of 800 W, and the mixture was allowed to cool naturally. The recycled product was then washed with deionized until the pH matched that of the deionized water. Finally, the product was analyzed using FTIR and ^{1}H-NMR.

2.3 *Chemical Recycling of Nylon 6,6*

The process began by adding 1 g of nylon 6,6 or nylon cable tie to a 100 mL Teflon-lined pressure vessel containing 50 mL of 25% v/v HCl. The reaction took place in a microwave reactor with an input power of 800 W at 150 °C for 30 min, followed by natural cooling. The digested sample was then transferred to a 100 mL beaker, where 20 mL of deionized water was added and stirred for 30 min. During this process, a white solid of adipic acid precipitated out of the solution. After stirring, the solution was left at 0 °C overnight. The sample was then centrifuged to separate the adipic acid solids, which were washed with deionized water and dried at 90 °C overnight. The remaining solution contained hexamethylenediamine (HMD). All samples were characterized using FTIR and ^{1}H-NMR.

2.4 *MIL-88B(Fe) Synthesis*

To synthesize MOFs using pure chemicals, 5 mmol of TPA (0.8306 g) was dissolved in 50 mL of DMF and stirred until fully dissolved. Next, 5 mmol of $FeCl_3 \cdot 6H_2O$ (1.3516 g) was added to the solution, which was then transferred to a Teflon-lined microwave vessel and heated at 150 °C for 20 min. The temperature was increased from room temperature to 150 °C over 10 min, after which the reactor was allowed to cool naturally. The supernatant was centrifuged to separate the solvent from the orange solid, which was then washed with DMF to remove any unreacted TPA, followed by ethanol (ETOH) for solvent exchange. The solid was then dried at 90 °C for 2 h. Finally, the MOFs were activated by heating them at 150 °C for 8 h in a vacuum oven.

2.5 *Hexamethylenediamine Functionalization of MIL-88B(Fe)*

Three hundred milligrams of MIL-88B(Fe) was added to 50 mL of HMD solution. The mixture was stirred for 6 h at room temperature. Afterward, the suspension was centrifuged to separate the MIL-88B(Fe)-HMD, which was then thoroughly washed multiple times with ethanol.

3 Experimental

3.1 Chemical Recycling of PET

The FTIR spectra confirm the successful recycling of PET using 25% v/v HCl at 200 °C for 30 min. Characteristic peaks, such as the ν C=O stretching of the carboxylic acid group at 1673 and 1712 cm^{-1}, ν C–O stretching at 1280 and 1240 cm^{-1}, and the δ O–H bending at 1421 and 1408 cm^{-1}, were evident in both the PET-derived TPA and commercial TPA. Additionally, prominent peaks at 927 cm^{-1} (δ O–H) and 722 cm^{-1} (δ C–H) in the spectra further supported the successful recovery of TPA from PET (Fig. 1a).

The chemical recycling of PET to TPA was further analyzed using ^{1}H-NMR spectroscopy by dissolving the product in DMSO-d_6. The ^{1}H-NMR spectrum of PET-derived TPA, obtained under the same reaction conditions, showed a peak at 8.044 ppm corresponding to the four protons of the benzene ring in TPA, and a peak at 13.290 ppm corresponding to the two protons of the carboxylic acid group in TPA. The integration ratio for commercial TPA aligned with the expected proton count in TPA, but the PET depolymerization sample displayed a different ratio, indicating potential challenges in detecting acidic protons due to factors such as hydrogen bonding or rapid exchange. The solubility of TPA in DMF confirmed the complete digestion of PET by HCl into TPA [4] (Fig. 1b).

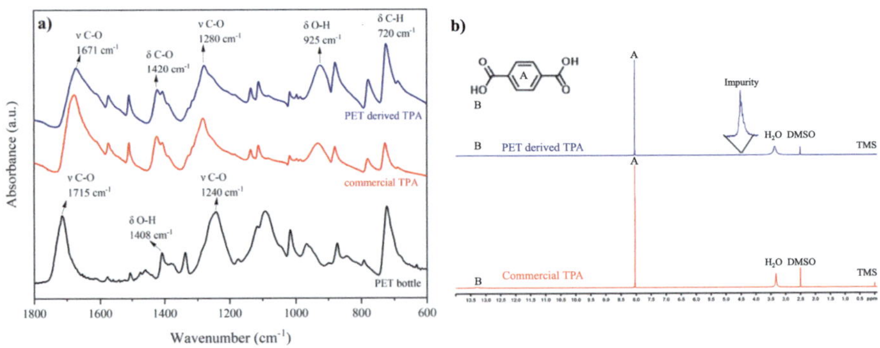

Fig. 1 **a** FTIR spectra and **b** ^{1}H-NMR spectra of commercial TPA and derived TPA

3.2 *Chemical Recycling of Nylon 6,6*

The FTIR spectra in Fig. 2 provide a detailed comparison of the acidic hydrolysis products of nylon 6,6 (Ny 66) and nylon cable. In Fig. 2a, the FTIR spectra of Ny 66 and the nylon cable exhibit nearly identical peaks, confirming that the two materials undergo similar chemical changes during acidic hydrolysis. The prominent peaks observed at approximately 3298 and 3084 cm^{-1} correspond to the ν N–H vibrations, while the peak at around 1642 cm^{-1} corresponds to the ν C=O vibrations, indicating the presence of amide groups characteristic of nylon structures. The spectra also show peaks at 2934 and 2854 cm^{-1}, which correspond to ν C–H vibrations, further supporting the presence of the nylon backbone.

In Fig. 2b, the focus shifts to the adipic acid, the desired product of nylon hydrolysis. The FTIR spectra of both adipic acid and the hydrolyzed products of Ny 66 and the nylon cable exhibit similar peaks, particularly the strong absorption at approximately 1707 cm^{-1} and 1413 cm^{-1}, which correspond to the ν C=O and δ O–H

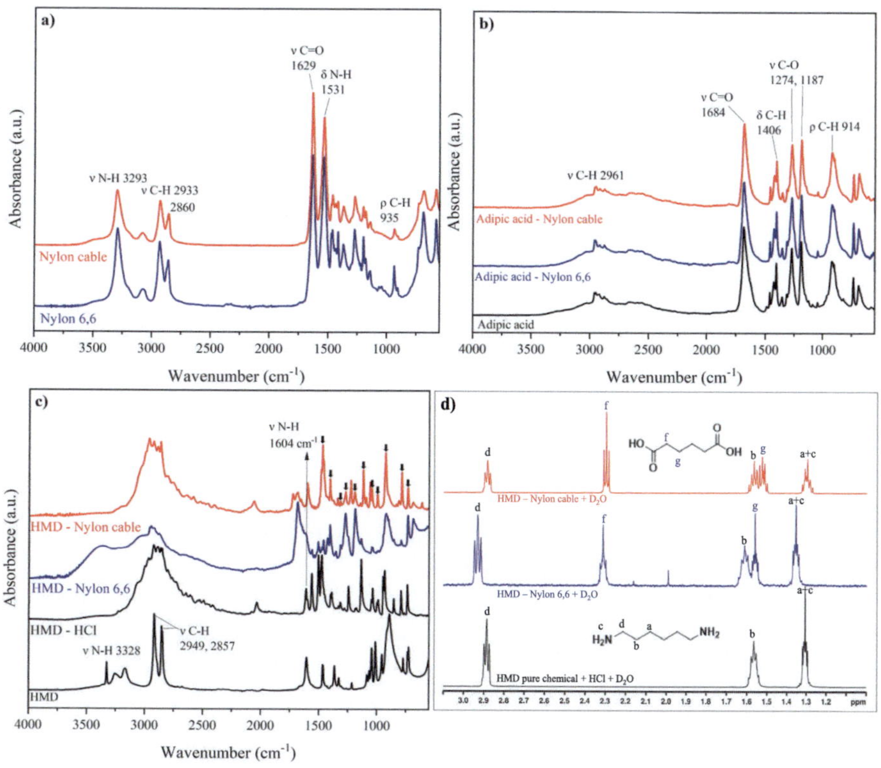

Fig. 2 FTIR Spectra of **a** nylon 6,6 and nylon cable, **b** adipic acid, **c** HMD and **d** ^{1}H-NMR of HMD

vibrations of the carboxylic acid group, respectively. This confirms the successful formation of adipic acid from the hydrolysis of both nylon sources.

In Fig. 2c, the analysis highlights hexamethylenediamine (HMD), where the FTIR spectra show that both pure HMD and HMD mixed with HCl have quite similar characteristic peaks. Notably, the N–H stretching vibrations around 3320 cm^{-1} and several other peaks, marked by black arrows, confirm that the HMD produced from hydrolysis retains structural similarities to pure HMD, even when mixed with HCl. Additionally, the FTIR spectra for HMD derived from Ny 66 and nylon cable reveal peaks that correspond to adipic acid, suggesting its presence as a byproduct.

The 1H-NMR spectrum in Fig. 2d further elucidates the chemical composition of the hydrolysis products. NMR analysis indicates that both HMD and adipic acid are present in the final product, pointing to some contamination of HMD with adipic acid. The process involves dissolving Ny 66 in HCl, heating, and then freezing the solution overnight at 0 °C, which causes adipic acid to precipitate while HMD remains in the solution. However, the detection of adipic acid in the NMR spectrum of the HMD sample suggests that the separation process was incomplete, necessitating further purification steps to obtain pure HMD. The NMR peaks for HMD and adipic acid confirm their co-existence, complicate the analysis of the final material.

3.3 MIL-88B(Fe)-HMD

In this study, FTIR spectra analysis of both samples reveals characteristic peaks of MIL-88B(Fe) and MIL-88B(Fe)-HMD, confirming successful synthesis. Key observations include the 1383 cm^{-1} peak, corresponding to the asymmetric and symmetric stretching of O–C–O in coordinated carboxylates with Fe metal centers [5]. The 1660 cm^{-1} peak, present in both MIL-88B(Fe) and waste-derived MIL-88B(Fe), indicates the stretching vibration of the carbonyl (C=O) group, suggesting the presence of unreacted TPA and a significant amount of DMF, likely due to insufficient activation time and temperature, as DMF has a boiling point of 153 °C, and the material was only activated at 150 °C for 8 h [6, 7]. The ν_{as} C=O peak at 1590 cm^{-1} in MIL-88B(Fe) confirms proper coordination between metal ions and the organic linker, while the δ C–H peak at 770 cm^{-1} corresponds to the bending vibrations of the C-H bonds in the organic linker [5, 8].

Additionally, for MIL-88B(Fe)-HMD, a distinct difference is observed in the peak around 1500–1600 cm^{-1}, which suggests some interaction involving the C–O group of the carboxylic acid, causing the peak to shift and broaden. The FTIR spectra of MIL-88B(Fe) depicted in Fig. 3a align with these findings, further confirming the successful synthesis of MIL-88B(Fe) and MIL-88B(Fe)-HMD.

The PXRD pattern spectra of MIL-88B(Fe) shown in Fig. 3b confirm the successful synthesis of MIL-88B(Fe) under specific conditions, with results aligning closely with the simulated pattern obtained from the Cambridge Crystallographic Data Centre (CCDC) under identifier YEDKOI 285810. The PXRD patterns for MIL-88B(Fe) synthesized using pure chemical reagents exhibit strong agreement with the

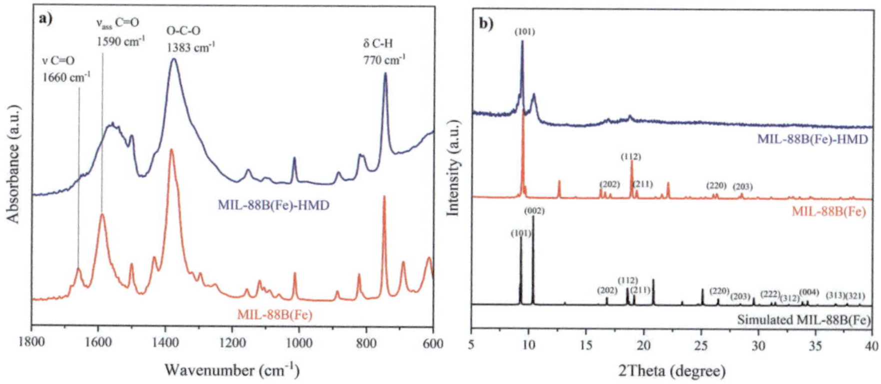

Fig. 3 **a** FTIR and **b** XRD spectra of MIL-88B(Fe) and MIL-88B(Fe)-HMD

simulated data, with key diffraction peaks at 2θ values of approximately 9.33° (101), 13.15° (002), 16.79° (202), 18.56° (112), 19.15° (211), 26.48° (220), 28.41° (203), 31.11° (222), 33.80° (312), 34.27° (004), 36.82° (313), and 37.79° (321), confirming the formation of the desired crystalline structure. Similarly, the PXRD pattern of MIL-88B(Fe)-HMD synthesized using waste materials also shows characteristic peaks of MIL-88B(Fe), indicating the successful formation of the MOF structure via the rapid microwave synthesis route. However, some peaks observed in the MIL-88B(Fe)-HMD sample seem to disappear after the post-synthesis process, likely influenced by the solvents used during synthesis. Incomplete removal of residual solvents, such as dimethylformamide (DMF), during activation at 150 °C for 8 h, may have led to residual solvent molecules within the MOF pores, affecting its crystal structure and generating additional peaks in the PXRD pattern [9, 10]. This phenomenon is further supported by the carbonyl stretching vibrations observed in the FTIR spectra at 1500–1670 cm^{-1}.

The FE-SEM analysis was conducted to examine the morphological structure of the MOFs. The image of MIL-88B(Fe), shown in Fig. 4, homogeneous octahedral crystal structure. This sample was formed by combining equal amounts of metals and organic ligands in a DMF solvent. The octahedral crystal structure of MIL-88B(Fe) and MIL-88B(Fe)-HMD is similar, with the only noticeable difference being a small crack observed on the MIL-88B(Fe)-HMD crystals.

The N_2 adsorption–desorption isotherms for MIL-88B(Fe) and MIL-88B(Fe)-HMD are illustrated in Fig. 5, showing the differences in adsorption capacity between the two materials. The isotherms of both samples exhibit a typical Type-IV pattern, which is indicative of mesoporous materials. However, MIL-88B(Fe)-HMD shows a significantly higher adsorption capacity compared to MIL-88B(Fe) across the entire range of relative pressures. This enhanced adsorption is particularly noticeable at higher relative pressures (p/p°), where the MIL-88B(Fe)-HMD sample exhibits a pronounced uptick, indicating larger pore volumes and greater surface area. The

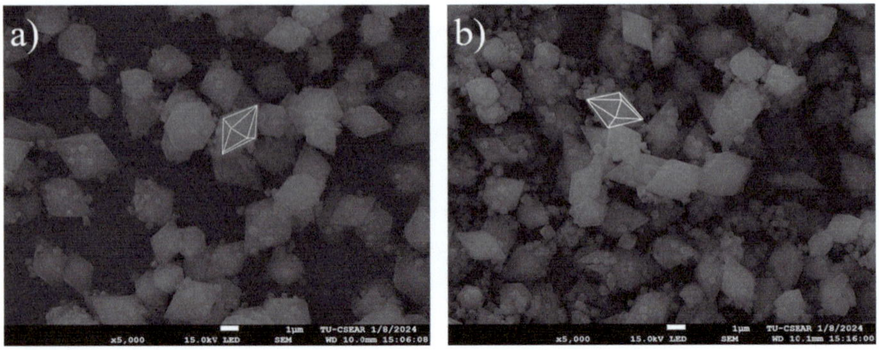

Fig. 4 SEM of **a** MIL-88B(Fe) and **b** MIL-88B(Fe)-HMD at 5000x

increased adsorption capacity of MIL-88B(Fe)-HMD can be attributed to the structural modifications induced by the incorporation of HMD, which likely results in more accessible pore structures and higher surface area, as evidenced by the BET surface area measurements.

Further analysis of the N_2 surface area data presented in Table 1 supports these observations. MIL-88B(Fe)-HMD demonstrates a lower BET surface area (25.42 m^2/g) compared to MIL-88B(Fe) (27.44 m^2/g). Additionally, the pore volume of MIL-88B(Fe)-HMD (0.027 cm^3/g) is notably higher than that of MIL-88B(Fe) (0.007 cm^3/g), indicating more extensive porosity within the modified material.

The CO_2 adsorption isotherms for MIL-88B(Fe) and MIL-88B(Fe)-HMD at 298.15 K and up to 1 bar demonstrate an unexpected trend, where MIL-88B(Fe) exhibits a higher CO_2 uptake capacity of 0.997 mmol/g compared to MIL-88B(Fe)-HMD, which adsorbs only 0.353 mmol/g. Typically, the presence of amino groups (NH_2) in MIL-88B(Fe)-HMD would be expected to enhance CO_2 adsorption due to

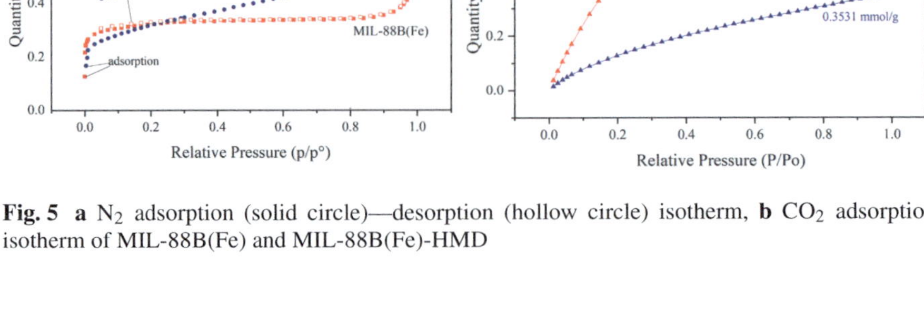

Fig. 5 **a** N_2 adsorption (solid circle)—desorption (hollow circle) isotherm, **b** CO_2 adsorption isotherm of MIL-88B(Fe) and MIL-88B(Fe)-HMD

Table 1 Surface area, pore volume and pore size of MIL-88B(Fe) and MIL-88B(Fe)-HMD

Sample	Surface area	Pore volume	Pore size
Models-Units	(BET) m^2/g	(BJH) cm^3/g	(BJH) nm
MIL-88B(Fe)	27.436	0.007	7.819
MIL-88B(Fe)-HMD	25.415	0.027	8.102

chemisorption interactions between -NH_2 and CO_2. However, the observed reduction in CO_2 adsorption for MIL-88B(Fe)-HMD suggests that the hexamethylenediamine has blocked the pores, thereby reducing the available surface area for physical sorption of CO_2. This pore blocking limits the material's capacity to adsorb CO_2, overriding the potential benefits of chemisorption, and leading to lower overall CO_2 uptake in MIL-88B(Fe)-HMD compared to MIL-88B(Fe).

4 Conclusion

The experimental findings demonstrate the successful chemical recycling of PET and nylon 6,6, yielding TPA, adipic acid, and HMD. The FTIR and 1H-NMR analyses confirm the efficient recovery of these chemical components, with notable similarities between the recycled and commercial counterparts. The study also highlights the synthesis and functionalization of MIL-88B(Fe) with HMD, showing significant modifications in its textural properties. The N_2 BET surface area and pore volume analyses reveal that MIL-88B(Fe)-HMD exhibits a higher surface area and porosity compared to the unmodified MIL-88B(Fe). However, the CO_2 adsorption isotherms indicate an unexpected decrease in CO_2 uptake for MIL-88B(Fe)-HMD, likely due to pore blockage by HMD, which restricts physical adsorption despite the presence of –NH_2 groups that could enhance chemisorption.

These findings highlight the complexities involved in MOFs for applications, where elements like pore structure and interactions between functional groups significantly influence adsorption performance. The research offers important insights into the recycling of synthetic polymers into functional MOFs for sustainable material design and environmental conservation.

Acknowledgements This research received funding support from the NSRF via the Program Management Unit for Human Resources & Institutional Development, Research, and Innovation grant no. B40G660035. CJ acknowledges the scholarship under the Thailand Advanced Institute of Science and Technology and Tokyo Institute of Technology (TAIST-Tokyo Tech) program, awarded by Sirindhorn International Institute of Technology, Thammasat University, funded by the National Research Council of Thailand (NRCT) and the National Science and Technology Development Agency (NSTDA). WC acknowledged support from Thailand Science Research and Innovation (TSRI) Fundamental Fund, fiscal year 2025/2568. CR gratefully acknowledged support from Program Management Unit for National Competitiveness Enhancement (PMU-C), Research & Innovation for Sustainability Center (RISC by MQDC) and National Science and Technology Development Agency (NSTDA) Grant no. C10F650194. TK acknowledges the TU-NSTDA Excellent Research Graduate Scholarship (No. T2.1-6.-01) from the National Science and Technology

Development Agency (NSTDA), Thammasat University, and the EFS Scholarship from Sirindhorn International Institute of Technology. Partial support from the Center of Excellence in Digital Earth and Emerging Technology (CoE: DEET), Thammasat University is highly appreciated.

References

1. Nayanathara Thathsarani Pilapitiya, P.G.C., Ratnayake, A.S.: The world of plastic waste: a review. Elsevier Ltd. (2024). https://doi.org/10.1016/j.clema.2024.100220
2. Hou, S., et al.: Green synthesis and evaluation of an iron-based metal-organic framework MIL-88B for efficient decontamination of arsenate from water. Dalton Trans. **47**(7), 2222–2231 (2018). https://doi.org/10.1039/c7dt03775a
3. Asghar, A., Iqbal, N., Noor, T., Khan, J.: Ethylendiamine (EDA) loading on MOF-5 for enhanced carbon dioxide capture applications. In: IOP Conference Series: Earth and Environmental Science. Institute of Physics Publishing (2020). https://doi.org/10.1088/1755-1315/471/1/012009
4. Bä, J.-E., Baldwin, J.E., Williams, R.M. (Eds.): Tetrahedron Organic Chemistry Series: High-Resolution NMR Techniques in Organic Chemistry, 2nd edn (Online). Available: www.elsevierdirect.com
5. Nguyen, D.T.C., et al.: Metal-organic framework MIL-53(Fe) as an adsorbent for ibuprofen drug removal from aqueous solutions: response surface modeling and optimization. J. Chem. **2019** (2019). https://doi.org/10.1155/2019/5602957
6. Nurani, D.A., Anisa, N., Khatrin, I., Yasmine, Kadja, G.T.M., Krisnandi, Y.K.: Breathable iron-based MIL-88 framework as dye adsorbent in aqueous solution. Chemistry (Switzerland) **6**(2), 283–298 (2024). https://doi.org/10.3390/chemistry6020015
7. Horcajada, P., Serre, C., Vallet-Regí, M., Sebban, M., Taulelle, F., Férey, G.: Metal-organic frameworks as efficient materials for drug delivery. Angew. Chem. Int. Ed. **45**(36), 5974–5978 (2006). https://doi.org/10.1002/anie.200601878
8. Akbar, M.U., et al.: A pH-responsive bi-MIL-88B MOF coated with folic acid-conjugated chitosan as a promising nanocarrier for targeted drug delivery of 5-Fluorouracil. Front. Pharmacol. **14** (2023). https://doi.org/10.3389/fphar.2023.1265440
9. Jun, H., Oh, S., Lee, G., Oh, M.: Enhanced catalytic activity of MOF-74 via providing additional open metal sites for cyanosilylation of aldehydes. Sci. Rep. **12**(1) (2022). https://doi.org/10.1038/s41598-022-18932-z
10. Villajos, J.A., Jagorel, N., Reinsch, S., Emmerling, F.: Increasing exposed metal site accessibility in a Co-MOF-74 material with induced structure-defects. Front. Mater. **6** (2019). https://doi.org/10.3389/fmats.2019.00230

Study on Carbon Emission Accounting and Carbon Reduction Measures for a Secondary Pumping Station in Jiashi County, Xinjiang

Wan-Ying Xie, Hong-Xin Zhao, Xin-Ran Li, Xi-Dong Wu, Jia-Rui Zhuang, Qiu-Ying Zhang, Zhuang Ma, Teng Chen, and Shang-Ye Yang

Abstract China is highly committed to tackling climate change caused by carbon emissions to complete the 3060 goals. However, the work of carbon emission verification and accounting in urban water system needs to be improved. In this study, the carbon emission of a secondary pumping station in Jiashi County is calculated. In order to meet the demand of annual water supply (601.92×10^4 m^3), the annual electricity consumption of the pumping station was 108.2448×10^4 kWh, and the annual carbon emission was 81.0756×10^4 kg CO_2, indicating a severe situation of carbon emission reduction. Two energy-saving and carbon reduction schemes are proposed: the fixed-frequency pump is optimized as a variable frequency speed regulating pump (Scheme 1), and on this basis, the three-phase asynchronous motor (TPAMs) is optimized as a permanent magnet synchronous motor (Scheme 2). The comprehensive analysis shows that Scheme 1 is expected to reduce the annual electricity consumption (E_d) and annual carbon emissions (E_{GHG_CO2}) of the pumping station by about 11.55%, and reduce the carbon emission intensity (CES_d) of 0.0456 kg CO_2-eq/m^3 when transporting each ton of water. The variable frequency speed regulating pumps (VFSRPs) is expected to achieve the double goals of energy saving and carbon reduction of the secondary pumping station in Jiashi County. The comparative analysis shows that the energy saving and carbon reduction effects of scheme 1 and scheme 2 are similar, but Scheme 1 has lower cost and is more feasible.

Keywords Secondary pumping station · Carbon emission accounting · E_d · E_{GHG_CO2} · CES_d · VFSRPs

W.-Y. Xie · H.-X. Zhao · X.-R. Li · X.-D. Wu · J.-R. Zhuang · Q.-Y. Zhang · Z. Ma · T. Chen · S.-Y. Yang (✉)
College of Civil Engineering and Architecture, Xinjiang University, Urumqi, China
e-mail: yangshangye@xju.edu.cn

Y. Zhu (ed.), *Proceedings of the 4th International Symposium in Environmental Science and Industrial Ecology*, https://doi.org/10.1007/978-981-96-1578-0_8

1 Introduction

Climate change is a key global concern, and reducing carbon emissions has become an international consensus. The United Nations Framework Convention on Climate Change (UNFCCC) and the Kyoto Protocol identify six greenhouse gases (CO_2, CH_4, N_2O, HFCs, PFCs, and SF_6) caused by human activities that require control and deserve special attention [1]. In recent years, many scholars both domestically and internationally have begun focusing on carbon emission research, including carbon emission accounting across various industries [2], such as electricity [3], coal [4], construction [5], chemicals [6], non-ferrous metals [7], papermaking [8], steel [9], and petrochemicals [10]. Additionally, numerous studies have focused on the establishment of carbon emission prediction-models [11] and the development of measures such as carbon reduction [12], carbon capture [13], and enhancing carbon sink potential [14]. China places great emphasis on addressing climate change caused by carbon emissions. In 2020, President Xi proposed that China would strive to peak its carbon emissions by 2030 and achieve carbon neutrality by 2060, known as the 3060 Goals [15]. As China accounts for 28.8% of the world's total energy-related carbon emissions [16], it plays a pivotal role in realizing global carbon peaking and neutrality. Subsequently, the Central Committee and the State Council introduced relevant policies [17] to support carbon reduction efforts across all sectors. In 2022, the Ministry of Ecology and Environment, along with several other departments, jointly issued the "Implementation Plan for Synergistic Pollution Reduction and Carbon Emission Mitigation" [18], which requires the coordination of carbon peaking, carbon neutrality, and ecological environmental protection efforts, as well as the coordination of emission reduction requirements across water, air, soil, solid waste, and greenhouse gas sectors. It also emphasizes promoting energy conservation and consumption reduction in sewage treatment plants, conducting carbon emission calculations for urban sewage treatment and resource utilization, and optimizing energy consumption and carbon emission management in sewage treatment facilities. Currently, the carbon emission accounting in China's water supply and drainage industry is not systematic, comprehensive, or thorough.

As a critical infrastructure for ensuring the normal life of residents and social development, the urban water system has been continuously developing and expanding since the reform and opening-up policy. With the continuous improvement of facilities in water supply, drainage, and water treatment plants, the capacity for water supply and drainage has also been continuously enhanced. However, it still faces significant challenges. This is because urban water supply consumes a large amount of electric energy, accounting for about 1.5% of the total electricity consumption in the country. Moreover, the efficiency of urban sewage treatment systems is relatively low, requiring high electric energy and chemical consumption to meet effluent standards, which leads to substantial greenhouse gas emissions [19]. To achieve the 3060 Goals early, the urban water system needs to take action as soon as possible. Carbon emission accounting and verification are the first step for the urban water industry to move towards a low-carbon future. In particular, carbon emissions in urban water systems are related to multiple factors such as treatment processes [20], influent water quality [21], effluent requirements [22], and water intake and transmission facilities [23]. The complexity of urban water systems makes carbon emission accounting and verification challenging. Carbon emission verification in urban water systems mainly includes four aspects: municipal water supply systems [23], sewage systems [24], reclaimed water systems [25], and rainwater systems [26]. Municipal water supply systems mainly consist of three key components: water intake (including intake facilities and pumping stations), water treatment (including treatment equipment and water treatment plants), and water transmission and distribution (including secondary pumping stations and transmission pipelines). Among them, raw water is transported from the water source to water purification structures or treatment plants during the water intake process, and the treated water needs to be transported to the pipe network through secondary pumping stations, which fully demonstrates that water pumps are very important and indispensable in the water supply system. The municipal water supply system uses a large number of water pumps, with electric motor pumps being the most common. When current passes through the motor coils, it generates a magnetic field, which produces torque to rotate the motor. The mechanical energy converts the power generated by the motor into the energy for water flow, achieving water pressure boosting and transmission. The electric energy consumed by electric motor pumps and the greenhouse gas emissions resulting from electricity purchases are considered indirect emissions.

Currently, there is relatively little research on carbon emission accounting for pumping stations in China's urban water systems, making it necessary to conduct accounting research on carbon emissions from pumping stations in water supply systems. In particular, the boundaries for carbon emission accounting for water pumps and pumping stations in water supply systems are not yet clear. Coupled with significant differences in energy intensity across different regions, there is a lack of accounting and statistical analysis for data such as carbon emissions generated by motors and water pumps. This study conducted research on a secondary pumping station in Jiashi County, Xinjiang. In the context where effective data on electricity consumption activities could not be obtained, an alternative assessment and quantification method based on the operational data of the pumping station was developed. Additionally, a comparative analysis was conducted to assess the impact of energy-saving measures on the carbon reduction efficiency of the pumping station, and the feasibility of these measures was evaluated through an economic cost analysis. It is hoped that this study will provide theoretical support and engineering guidance for carbon reduction efforts in China's urban water supply systems.

2 Research Contents and Methods

2.1 Research Contents

According to the policy documents issued by China regarding energy conservation, emission reduction, carbon peaking, and carbon neutrality, as well as the national requirements for various industries to actively and steadily promote carbon peaking and carbon neutrality, this project takes a secondary pumping station in Jiashi County (Kashgar Prefecture, Xinjiang) as the research object to carry out the following studies:

(1) Investigate the operation data of the pumping station, including operation schemes, pump types, and pump operation parameters (such as flow rate, head, efficiency, etc.), to calculate the electricity consumption of the pumping station. Account for the electricity consumption and carbon emissions of the pumping station to assess the current carbon emissions situation. Determine the daily maximum water consumption variation curve and average daily water consumption of users.

(2) This study proposes two energy-saving and emission reduction schemes. Scheme 1 involves optimizing centrifugal pumps to variable-frequency speed-regulating pumps. Scheme 2 involves the combination of VFSRPs and permanent magnet synchronous motor (PMSMs). Finally, compare and analyze the impacts of these two schemes on the energy consumption, CO_2 emissions, and economic costs of the pumping station operation.

2.2 Research Methods

Calculation of Water Pump Electricity Consumption. Generally, centrifugal pumps and axial flow pumps are commonly used in water supply and drainage projects in urban areas and industrial enterprises. The pumps used in a secondary pumping station in Jiashi County in this study are centrifugal pumps. These pumps belong to the category of vane pumps, and their performance is related to factors such as flow rate, head, shaft power, and efficiency. Therefore, the calculation of electricity consumption for the water pump is shown in formula (1) [27]:

$$W = \frac{\rho g Q H}{1000\,\eta_1 \eta_2} t \tag{1}$$

where, W is the electricity consumption of pump, kWh; ρ is the density of water, 1000 kg/m^3; g is gravitational acceleration, 9.81 m/s^2; Q is the flow rate, m^3/s; t is the pump operation time, h; η_1 is the pump efficiency, %; η_2 is the efficiency of electric motor, %; H is the pump head.

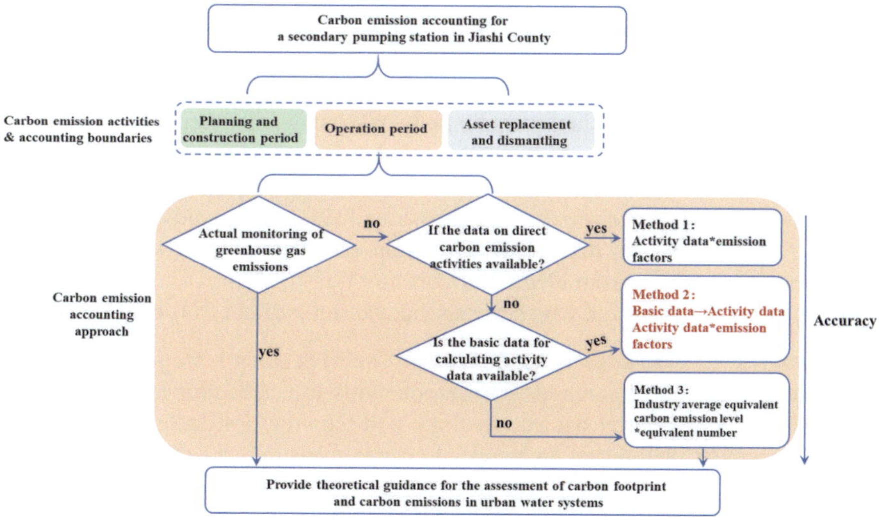

Fig. 1 The thought diagram of carbon emission accounting for the secondary pump station in Jiashi County, Xinjiang

Calculation of Carbon Emissions. The carbon emission accounting approach for a secondary pumping station in Jiashi County is illustrated in Fig. 1, which includes identifying carbon emission activities, defining accounting boundaries, selecting accounting methods, and calculating carbon emissions. Carbon emission accounting refers to the assessment and calculation of greenhouse gas emissions across different industries from national or regional levels, corporate or organizational levels, product or service levels, and individual levels [27]. The carbon emissions of urban water systems basically cover the aforementioned aspects. When accounting for the carbon emission intensity of a secondary pumping station in Jiashi County, Xinjiang, its water supply target and water quality requirements are relatively stable, making it a carbon emission accounting task at the product or service level.

Boundaries of Carbon Emission Accounting. This secondary pumping station is divided into three phases: design and construction, operation, and asset replacement and dismantling. The design phase in the design and construction period, as well as the asset replacement and dismantling phase, are relatively short, resulting in lower overall carbon emissions from the pumping station. Compared to the construction phase in the design and construction period, the operation phase spans a longer time frame, ranging from several years to over a decade. Therefore, the boundary for carbon emission accounting for the secondary pumping station in this study is the annual carbon emission intensity during its operation phase.

Carbon Emission Activities. Referring to the Revised Guidelines for National Greenhouse Gas Inventories issued by the Intergovernmental Panel on Climate Change (IPCC) and the Research on China's Greenhouse Gas Inventor, all greenhouse gases generated from carbon emission activities within the accounting boundaries should be calculated. Carbon emission activities in urban water systems include two forms, direct and indirect carbon emissions [27]. Among them, greenhouse gases directly produced by microbial biochemical reactions or fossil fuel combustion are considered direct carbon emissions. Otherwise, greenhouse gases generated by electricity consumption, consumption of various chemical agents or materials, or transportation are considered as indirect carbon emissions [19]. In this study, the secondary pumping station does not involve direct carbon emissions. Only the carbon emission activities resulting from electricity consumption by pumps and motors in the pumping station are studied, which means calculating indirect CO_2 emissions.

Calculation of CO_2 Emissions by Using the Emission Factor Method. The IPCC [28] recommends three methods for carbon emission accounting: the measurement method, the material balance method, and the emission factor method [29]. The measurement method involves actual on-site measurement of greenhouse gas emissions and is suitable for accounting direct carbon emissions such as CH_4 and N_2O. The material balance method calculates based on the principle of CO_2 input–output balance. For example, there is research that calculates the net production of CO_2 through the material balance of carbon absorption and emission by photosynthetic bacteria. Compared with the first two methods, the emission factor method is currently a widely used carbon emission accounting method and has been extensively applied in the field of carbon emission accounting for urban water systems both domestically and internationally. In this study, the main carbon emission activity during the operation of the secondary pumping station is indirect carbon emissions generated by electricity consumption of pumps and motors. Both the measurement method and the material balance method are not applicable, making the emission factor method more suitable. Its principle is to calculate carbon emission activity data based on the operation data of pumps and motors, and then calculate carbon emissions using the emission factor method (Method 2 in Fig. 1). The formula for calculating carbon emissions [30] is shown in Eq. (2). The carbon emission intensity resulting from electricity consumption at the secondary pumping station can be accounted for based on the actual total electricity consumption (Eq. (3)).

$$\mathrm{E_{GHGi}} = \mathrm{AD_i} \times \mathrm{EF_i} \tag{2}$$

where, i is the greenhouse gas; E_{GHGi} is the emissions of the ith greenhouse gas; AD is the activity level data; EF is the emission factors.

$$\mathrm{CES_d} = (\mathrm{E_d} \cdot \mathrm{EF_d})/Q' \tag{3}$$

where, CES_d is the carbon emission intensity generated by purchasing electricity for operational consumption, kg CO_2-eq/m^3, and only CO_2 is present in this study; E_d is the total electricity consumption during one year, kWh/a; EF_d is the carbon emission factor in Xinjiang region and is 0.749 kg CO_2/kWh in this study, according to the Research on Carbon Dioxide Emission Factors of China Regional Power Grid (2023); Q' is the total volume of treated water per year, m^3/a.

3 Carbon Emission Assessment of Secondary Pumping Station

3.1 Operation Status of Secondary Pump Station

The annual water supply volume of this pumping station in 2023 is 601.92 × 104 m^3, with a maximum daily water supply volume of 19,500 m^3/d. The layout of the pumping station units is arranged horizontally, as shown in Fig. 2. There are three sets of pumps: Pump A is pipeline centrifugal pump, Pump B is single-stage vertical centrifugal pump, and Pump C is double-suction pump. Each set consists of two identical pumps, which operating in a mode of one in use and one in standby. Each type of pump operates for 12 h, alternating in the sequence of Pump A to B, then to C. When a pump requires maintenance, the standby pump in that set is activated.

Pump A is an LSG250-400 pipeline centrifugal pump (Shanghai Xizhiquan Pump and Valve Co., Ltd.), Pump B is an SLS300-450 single-stage centrifugal pump (Shanghai Liancheng Co., Ltd.), and Pump C is a DFSS300-8/4A double-suction pump (Shanghai East Pump Co., Ltd.). Motor A is a YE2-315S TPAMs (Shanghai General Electric Motor Manufacturing Co., Ltd.), Motor B is a YE2-315 M-4 high-efficiency TPAMs (Shanghai Tongcheng Electric Motor Co., Ltd.), and Motor C is a YE2-315L1-4 high-efficiency TPAMs (Hengshui Electric Motor Co., Ltd.) (Tables 1 and 2).

Due to factors such as living habits, living standards, housing sanitation facilities, and social production activities, domestic water consumption varies every hour. The variation curve of water consumption on the peak day for residents in Jiashi County is illustrated in Fig. 3. The daily variation coefficient of water consumption, K_d is

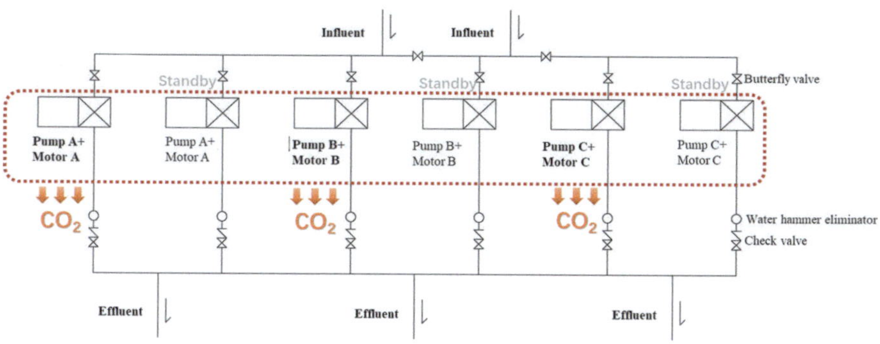

Fig. 2 The layout diagram of one secondary pump station in Jiashi County, Xinjiang

Table 1 Parameter table of water pump used in the secondary pump station

Pump	Model	Q (m^3/h)	H (m)	Power (kW)	Rotation speed (r/min)	η_1 (%)	Operation time per cycle (h)
Pump A	ISG250-400	600	50	110	1450	80	12
Pump B	SLS300-450	720	50	132	1480	83	12
Pump C	DFSS300-8/ 4A	770	53	160	1480	82	12

Table 2 Parameter table of electric motor used in the secondary pump station

Motor	Model	Rated power (m^3/h)	Rated voltage (m)	Rated current (kW)	Operating frequency (r/min)	Rotation speed (r/ min)	H_2 (%)
Motor A	YE2-315S	110	380	201	50	1480	94.5
Motor B	YE2-315 M-4	132	380	237.9	50	1480	94.7
Motor C	YE2-315L1-4	160	380	285	50	1480	94.9

1.17. The Maximum Hourly Water Consumption accounts for 6% of the Maximum Daily Water Consumption. The peak periods concentrated from 8:00 to 10:00 in the morning and from 16:00 to 21:00 in the afternoon.

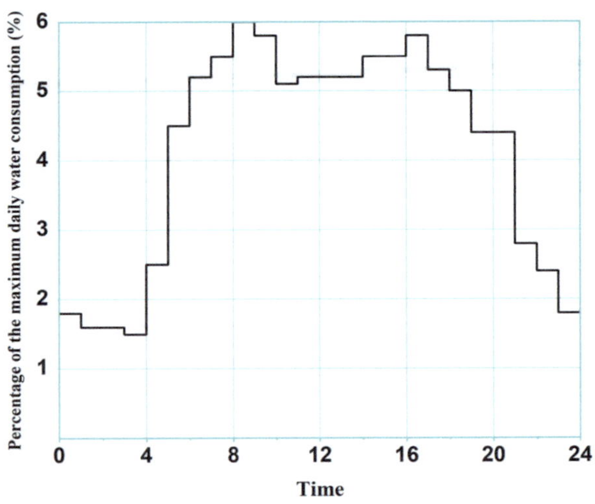

Fig. 3 The highest daily water consumption variation curve of the secondary pumping station in Jiashi County, Xinjiang

3.2 *Electricity Consumption and Carbon Emissions of the Secondary Pumping Station*

The electricity consumption and carbon emissions accounting for a secondary pumping station in Jiashi County in 2023 are shown in Table 3. Based on the monthly operation frequency (20 times per month) of Pump A, B, and C, the average monthly water supply of this secondary pumping station is calculated to be 50.16×10^4 m^3, with an annual water supply of 601.92 × 104 m^3. According to the efficiency of the pumps and motors, the monthly electricity consumption of Group A (Pump A + Motor A), Group A (Pump B + Motor B), and Group C (Pump C + Motor C) is calculated to be 25,952, 29,954, and 34,298 kWh, respectively. Using Eq. (2), the monthly CO_2 emissions from the three pumps are determined to be 19,438, 22,435, and 25,689 kg CO_2, respectively. Assuming no pump maintenance or repairs during the month, the monthly electricity consumption of this secondary pumping station is 90204 kWh, with monthly carbon emissions of 67,563 kWh. It is estimated that the annual electricity consumption (E_d) of this secondary pumping station in 2023 is 108.2448×10^4 kWh, and the annual carbon emissions (E_{GHG_CO2}) are 81.0756×10^4 kg CO_2.

Table 3 Summary table of annual electricity consumption and annual carbon emissions of the secondary pumping station

Annual water supply of the pumping station(m^3)	Group	Q (m^3/h)	H (m)	η_1 (%)	η_2 (%)	W (kWh)	Carbon emissions (kg CO_2)
601.92×10^4	Group A	600	50	80	94.5	25,952	19,438
	Group B	720	50	83	94.7	29,954	22,435
	Group C	770	53	82	94.9	34,298	25,689
Total for a single month						90,204	67,562
Total for the whole year						108.2448×10^4	81.0744×10^4

4 Impact of Energy-Saving Schemes on Carbon Reduction Effectiveness and Economic Costs at the Secondary Pumping Station

4.1 Energy-Saving Schemes

This study proposes two energy-saving and carbon-reduction schemes in Table 4 for the secondary pumping station. Scheme 1 is keeping the original motors unchanged and adjusting the original fixed-frequency pumps to VFSRPs, specifically Pump A to D, Pump B to E, Pump C to F. Each pump operates for 12 h before switching to the next one. The VFSRPs are controlled by a control box that adjusts the motor frequency to control the rotational speed, achieving automatic adjustment of flow rate and head, and avoiding unnecessary waste of electrical energy. Moreover, on the basis of Scheme 1, Scheme 2 is replacing the current motors with more efficient models, specifically Pump D + Motor D (Group D), Pump E + Motor E (Group E), Pump F + Motor F (Group F). The Motor D, E, F are PMSMs. PMSMs have numerous advantages, including simple structure, reliable operation, low losses, high efficiency, and flexible shapes and sizes. Additionally, PMSMs do not require reactive excitation current, which reduces stator current and stator resistance losses, significantly improving the power factor.

With the goal of meeting the maximum daily water consumption in Jiashi County, this study simplifies Fig. 3 into Fig. 4 based on the convenience of carbon emission calculation and the operability for workers during operation. According to Fig. 3, daily water supply is divided into six periods, and the principles of simplification are outlined as follows. Period 1 (21:00–4:00) is the off-peak period for water use when residents are resting, and water demand is low. In Period 2 (4:00–5:00), there is a slight increase in water use, which may be due to early morning farming or night shift production requiring water. During Period 3 (5:00–8:00), water use increases significantly, due to increased demand from daily activities, such as schooling, work, and farming. In Period 4 (8:00–10:00), most residents start getting up, washing up, or going to work in the fields, and all production activities begin, leading to the first peak in water demand. Period 5 (10:00–19:00) covers daytime production and life activities, with high water demand. Period 6 (19:00–21:00) is the rest time after production activities, with water demand for activities such as cooking, showering, and laundry. The cumulative water consumption in these six periods from Fig. 3, is then averaged out across the respective time periods to obtain Fig. 4.

Table 4 Parameter table of variable frequency speed control pump used in this study

Group		Q (m^3/h)	H (m)	Power and η_1 of pump (kW)/(%)	Power and η_1 of motor (kW)/(%)	Information
Group D	TD250-50/4 + DL355M-132-7R52W	600	50	132/94	132/95.8	Pump body: cast iron HT200; Impeller: cast iron/stainless steel; Coupling: cast iron; Pump shaft: stainless steel; Mechanical seal: graphite/ tungsten carbide; Air release assembly: stainless steel; The water pump is equipped with a variable frequency speed control box
Group E	TD300-44/4 + DL355M-132-7R52W	720	50	160/94	132/95.8	
Group F	TD300-55/4 + DL355M-132-7R52W	770	53	200/94	132/95.8	

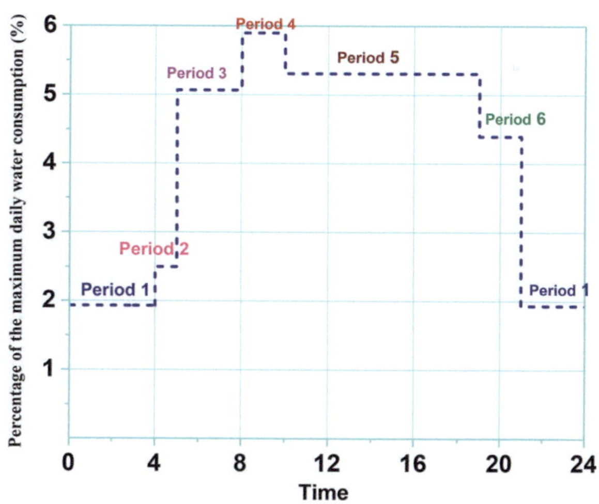

Fig. 4 The simplified water consumption variation curve according to different water supply periods

4.2 The Impact of Variable Frequency Speed Control Pump on Energy Saving and Carbon Reduction in Secondary Pumping Stations

It takes 3 days for pumps D, E, and F to complete two sets of small cycles. During these 3 days, each of the pumps D, E, and F operates once within the period from 0:00 to 12:00. In one month, each of the pumps D, E, and F operates 10 times during the period from 0:00 to 12:00 and another 10 times during the period from 12:00 to 24:00. With the goal of meeting the maximum daily water supply of 19,500 m^3/d, to calculate the electricity consumption and carbon emissions for Scheme 1 during the maximum daily water supply, and the results are shown in Table 5. Subsequently, use the K_d to convert these values into the average daily electricity consumption and carbon emissions, under average daily water supply.

According to Table 5, based on meeting the maximum daily water supply of 19,500 m^3/d, the monthly electricity consumption of pumps D, E, and F is 30,577, 30,492, and 32,284 kWh, respectively, with corresponding monthly carbon emissions of 22,902, 22,839, and 24,181 kg CO_2. This reveals that the monthly water supply volume of this secondary pumping station is 587,640 m^3. While meeting the water supply demand, the monthly electricity consumption and carbon emissions are 93,353 kWh and 69,922 kg CO_2, respectively. During daily operation, the regulating structures in the water transmission and distribution network can adjust the difference between the water supply and demand of the secondary pumping station. Therefore, the Q of the secondary pumping station is basically maintained around the average daily water consumption. To compare the results with those in Sect. 3.2 under the same water supply volume baseline, the monthly electricity consumption and carbon

Table 5 Estimation table of electricity consumption and carbon emissions of the secondary pumping station under the highest daily water supply, when operating under the Scheme 1

Period	Percentage of highest daily water usage (%)	Q (m^3/h)	Monthly electricity consumption (kWh)			Monthly CO_2 emissions (kg CO_2)		
			Pump D + Motor A	Pump E + Motor B	Pump F + Motor C	Pump D + Motor A	Pump E + Motor B	Pump F + Motor C
4:00–5:00	2.50	500	787	765	840	589	573	629
5:00–8:00	5.07	1014	4666	4656	4925	3495	3487	3689
8:00–10:00	5.90	1180	3620	3612	3821	2711	2706	2862
10:00–19:00	5.31	1062	14,660	14,629	15,474	10,981	10,957	11,590
19:00–21:00	4.40	880	2700	2694	2849	2022	2018	2134
21:00–4:00	1.93	386	4144	4136	4375	3104	3098	3277
Total for a single month by one pump			30,577	30,492	32,284	22,902	22,839	24,181
Total for a single month by three pumps			93,353			69,922		

emissions in Table 5 need to be adjusted based on the K_d. Thus, under Scheme 1, the monthly water supply volume of the pumping station is 502,256 m^3, and the annual water supply volume is 6,027,072 m^3. The monthly electricity consumption and carbon emissions of the pumping station are 79,789 kWh and 59,762 kg CO_2, respectively, resulting in E_d and E_{GHG_CO2} of 957,467 kWh and 717,149 kg CO_2, respectively.

In Scheme 2, the efficiency of the PMSM is 95.8%. Based on this, the estimated electricity consumption and carbon emissions of the secondary pumping station under maximum daily water supply are calculated and presented in Table 6. Both Scheme 2 and Scheme 1 have the same annual water supply capacity, meeting the water demand of Jiashi County. At this point, the monthly electricity consumption of Groups D, E, and F is 30,143, 30,143, and 31,951 kWh, respectively, with corresponding monthly carbon emissions of 22,577, 22,577, and 23,931 kg CO_2. Therefore, the monthly and annual electricity consumption of the secondary pumping station is 92,237 kWh and 1,106,844 kWh, respectively, with monthly and annual carbon emissions of 69,085 kg CO_2 and 829,020 kg CO_2, respectively. After adjusting via K_d, when the pumping station meets a monthly water supply volume of 502,256 m^3, the monthly electricity consumption of Scheme 2 is 78,835 kWh, and the monthly carbon emissions are 59,047 kg CO_2. Thus, according the pump station operated under Scheme 2, E_d is 946,020 kWh, and E_{GHG_CO2} is 708,564 kg CO_2.

Table 6 Estimation table of electricity consumption and carbon emissions of the secondary pumping station under the highest daily water supply, when operating under the Scheme 2

Period	Percentage of highest daily water usage (%)	Q (m^3/h)	Monthly electricity consumption (kWh)			Monthly CO_2 emissions (kg CO_2)		
			Pump D + Motor D	Pump E + Motor E	Pump F + Motor F	Pump D + Motor D	Pump E + Motor E	Pump F + Motor F
4:00–5:00	2.50	500	757	757	802	567	567	601
5:00–8:00	5.07	1014	4603	4603	4879	3447	3447	3654
8:00–10:00	5.90	1180	3571	3571	3785	2674	2674	2835
10:00–19:00	5.31	1062	14,461	14,461	15,329	10,832	10,832	11,481
19:00–21:00	4.40	880	2663	2663	2823	1995	1995	2114
21:00–4:00	1.93	386	4088	4088	4333	3062	3062	3246
Total for a single month by one pump			30,143	30,143	31,951	22,577	22,577	23,931
Total for a single month by three pumps			92,237			69,085		

4.3 Comparative Analysis of Carbon Reduction Effect and Economic Cost of Energy-Saving Schemes

For the pumping station in Jiashi County, a comparison was made between the current operation scheme and two energy-saving schemes, including annual water supply, annual electricity consumption, annual carbon emissions, and annual carbon emission intensity, as shown in Fig. 5. According to Fig. 5a, the annual water supply of the two energy-saving schemes is close to that of the current scheme, indicating that two schemes can meet the domestic water demand of Jiashi County residents. Scheme 1 reduces the E_d of the pumping station from 108.2448×10^4 to 95.7467×10^4 kWh and reduces E_{GHG_CO2} from 81.0744×10^4 to 71.7149×10^4 kg CO_2. The E_{GHG_CO2} of Scheme 1 are 88.46% of the current scheme, indicating that the VFSRPs can achieve the dual goals of energy saving and carbon reduction in the secondary pumping station. Comparative analysis reveals that Scheme 2 can reduce the E_{GHG_CO2} of the pumping station to 87.40%, indicating that the synergistic effect of PMSMs in energy saving and carbon reduction is less significant than that of VFSRPs.

Additionally, the carbon emission intensity (CES_d) of the pumping station under the three schemes was calculated according to Eq. (3), and the results are shown in Fig. 5b. When the pumping station operates under the current scheme, Scheme 1, and Scheme 2, the CES_d generated per ton of water transported is 0.1347, 0.1190, and 0.1176 kg CO_2-eq/m^3, respectively. Scheme 1 and Scheme 2 reduce the CES_d of the pumping station by 11.66% and 12.71%, respectively, indicating that the carbon emission intensity of the secondary pumping station is the highest under the current scheme, and the two energy-saving schemes have significant carbon reduction effects.

An estimation of the economic costs under the three operation schemes of the secondary pumping station in Jiashi County is conducted. According to Table 7, under the pump layout scheme of one for use and one for standby, the pump cost of the current scheme is 171,200 yuan, and the cost of the TPAMs is 90,200 yuan, with a total cost of 261,400 yuan. According to Table 8, the total cost of Scheme 1 is 630,800 yuan, which is 2.4 times that of the current scheme. The total cost of Scheme 2 is 735,700 yuan, which is 2.8 times that of the current scheme. Through comprehensive comparison, it is concluded that Scheme 1, with an increase of 1.4 times in investment cost, is expected to reduce electricity consumption and carbon emissions by approximately 11.55% annually and decrease carbon emission intensity by 0.0157 kg CO_2-eq/m^3. Therefore, Scheme 1 is more economical and easier to implement.

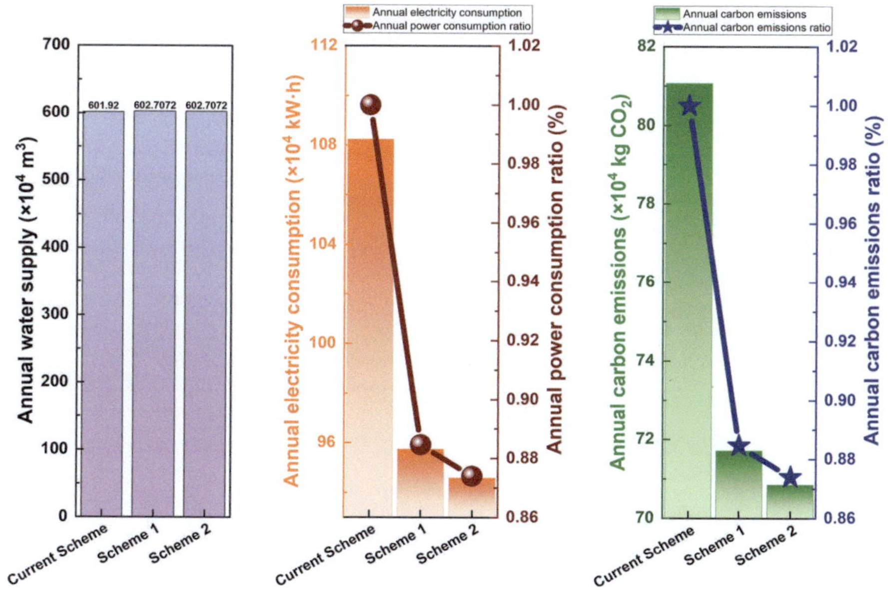

(A)

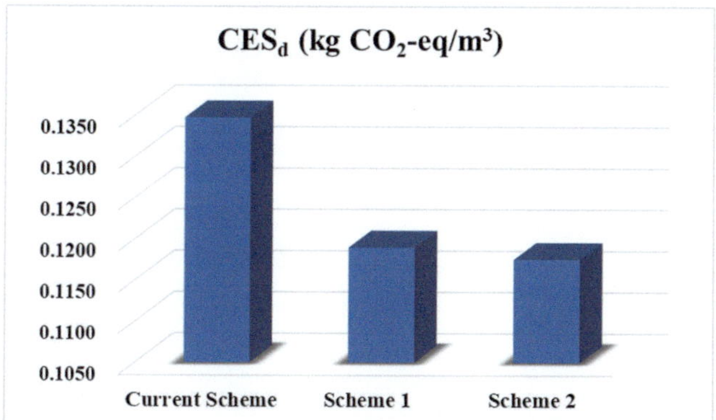

(B)

Fig. 5 The impact of energy-saving and emission reduction plans on the annual water supply, annual electricity consumption, annual carbon emissions, and CES_d of the secondary pumping station in Jiashi County, Xinjiang

Table 7 Economic cost analysis table for the current operating plan of the pumping station

Pump models	Operation scheme	Number of devices	Costs
Pump A (ISG250-400) + Motor A (YE2-315S-4)	One for use and one for standby	2/2	28,300 yuan per pump, 13,000 yuan per motor;
Pump B (SLS300-450) + Motor B (YE2-315 M-4)		2/2	30,300 yuan per pump, 16,500 yuan per motor;
Pump C (DFSS300-8/4A) + Motor C (YE2-315L1-4)		2/2	27,000 yuan per pump, 15,600 yuan per motor;
Total costs for pumps or motors			171,200 yuan/90,200 yuan
Total costs for the current operation scheme			261,400 yuan

Table 8 Economic cost analysis table for the Scheme 1 of the pumping station

Pump models	Operation scheme	Number of devices	Costs (yuan per device)
Pump A (ISG250-400)	One for use and one for standby. Pumps D-F are commonly used pumps, and pumps A-C are standby pumps	1	28,300
Pump B (SLS300-450)		1	30,300
Pump C (DFSS300-8/4A)		1	27,000
Pump D (TD250-50/4)		1	130,000
Pump E (TD300-44/4)		1	155,000
Pump F (TD300-55/4)		1	170,000
Total costs for pumps in Scheme 1 or Scheme 2			540,600
Motor models	Operation scheme	Number of devices	Costs (yuan per device)
Motor A (YE2-315S-4)	One for use and one for standby	2	13,000
Motor B (YE2-315 M-4)		2	16,500
Motor C (YE2-315L1-4)		2	15,600
Total costs for motors in Scheme 1			90,200

(continued)

Table 8 (continued)

Motor models	Operation scheme	Number of devices	Costs (yuan per device)
Motor A (YE2-315S-4)	One for use and one for standby. Motors D-F are commonly used pumps, and motors A-C are standby pumps	1	13,000
Motor B (YE2-315 M-4)		1	16,500
Motor C (YE2-315L1-4)		1	15,600
Motor D (YE2-315S-4)		1	50,000
Motor E (YE2-315 M-4)		1	50,000
Motor F (YE2-315 M-4)		1	50,000
Total costs for motors in Scheme 2			195,100

5 Conclusion

Based on the survey of a secondary pumping station in Jiashi County, the emission factor method was used to calculate the E_d and E_{GHG_CO2} of the current operation scheme of the pumping station. In response to the high annual carbon emissions of the pumping station, two energy-saving schemes were proposed. A comprehensive analysis of their carbon reduction effects and economic costs led to the following conclusions:

(1) The secondary pumping station includes three types of centrifugal pumps: pipeline centrifugal pumps, single-stage vertical centrifugal pumps, and double-suction pumps. These pumps are paired with TPAMs to form the water pump station. The pumping station adopts an alternating operation mode of these three types of pumps to supply water to the urban residents of Jiashi County. While meeting the annual water supply demand of 601.92×10^4 m^3, the E_d of the secondary pumping station is 108.2448×10^4 kWh, and the E_{GHG_CO2} is 81.0756×10^4 kg CO_2, indicating high carbon emissions and a severe situation.

(2) To address the issue of high carbon emissions at the pumping station, two energy-saving schemes are proposed. While meeting the monthly water supply demand of 502,256 m^3, under Scheme 1, the E_d and E_{GHG_CO2} of pumping station are respectively 95.7467×10^4 kWh and 71.7149×10^4 kg CO_2. By replacing the TPAMs to the more efficient PMSMs, when pumping station operates under Scheme 2, the E_d and E_{GHG_CO2} are separately reduced to 94.602×10^4 kWh and 70.8564×10^4 kg CO_2. This result reflects that Scheme 1 is expected to reduce the E_d and E_{GHG_CO2} of the secondary pumping station by approximately 11.55% annually, while the reduction of Scheme 2 is 12.71%.

(3) Through CES_d and economic cost analysis, it is evident that with an increase of 1.4 times in cost to decrease the CES_d by 0.0157 kg CO_2-eq/m^3 for each ton of water transported. The aforementioned research results indicate that adjusting fixed-speed pumps to VFSRPs is likely to achieve the dual goals of energy conservation and carbon reduction in secondary pumping stations. However, after adding PMSMs, the energy-saving and carbon-reduction effects of Scheme 2 are not as significant as those of Scheme 1.

Acknowledgements This research was funded by the Tianchi Talent Program—Young Doctoral program (Xinjiang, China), and really appreciate for the support from Xinjiang University and Department of Human Resources and Social Security, Xinjiang Uyghur Autonomous Region.

References

1. Liu, J., Wang, G., Zhang, J., et al.: Introduction of UNFCCC and Kyoto Protocol/Research Center for Climate change, Ministry of water resources. China Water Resour. **2**, 65–80 (2008)
2. Jin, X., Zhao, M., Yang, J., et al.: Methodology for greenhouse gas inventory accounting at the county level and analysis of emission characteristics—a case study of Huairou district, Beijing. Environ. Prot. Sci. https://doi.org/10.16803/j.cnki.issn.1004−6216.202310019
3. Ding, N., Guo, P., Xi, Y., et al.: Low-carbon development in power systems based on carbon emission analysis models: a comprehensive review. Sustain. Energy Technol. Assess. **65**, 103774 (2024)
4. Su, H., Zhang, C., Wang, C., et al.: Analysis of the impact path of green technology spillovers on the carbon emission intensity of the coal industry in Shanxi, Shaanxi and Inner Mongolia. China Coal **50**(3), 27–38 (2024)
5. Wen, L., Yan, X.: Research on the distribution characteristics and regional impact of carbon emissions in China's construction industry. J. Eng. Manage. **38**(2), 7–12 (2024)
6. Deepti, A., Kelvin, A., Aeyed, N.A., et al.: Carbon emissions and decarbonisation: the role and relevance of fermentation industry in chemical sector. Chem. Eng. J. **475**, 146308 (2023)
7. Cao, Y., Guo, L., Qu, Y.: Evaluating the dynamic effects of mitigation instruments on CO_2 emissions in China's nonferrous metal industry: a vector autoregression analysis. Sci. Total. Environ. **853**, 158409 (2022)
8. Song, J., Cheng, Z., Liu, L., et al.: Analysis on greenhouse gas emission standards and accounting methods in pulp and paper industry. China Pulp Paper **42**(8), 11–18 (2023)
9. Zhu, T., Liu, X., Wang, X., et al.: Technical development and prospect for collaborative reduction of pollution and carbon emissions from iron and steel industry in China. Engineering **31**, 37–49 (2023)
10. Zhang, P., Li, W.: Research on the implementation path of carbon reduction in China's petrochemical industry. Price Theory Pract. **12**, 116–121 (2023)
11. Daseung, K., Dongwook, K., Nosuk, P., et al.: Estimation of CO_2 emission from water treatment plant e model development and application. J. Environ. Manage. **13**, 74–81 (2013)
12. Cao, J.X., Zhang, J., Chen, Y., et al.: Current status, future prediction and offset potential of fossil fuel CO_2 emissions in China. J. Clean. Prod. **426**, 139207 (2023)

13. Gimun, G., David, I.K., Jungwon, K., et al.: An integrated system for CO_2 capture and water treatment by forward osmosis driven by an amine-based draw solution. J. Membr. Sci. **581**, 9–17 (2019)
14. Zeng, P., Liang, L., Zeng, F.: Differences in carbon sink potential between urban agglomerations are decreasing: evidence from China. Ecol. Ind. **163**, 112062 (2024)
15. Hu, A.: China's goal of achieving carbon peak by 2030 and its main approaches. J. Beijing Univ. Technol. (Soc. Sci. Ed.) **21**(3), 1–15 (2021)
16. Chai, Q.: Policy outlook on China's new goal of peaking carbon dioxide emissions and vision of carbon neutrality. World Environ. **1**, 20–22 (2021)
17. Options of the Central Committee of the Communist Party of China and the State Council on the complete, accurate and comprehensive implementation of the new development concept to do a good job in carbon peak and carbon neutrality. Resour. Recycl. **10**, 52–56 (2021)
18. China's Ministry of Ecology and Environment, National Development and Reform Commission, Ministry of Industry and Information Technology, Ministry of Housing and Urban-Rural Development, Ministry of Transport, Ministry of Agriculture and Rural Affairs, and National Energy Administration: Implementation plan for synergistic effect of pollution reduction and carbon reduction. Resour. Recycl. **6**, 44–49 (2022)
19. China Urban Water Supply and Drainage Association: Technical Guidelines for Carbon Accounting and Emission Reduction Pathways in China's Urban Water Systems, 1st edn. China Architecture & Building Press, Beijing (2022)
20. Wang, Z.: Efficiency and carbon emission analysis of step-feed multistage AO process in low C/N wastewater treatment. Water Purif. Technol. **42**(s1), 146–153 (2023)
21. Wang, J., Sun, Y., Zhang, X., et al.: Carbon-efficient nutrients removal from real municipal wastewater under conditions of highly variable influent quality and low temperature. Chem. Eng. J. 155268 (2024)
22. Lee, S., Choi, J., Choi, H., et al.: Assessment and optimization of wastewater treatment plant in terms of effluent quality, energy footprint, and greenhouse gas emissions: an integrated modeling approach. Ecotoxicol. Environ. Saf. **283**, 116820 (2024)
23. Liu, S., Qiu, J., Bai, L., et al.: Carbon emissions of water supply systems in China: characteristic, right, right balance and reduction potential assessment. J. Clean. Prod. **428**, 139446 (2023)
24. Liu, Y., Shi, Z., Wen, H., et al.: Research on carbon accounting method of urban sewage system and analysis of typical process carbon reduction path. Water Wastewater Eng. **50**(1), 37–45 (2024)
25. Jiang, T., Duan, X., Cai, Y.: Research on carbon emission behavior in waster saving process and its benefit evaluation method. China Pulp Paper Ind. **44**(8), 1–5 (2023)
26. Liu, J., Wang, J., Ding, X., et al.: Assessing the mitigation of greenhouse gas emissions from a green infrastructure-based urban drainage system. Appl. Energy **278**, 115686 (2020)
27. Qin, H., Yuan, H., et al.: Urban water system and carbon emission, 1st edn. Science Press, Beijing (2022)
28. Singh, A.K., Singh, U., Panigrahi, D.C., et al.: Updated greenhouse gas inventory estimates for Indian underground coal mining based on the 2019 IPCC refinements. iScience **25**(9), 104946 (2022)
29. Li, C., Ma, S., He, X., et al.: Carbon emission calculation and carbon emission reduction path of typical waterworks operation. Water Wastewater Eng. **49**(7), 1–7 (2023)
30. Jiang, X., Wang, D., Yang, Q.: Carbon Emission Accounting Methodology, 1st edn. China Social Sciences Press, Beijing (2021)

Research Progress of Phosphogypsum Pretreatment and Comprehensive Utilization

Jing Tan, Guodong Li, Xinghua Peng, Dahai You, Jin Zhang, Chaohong Zhang, Li Zhang, and Xiuquan Hu

Abstract Phosphogypsum (PG) is the main by-product of phosphoric acid production, and its large amount of storage has caused serious burden to the environment. In view of the serious environmental pollution caused by PG accumulation, it is necessary to speed up the research on PG pretreatment and comprehensive utilization. In this paper, the pretreatment methods and comprehensive utilization status of PG are reviewed, aiming to provide reference for improving the utilization rate of PG as resources. PG contains many impurities, such as phosphorus, fluorine, organic matter, heavy metals and radionuclides, which limit its application in construction, chemical industry, agriculture and environment. Pretreatment methods include physical method, chemical method and heat treatment method. In terms of comprehensive utilization, PG can be used in the construction field to prepare cement, cement retarder and construction gypsum, in the chemical field to prepare ammonium sulfate, high-purity calcium carbonate and calcium sulfate whisker, in the agricultural field as a soil conditioner and preparation of compound fertilizer, and in the environmental field, PG can be used to adsorb pollutants and store CO_2. The comprehensive utilization of PG has broad prospects, but there are still many aspects that need to be made efforts. Future research needs to focus on the development of low-cost, low-energy and environment-friendly PG treatment technology, accelerate the rational utilization of PG resources, and achieve a win–win situation of economic benefits and environmental protection.

Keywords Industrial solid waste · Phosphogypsum · Pretreatment · Impurity removal

J. Tan (✉) · G. Li · X. Peng · D. You · J. Zhang · C. Zhang · L. Zhang · X. Hu
Hubei Institute of Metallurgical Geology (Central South Institute of Metallurgical Geology), Yichang, China
e-mail: tanjing0809@outlook.com

Mineral Materials and Applied Engineering Technology Research Center of Hubei Province, Yichang, China

Yichang Key Laboratory of Comprehensive Utilization of Mineral Resources, Yichang, China

Y. Zhu (ed.), *Proceedings of the 4th International Symposium in Environmental Science and Industrial Ecology*, https://doi.org/10.1007/978-981-96-1578-0_9

1 Introduction

Phosphogypsum (PG) is a by-product in the process of decomposition of phosphate rock by sulfuric acid to produce phosphoric acid. The main component of PG is calcium sulfate dihydrate, which is an important recycled gypsum resource. According to statistics, for every 1 t of phosphoric acid produced, 4–5 t of PG will be by-produced. At present, China's PG stockpile has exceeded 800 million tons, and the annual output of PG has exceeded 80 million tons/year.

PG is a fine powdery solid, acidic, slightly soluble in water, soluble in ammonium salt, acid and glycerin, with a particle diameter of about 10 um and a density of 2.2–2.3 t/m^3. Due to impurities, its surface luster is dim, generally light gray, yellow white or black gray. Due to the different characteristics and production processes of phosphate ores, the impurities in PG are also different, mainly including phosphorus-containing impurities, fluorine-containing impurities and organic impurities, and some PG also contains heavy metals and radioactive impurities. The impurity composition of PG is complex, but the pretreatment and impurity removal technology of PG is immature, resulting in a long-term low utilization rate of PG, and most of them use stacking treatment. However, the accumulation of a large amount of PG not only occupies land and causes pollution to local air, water and other environments, but also affects regional hydrodynamic conditions, leading to changes in groundwater level, and even changes in local natural landscape and ecosystem. In order to solve the environmental pollution caused by PG accumulation, it is necessary to speed up the research on PG pretreatment and comprehensive utilization. Therefore, this paper summarizes the research progress and comprehensive utilization of PG pretreatment methods in order to provide reference for the comprehensive utilization of PG.

2 Pretreatment of PG

In the comprehensive utilization of PG, the impurities contained in it will affect the effect and performance of the product, thus reducing the application value of PG product and limiting its practical application. Therefore, to clarify the characteristics and migration and transformation rules of various impurities in PG, and explore the pretreatment method with high efficiency and low cost, is the premise of realizing the efficient comprehensive utilization of PG.

2.1 Types of Impurities in PG

Impurities in PG mainly include phosphorus impurities, fluorine impurities, organic impurities, heavy metal impurities, radionuclides and so on. Phosphorus impurity and fluorine impurity are the main impurity components, which are caused by the incomplete decomposition of phosphate rock by sulfuric acid. Organic impurities are produced by the organic matter contained in the phosphate ore or the organic reagents added during the process production. Compared with other countries, the content of heavy metals and radioactive elements in phosphate ores in China is low, so the content of heavy metals and radioactive impurities in PG is also low. The main forms of various impurities in PG are shown in Table 1.

The existence of phosphorus impurities can be roughly divided into soluble phosphorus, eutectic phosphorus, insoluble phosphorus three categories. Soluble phosphorus mainly exists in solution in the form of ions, and is easy to combine with calcium ions to form insoluble calcium salts that hinder the hydrolysis of gypsum dihydrate, thus extending the setting time of PG and cement products, loosening the structure and reducing the strength. Eutectic phosphorus exists in the form of co-crystallization, and its effect is similar to that of soluble phosphorus. The properties of insoluble phosphorus are more stable and the distribution is more concentrated. The fluorine impurities can be divided into soluble fluorine and insoluble fluorine. The influence of soluble fluorine on PG is that the interaction between molecules is weakened, and the compressive folding strength of PG is reduced. Organic impurities mainly come from phosphate rock itself and flotation reagents added in flotation production, mainly ethylene glycol methyl ether acetate, isothiocyanomethane, 3-methoxy-n-pentane, 2-ethyl-1, 3-dioxopentane, etc. The mass fraction increases with the increase of PG particle size. The organic impurities attached to the surface of the gypsum crystal will reduce the whiteness and strength of the gypsum, and weaken the bonding force between the gypsum crystals, thus making the hardened structure loose. In addition, PG also contains trace radioactive elements and heavy metal elements, such as As, Cd, Zn, Ra, etc. These elements are seriously polluted to the environment, easy to be absorbed by plants and animals and affect human health.

Table 1 Main forms of impurities in PG

Impurity type	Impurity	Main form of existence
Phosphorus impurity	Soluble phosphorus	H_3PO_4, $H_2PO_4^-$, HPO_4^{2-}
	Insoluble phosphorus	$Ca_3(PO_4)_2$
	Eutectic phosphorus	$CaHPO_4 \cdot 2H_2O$
Fluorine impurity	Soluble fluorine	NaF, KF
	Insoluble fluorine	Na_3AlF_6, $CaSiF_6$, CaF_2
Organic impurity	Organic matter	Plant roots, humus
	Chemical reagent	Ethylene glycol encrypted acetate, isothiocyanomethane
Other impurities	Heavy metals	As, Hg, Pb, Cd
	Radioactive class	Ra-226, Th-232, K-40

2.2 *Hazards of Impurities in PG*

Impact of Impurities on the Environment

Phosphoric acid in PG impurities will produce a large amount of H+ and enhance the acidity of the environment, and through rain leaching and infiltration, make the surface water, groundwater and soil environment acidic, resulting in a large number of fish, shrimp and other organisms in the nearby surface water body. The phosphorus element in PG will be washed into rivers, lakes and seas with rain, causing water eutrophication. After the soluble fluoride in PG flows into water or soil, it is easily absorbed by plants, and then transmitted to humans through the food chain. When humans consume too much fluoride, they are prone to endemic fluorosis, tooth malformation and softening, osteoporosis and other symptoms.

Influence of Impurities on Properties of PG Products

The influence of impurities in PG on the properties of PG products includes two ways. One is the effect on the apparent properties of PG products, this adverse effect will be significantly reduced or disappeared after the removal of impurities, which contains P_2O_5 PG appearance is significantly different from natural gypsum, the higher the impurity content, the greater the difference in appearance. Na+ and K+ impurities can produce obvious crystals on the surface of gypsum products. Organic impurities will greatly deepen the color of PG surface. The other is the effect on the application performance of PG products. In the process of PG hydration, phosphate impurities will react with Ca^{2+} to form $Ca_3(PO_4)_2$. This insoluble substance will form a shell wrapped on the surface of gypsum, which will hinder the dissolution of gypsum and the water chemical reaction until its termination. As a result, the setting time is prolonged, the structure is loosened, and the product strength is reduced. Soluble fluorine impurities will weaken the intercrystalline molecular force, coarsing the crystals of the hydration products, and also lead to the loose structure of the final product and the reduction of compressive and folding strength. Organic impurities will increase the water requirement of PG and weaken the intercrystal bonding force of $CaSO_4 \cdot 2H_2O$, resulting in loose texture of the cementified material.

2.3 *Pretreatment Method of PG*

In order to eliminate the adverse effects of impurities on the properties of PG, PG needs to be pre-treated to remove impurities. At present, pretreatment methods mainly include physical method, chemical method and heat treatment method. Physical method includes water washing method, flotation method and ball milling method, chemical method includes lime neutralization method, acid leaching method, etc. Heat treatment method includes burning method and fast burning method. Among them, physical and chemical methods are the most widely used and have the best effect on impurity removal, and heat treatment is the main method to achieve PG dehydration and stimulate gelling activity. The advantages and disadvantages of main PG pretreatment methods are shown in Table 2.

Physical Method

Water Washing Method

The water washing method mainly removes soluble impurities, which will dissolve in water and separate from PG particles. After washing, PG crystals are clean and clear with clear outline. PG is added to water, impurities are removed by controlling liquid–solid mass ratio, washing time, washing temperature, etc., and almost all soluble impurities can be removed by washing until neutral. However, water washing can not remove eutectic phosphorus, which will precipitate from the crystal lattice during the hydration process of semi-aqueous gypsum, affecting the strength of the product. The advantage of washing is that the operation is simple and the impurity removal effect is obvious, the disadvantage is that the water consumption is large and the production line is a large investment. On the basis of the traditional water

Table 2 Comparison of common pretreatment methods of PG

Pretreatment method	Impurity removal category	Processing mode	Advantage	Shortage
Physical method	Soluble impurities, Organic impurities, Heavy metals	Water washing	High removal rate of soluble impurities	Insoluble impurities cannot be removed, resulting in a large amount of sewage
		Flotation	High removal rate of organic impurities	Flotation equipment is easy to be corroded, easy to cause secondary pollution
		Ball mill	The structure of phosphogypsum can be improved, no secondary pollution	Reaction conditions are demanding, Eutectic phosphorus cannot be removed
Chemical method	Slightly soluble impurities, insoluble impurities	Lime neutralization process	The process is simple and the cost is low	Pre-homogenization is required
		Acid leaching method	Can remove some insoluble impurities, simple operation process	The treatment of reaction residue is difficult
Heat treatment	Organic impurities, eutectic phosphorus	Calcination	Eutectic phosphorus can be removed	High energy consumption, acid impurities corrosion equipment, produce harmful gases

washing method to remove impurities, Liu et al. [1] discussed the influence of water washing treatment on PG backfilling in cement slurry, and the results showed that water washing could remove most soluble impurities in PG, significantly promote the hydration process of slurry, and significantly improve the mechanical properties of gypsum after treatment.

Flotation Method

The flotation method, also known as the semi-washing method, can effectively remove organic matter in PG, eliminate the influence of organic impurities in building gypsum on the cementiousness of semi-aqueous gypsum, can also remove silicon dioxide in PG, improve the purity of calcium sulfate, and solve the problem of poor quality and low whiteness of PG. By reverse flotation of PG, Wang et al. [2] made the whiteness of PG increase from 31.5 to 58.4%, the mass fraction of P_2O_5 decrease from 1.78 to 0.89%, and the mass fraction of calcium sulfate reach 96.6%, and the obtained products can be used as building materials. Zhang et al. [3] added a surfactant (TDBAC) when treating PG by flotation method. TDBAC can successfully adsorb on the surface of quartz and has good selectivity and trapping ability for quartz, which is expected to become a trapping agent that can efficiently remove quartz impurities.

Ball Mill Method

After the water washing method and flotation method, the water content of PG will increase, but the ball milling method will not change its water content, and the grinding treatment will make the PG particles smaller and more uniform particle size, which can fundamentally reduce the porosity of PG and make the structure of the product more dense. Therefore, when the particle size structure of PG particles needs to be changed or the impurity content is not high, it should be treated by ball milling method. Xu et al. [4] found that ball milling can shorten the setting time of PG based materials and improve the mechanical strength and water resistance. Wu et al. used ball milling method to prepare building gypsum, and explored the influence of ball milling time on the particle size of gypsum, physical and mechanical properties of slurry. It is found that with the extension of milling time, the particle size of gypsum decreases until it becomes stable, the water requirement of slurry to reach the standard consistency decreases, the coagulation time is shortened, and the strength of cured gypsum products is improved [5].

Chemical Method

Lime Neutralization Method

Lime neutralization method is to add lime materials such as calcium carbonate, quicklime and hydrated lime to PG to remove impurities. This method can improve the acidic environment caused by PG and transform impurities into inert substances for removal. In addition, adding lime substances can also effectively reduce the water content of PG. When dealing with PG with high water content or changing the acidic environment of PG, the lime neutralization method should be preferred. Wu et al.

developed an effective method for curing phosphorus and fluorine in PG by using calcium carbide slag and polymer. The test results show that in the calcium carbide slag modified PG system, soluble phosphorus and fluorine are fixed and converted into stable precipitates such as $Ca_3(PO_4)_2$ and CaF_2, and the addition of polymer can promote the curing of soluble phosphorus and fluorine in PG [6].

Acid Leaching Method

Acid leaching is a common method in PG pretreatment, which can effectively remove insoluble impurities in PG. The common acids are sulfuric acid and citric acid. Among them, sulfuric acid is the most widely used in the research and practical engineering of PG pretreatment. This method has a good effect on the removal of phosphorus and fluorine impurities in PG, especially the removal of eutectic phosphorus. In the process of leaching PG with citric acid, citric acid will react with soluble fluorine and phosphorus impurities in PG to form water-soluble substances. When treating PG with higher eutectic phosphorus content or requiring more strict impurity content, acid leaching method should be preferred. Mashifana used citric acid leaching to remove 34.7% of P_2O_5 from PG, while increasing CaO purity by 2.3% and reducing radioactive material by 72.3%. The study also compared the effects of citric acid, oxalic acid, sodium carbonate and sodium bicarbonate on PG leaching to remove impurities, and found that lemon acid leaching had the best effect, reducing the content of heavy metals and radioactive elements in PG [7].

Heat Treatment Method

When the content of organic matter and volatile impurities in PG is high or the required water content is low, the heat treatment method can be used to treat PG. Under the condition of high temperature, the impurities such as organic matter and volatile fluorine in PG are volatilized by heat, and the soluble phosphorus and eutectic phosphorus impurities will be transformed into inert substances such as pyrophosphate. In addition, heat treatment can also convert dihydrous gypsum, the main component of PG, into anhydrous gypsum, which has more stable properties, stronger hydration capacity, faster condensation rate, and greater compressive and flexion strength. Heat treatment method is divided into fast burning method and burning method, the difference between fast burning method and burning method is faster, more efficient, but it requires higher temperature and energy consumption.

Heat treatment method does not need admixtures and has less secondary pollution to the environment. Generally, the water-soluble phosphorus can be removed after roasting at 450 °C, and the fluorine content in PG can be effectively reduced when the temperature reaches 750 °C.

3 Comprehensive Utilization of PG

The comprehensive utilization of PG after pretreatment can not only solve the environmental harm caused by PG stacking, but also make it turn waste into treasure and generate economic value, which is the best way to solve the problem of PG stacking. At present, PG is mainly used in four major fields: construction, chemical industry, agriculture and environment.

3.1 Architectural Field

Preparation Cement

PG is mixed with sandstone, clay, coke and other materials in proportion, and the mixture can be processed to prepare Portland cement, super sulfate cement, calcium sulfoaluminate cement, etc. Chen et al. mixed PG with cement, fly ash, quicklime and foam in proportion to prepare foamed concrete, which has light texture, excellent mechanical properties and thermal insulation properties. Most of PG acts as fillers, and a small amount of sulfate stimulates the formation of calcium-silicon materials into calcium-vanadite and improves compressive strength [8]. Shen et al. prepared calcium-sulfoaluminate cement by mixing PG with fly ash. The cement was insensitive to impurities in PG and its performance was almost unaffected by impurities. It was also found that the decomposition of PG at high temperature was closely related to the sintering temperature, and its decomposition behavior would reduce the compressive strength of cement products [9].

Preparation of Cement Retarder

The main raw materials of cement retarder are natural gypsum and desulfurized gypsum. With the government's attention to the comprehensive utilization of PG and the protection of high-quality natural gypsum resources, PG has gradually become the raw materials for the preparation of cement retarder. Hu Yong et al. used autoclaved method to remove impurities such as fluorine and phosphorus from PG and then obtained autoclaved modified PG, whose performance when used as cement retarding agent was the same as that of natural gypsum as raw material. Li et al. modified PG by adding an appropriate amount of calcium carbide slag and fly ash, which significantly reduced the content of soluble phosphorus and fluorine in PG, significantly increased the strength of PG, and significantly shortened the setting time of the Portland cement produced than the original PG [10].

Preparation of Building Gypsum

Gypsum board and gypsum block are all building wall materials. With the progress of technology, more and more evidence shows that gypsum board produced by PG and gypsum board produced by natural gypsum have the same physical properties. Zhou et al. prepared paperless and fiberless high-strength gypsum board with PG as the only material. As it does not contain paper, fiber or additives, this gypsum board can be effectively recycled and used as a new wall material with good fire resistance, high efficiency and environmental friendliness. Wang et al. modified PG by adding an appropriate amount of calcium sulfaluminate cement and blast furnace slag, and the strength of the modified sample was significantly increased, and the smaller the particle size of the slag added, the denser the microstructure and the greater the strength of the prepared building gypsum [11]. Wu et al. used PG as raw material to prepare PG blocks by conventional mechanical pressing method, and found that when the pressing pressure was 300 MPa, the mass fraction of water in the blocks was 5%, and the mass fraction of elemental iron or aluminum was 1%, the 3d flexural strength of PG blocks could reach 8.0 MPa [12].

3.2 Chemical Industry

Preparation of Ammonium Sulfate and High-Purity Calcium Carbonate

PG is mixed with ammonia and CO_2 gas, or directly reacts with ammonium carbonate solution to produce ammonium sulfate and calcium carbonate residue, which also contains a small amount of SiO_2, calcium sulfate and fluorapatite and other impurities. Calcium carbonate residue can be purified to produce calcium carbonate with light quality and high purity, and can be used as raw materials for chemical production. Its potential utilization value is huge.

Ding et al. added ammonium acetate to PG to modify it, and under optimal conditions, 1000 kg modified PG could produce 510 kg of high-purity calcium carbonate [13]. Kandil et al. first pretreated PG with sulfuric acid to remove impurities, and then fully reacted the purified PG with excess ammonium carbonate, and the crystal purity of the obtained ammonium sulfate was as high as 95% [14].

Preparation of Calcium Sulfate Whisker

As an excellent additive for material synthesis, calcium sulfate whisker has a broad application prospect. The preparation of calcium sulfate whisker mainly uses natural gypsum as raw material, and natural gypsum consumption is large, which brings serious challenges to the sustainability of resources. However, the preparation of calcium sulfate whisker by PG provides a path for the comprehensive utilization of PG, and can also meet this challenge to a certain extent. Most of the calcium sulfate whisker products prepared by PG are hemihydrate calcium sulfate whiskers. Tan et al. used PG to prepare anhydrous calcium sulfate whiskers with smooth surface and large length-diameter ratio in the H_2O-H_2SO_4 high-pressure hydrothermal system, and found that adding maleic acid to the system would reduce the length-diameter ratio of crystals and increase the number of whiskers [15].

Recovery of Rare Earth Elements

Nowadays, rare earth elements have become a key strategic resource in the world, and the content of rare earth elements in PG is up to more than 2000 mg/kg, which is a potential rare earth secondary resource. The results show that the production process of phosphate rock has a significant effect on the content of rare earth elements in PG. Sulfuric acid reacts with phosphate rock to produce calcium sulfate which is not soluble in the liquid phase. According to the different hydration degree of calcium sulfate, it can be divided into dihydrate process and semi-hydrate process. Generally, about 70% of rare earth elements enter PG in the dihydrate process, and about 90% of rare earth elements enter PG in the semi-hydrate process.

Kijkowska et al. leach PG slurry with sulfuric acid solution with a mass fraction of 10–13% and hydrogen peroxide, stir at 60 °C for 60–120min, filter and wash, and evaporate and crystallize the leach solution at 100 °C, with a rare earth recovery rate of 50% [16]. Salo et al. first used sulfuric acid to leech rare earth elements from PG, and then placed the leech solution in a sulfate reduction bioreactor for treatment. The results showed that the bioreactor could operate reliably when the concentration of the leech solution was low, and mineral precipitation of enriched rare earth elements would be formed in the bioreactor, realizing efficient recovery of rare earth elements [17]. Wu et al. used phosphoric acid to leach rare earth elements from phosphate rock, and the maximum leaching rate reached 94.3%. After holding the leaching solution at 90 °C for 4 h, the selective precipitation of rare earth elements was achieved, and the recovery rate reached 81.3% [18].

3.3 Agricultural Field

PG can be used to prepare soil conditioner, special compound fertilizer and other agricultural products. SO_4^{2-} and Ca^{2+} in PG can replace substitutable aluminum and sodium in soil. PG with a grain size of about 900 um can enhance soil permeability, weaken surface runoff, and slow down soil fertility loss, so it can be used to improve saline-alkali land, acidic soil, polluted soil and erosive soil. Ca^{2+} in PG has a solidifying absorption effect on phosphorus substances in soil, which can reduce the loss of nutrient elements and fix soil fertility. PG is rich in calcium sulfate, and also contains sulfur, calcium, silicon, iron, phosphorus and other elements necessary for plant growth, which can be used to produce silica-calcium potassium fertilizer, compound fertilizer additives and other products.

Acid Soil Improvement

High content of substitution aluminum is more likely to lead to plant poisoning and soil acidification. The traditional method of improving acidic soil is to add lime, but a large amount of long-term application of lime will cause soil compaction, and the imbalance of calcium, potassium and magnesium in the soil. Calcium sulfate in PG can react with aluminum-containing substances in the soil to form $Al(OH)(SO_4)$, which can not only increase the exchangeable Ca^{2+} and available sulfur elements in the soil, but also increase the soil pH. PG is better than lime in improving acid soil. PG can also effectively improve the physical properties of soil. Panda et al. prepared a biochar—PG composite material by co-pyrolysis of banana flower stalks and PG at 700 °C, which can increase the sulfate content in soil and enhance soil fertility when applied to acidic red soil [19].

Saline-Alkali Soil Improvement

The harm of saline-alkali soil is mainly salt damage and alkali damage, too much salt and alkali in soil will affect the growth of plants, and too alkaline soil will seriously affect the physical and chemical properties of soil. The basic of saline-alkali land improvement is desalting. The addition of PG can reduce soil pH, improve soil structure and reduce alkaline damage to crops. Sagna et al. explored the conditioning effect of PG on saline-alkali soil, it was found that organic amendments combined with PG could significantly reduce the alkalinity of saline-alkali soil [20].

Preparation of Compound Fertilizer

PG contains some trace elements necessary for plants, such as iron and zinc, and PG is water-soluble to some extent, so the application of a certain amount of PG can promote the growth of plants. Melgaco et al. found that phosphogypse provided sulfate for sulfate-reducing bacteria and improved the richness, diversity and activity of soil bacteria, but it was unfavorable to the growth of fungi [21]. Zhantasov et al. combined with the mineralogy components of PG samples, proposed a process scheme of mixed slow-release fertilizer with phosphate fertilizer as fertilizer, vermiculite as water-retaining component and PG as neutralizing agent, which can effectively reduce the acidity and salinity of soil [22]. Samet et al. found that PG can improve the diversity of microbial population in compost. The compost with 30% PG added contains a large number of pseudomonas and penicillium, which have high tolerance to heavy metals and PG dissolution, and can also promote plant growth and control insect pests [23].

3.4 Environmental Field

The application of PG in the environment started late, the research is relatively few, and it is mainly used for adsorption of pollutants and storage of CO_2. PG has a good application prospect in the environmental field, but the research on the occurrence mechanism, migration and transformation of toxic and harmful substances in it is not enough, and the material preparation cost is high, there is secondary pollution, which limits the comprehensive utilization of PG in the environmental field.

Preparation of Adsorption Materials

PG can be used to prepare adsorbent materials. Zhao et al. modified PG with sodium dodecyl benzene sulfonate, and the adsorption rate of modified PG for Cu^{2+} in the solution was as high as 99.2% [24]. Bellanger et al. prepared a composite adsorption material using PG and clay. The maximum adsorption capacity of the material for U^{6+} reached 0.09 mol/kg, indicating that PG also has a good adsorption effect on radioactive elements [25]. Zhang et al. made hydroxyapatite nanoparticles from PG through microwave radiation technology, and the prepared nanoparticles have a good adsorption effect on fluorine in sewage [26]. Ma et al. made PG into porous materials for noise absorption, which can significantly reduce noise pollution [27].

Carbonation Storage of CO_2

The use of PG as a carbon sink material is an effective and energy-saving new technology. The use of PG-decomposed aqueous solution to capture CO_2 and make full use of calcium sources to prepare $CaCO_3$ materials with hard texture and excellent performance can simultaneously realize the comprehensive utilization of PG and the storage of greenhouse gas CO_2. Zhang et al. found that more than 90% of Ca^{2+} in PG aqueous solution can be converted into $CaCO_3$, and increasing temperature and CO_2 partial pressure can improve the adsorption efficiency of CO_2 [28]. Ding et al. used ammonium acetate to modify PG, and found that ammonium acetate could separate Ca^{2+} from PG and concentrate other impurity ions.1000 kg PG could chelate up to 224 kg CO_2, and the carbonization rate of PG reached 98.3% [13].

4 Conclusion

The pretreatment of PG is the key to the resource utilization of PG, and a single pretreatment method is difficult to achieve the deep purification of PG. Therefore, the development of low-cost, low-energy and environment-friendly coupling PG treatment process is the focus of future development. Although the comprehensive utilization of PG in China has achieved some stage results, the utilization rate has gradually increased, and the utilization way has gradually diversified, but due to the difference of the impurity content of PG and the cost and complexity of the treatment process, the current situation of low added value of products and large amounts of storage has not been significantly changed, and the resource utilization of PG has not been fully realized. In the future, it is necessary to increase research and development efforts, accelerate the structural adjustment and optimization of the phosphorus chemical industry, overcome the key difficulties in the resource utilization of PG, and achieve green and sustainable development of the phosphorus chemical industry.

Acknowledgements The work was supported by the Hubei Province Science and Technology Development Special Project (Project name: Key technology and application demonstration of self-leveling mortar with high pressure resistance and wear resistance of phosphogypsum base, Project code: 42000023205T000000218).

References

1. Liu, Y., Chen, Q., Wang, Y., Zhang, Q., Li, H., Jiang, C., Qi, C.: In situ remediation of phosphogypsum with water-washing pre-treatment using cemented paste backfill: rheology behavior and damage evolution. Materials **14**, 6993 (2021)
2. Wang, J., Dong, F., Wang, Z., Yang, F., Du, M., Fu, K., Wang, Z.: A novel method for purification of phosphogypsum. Physicochem. Probl. Min. Process. **56**, 975 (2020)
3. Zhang, H., Chai, W., Cao, Y.: Flotation separation of quartz from gypsum using benzyl quaternary ammonium salt as collector. Appl. Surf. Sci. **576**, 151834 (2022)
4. Xu, Q., Yang, D., Lv, W., Li, X., Qian, Y., Du, B.: Effect of grinding time on properties of phosphogypsum based cementitious materials. Inorgan. Chem. Ind. **54**, 101 (2022)
5. Wu, L., Tao, Z., Zhao, Z.-M., Ghafar Wahab, A., Tao, Y., Liao, S., Zhang, Y.: Effect of ball milling time on the performance of phosphorous building gypsum. Adv. Civ. Eng. 2022, 7670057 (2022)
6. Wu, F., Liu, S., Qu, G., Chen, B., Zhao, C., Liu, L., Li, J., Ren, Y.: Highly targeted solidification behavior of hazardous components in phosphogypsum. Chem. Eng. J. Adv. **9**, 100227 (2022)
7. Mashifana, T.P.: Chemical treatment of phosphogypsum and its potential application for building and construction. Procedia Manuf. **35**, 641 (2019)
8. Chen, M., Liu, P., Kong, D., Wang, Y., Wang, J., Huang, Y., Yu, K., Wu, N.: Influencing factors of mechanical and thermal conductivity of foamed phosphogypsum-based composite cementitious materials. Constr. Build. Mater. **346**, 128462 (2022)
9. Shen, Y., Qian, J., Chai, J., Fan, Y.: Calcium sulphoaluminate cements made with phosphogypsum: production issues and material properties. Cem. Concr. Compos. **48**, 67 (2014)
10. Li, B., Li, L., Chen, X., Ma, Y., Zhou, M.: Modification of phosphogypsum using circulating fluidized bed fly ash and carbide slag for use as cement retarder. Constr. Build. Mater. **338**, 127630 (2022)
11. Wang, Q., Jia, R.: A novel gypsum-based self-leveling mortar produced by phosphorus building gypsum. Constr. Build. Mater. **226**, 11 (2019)
12. Wu, F., Jin, C., Qu, G., Liu, Y., Wang, C., Chen, B., Liu, S., Li, J.: Enhancement of phosphogypsum mechanical block with the addition of iron and aluminum salts. J. Build. Eng. **52**, 104397 (2022)
13. Ding, W., Chen, Q., Sun, H., Peng, T.: Modified mineral carbonation of phosphogypsum for CO_2 sequestration. J. CO_2 Util. **34**, 507 (2019)
14. Kandil Abdel-Hakim, T., Cheira Mohamed, F., Gado Hady, S., Soliman Madiha, H., Akl Hesham, M.: Ammonium sulfate preparation from phosphogypsum waste. J. Rad. Res. Appl. Sci. **10**, 24 (2017)
15. Tan, H., Dong, F., Bian, L., He, X., Liu, J.: Preparation of anhydrous calcium sulfate whiskers from phosphogypsum in H_2O-H_2SO_4 autoclave-free hydrothermal system. Mater. Trans. **58**, 1111 (2017)
16. Kijkowska, R., Kowalczyk, J., Mazanek, Cz., Mikolajczyk, T., Pawlonska-Kozinska, D.: Rare Earth Elements from Kola Apatite Phosphogypsum (1989)
17. Marja, S., Oleg, K., Jarno, M., Xiaosheng, Y., Pertti, K.: Integrated acid leaching and biological sulfate reduction of phosphogypsum for REE recovery. Miner. Eng. **155**, 106408 (2020)
18. Wu, S., Zhao, L., Wang, L., Huang, X., Zhang, Y., Feng, Z., Cui, D.: Simultaneous recovery of rare earth elements and phosphorus from phosphate rock by phosphoric acid leaching and selective precipitation: towards green process. J. Rare Earth **37**, 652 (2019)
19. Panda, L., Kumar, M., Pradhan, A.: Leaching of sulphate from biochar and phosphogypsum-biochar for the treatment of acidic red soil. Asian J. Water Environ. Pollut. **19**, 23 (2022)
20. Sagna, Y.P., Diedhiou, S., Goudiaby, A.O.K., Diatta, Y., Diallo, M.D., Ndoye, I., Fall, S.: Do phosphogypsum combined with organic amendments improve rice growth in a saline environment? Curr. J. Appl. Sci. Technol. **42**, 52 (2023)
21. Antonio De Oliveira, M.L., Carolina, Q.N., Albis, L.V.: Phosphogypsum as sulfate source for sulphate-reducing bacteria in a continuous fluidized-bed reactor. Engenharia Sanitaria e Ambiental **25**, 157 (2020)

22. Kurmanbek, Z., Aitzhan, Z., Nursulu, S., Kamilya, K., Gani, I., Manap, Z., Guzaliya, S., Ayganym, A.: Ecologically friendly, slow-release granular fertilizers with phosphogypsum. Polish J. Environ. Stud. 31, 2935 (2022)
23. Mariem, S., Fatma, K., Najla, M., Lotfi, K., Sami, S., Radhia, G.-B.: Effect of phosphogypsum addition in the composting process on the physico-chemical proprieties and the microbial diversity of the resulting compost tea. Environ. Sci. Pollut. Res. **26**, 21404 (2019)
24. Zhao, L., Zhang, Q., Li, X., Ye, J., Chen, J.: Adsorption of Cu(II) by phosphogypsum modified with sodium dodecyl benzene sulfonate. J. Hazard. Mater. **387**, 121808 (2020)
25. Boris, B., Sylvain, H., Philipp, S., François, C., Fernando, V., Vincent, V., Kaspar, A., Norbert, C., André, M.: Oxic–anoxic conditions in the water column of a tropical freshwater reservoir (Pena-Larga dam, NW Venezuela). Appl. Geochem. **19**, 1295 (2004)
26. Zhang, D., Luo, H., Zheng, L., Wang, K., Li, H., Wang, Y., Feng, H.: Utilization of waste phosphogypsum to prepare hydroxyapatite nanoparticles and its application towards removal of fluoride from aqueous solution. J. Hazard. Mater. **241**, 418 (2012)
27. Ma, B., Jin, Z., Su, Y., Lu, W., Qi, H., Hu, P.: Utilization of hemihydrate phosphogypsum for the preparation of porous sound absorbing material. Constr. Build. Mater. **234**, 117346 (2020)
28. Zhang, W., Zhang, F., Ma, L., Yang, J., Wei, Y., Kong, D.: CO_2 capture and process reinforcement by hydrolysate of phosphogypsum decomposition products. J. CO_2 Util. 36, 253 (2020)

A Pathway Model for the Sustainable Development of Industrial Heritage in Old Industrial Cities: Cases of Zibo and Harbin, China

Zhihong Huang, Zijuan Zhang, Mengyi Chen, and Guansu Wang

Abstract The management of industrial heritage in old industrial cities has been a significant topic in both academia and government in the context of sustainable development. By adopting suitable strategies, specific types of industrial heritage can be adaptively reused and revitalized, thereby achieving sustainability and promoting the growth of other urban industries. The successful revitalization of such cities requires coordinated efforts among multiple stakeholders, particularly the government's role in mobilizing resources and formulating effective strategies. News reports, as a channel for presenting government action, provide insights into governmental strategies for industrial heritage revitalization. This study integrated industrial ecology and agenda-setting theory, employing the pre-trained language model RoBERTa to vectorize news report content, followed by the application of clustering algorithms for topic analysis. The results elucidated the successful strategies for sustainable industrial heritage development in two old industrial cities and constructed a pathway model for the sustainable reuse of the industrial heritages.

Keywords Old industrial city · Industrial heritage · Industrial ecology · Sustainable development · RoBERTa

1 Introduction

As many old industrial cities face decline amid economic transformation and social development, the question of how to leverage industrial heritages for revitalization has become a pressing issue. Effective repurposing and revitalization of industrial heritage can not only preserve its historical and cultural value but also stimulate economic growth, boost tourism, and foster community cohesion. Zibo and Harbin

Z. Huang · Z. Zhang · M. Chen · G. Wang (✉)
School of Humanities, Zhuhai College of Science and Technology, Zhuhai, China
e-mail: s2128778@siswa.um.edu.my

G. Wang
Asia-Europe Institute, Universiti Malaya, Kuala Lumpur, Malaysia

Y. Zhu (ed.), *Proceedings of the 4th International Symposium in Environmental Science and Industrial Ecology*, https://doi.org/10.1007/978-981-96-1578-0_10

were once major industrial cities in China, but with the decline of traditional industries, they have encountered significant challenges in sustaining economic vitality. However, after the COVID-19 pandemic, these cities swiftly repurposed their industrial heritage through tourism strategies and actions, which significantly boosted the cities' overall income and development. During this process, the local governments clearly formulated and communicated their strategy and action plans, using mass media to disseminate a public agenda. To study and present the strategic approaches of the two local governments, this study employed a method that integrates thematic analysis of news reports with industrial heritage, exploring agenda-setting practices in the repurposing of industrial heritage across different stages. The findings revealed that the strategic use of media coverage, combined with stakeholder collaboration, has played a critical role in achieving the sustainable development of industrial heritage in old industrial cities.

2 Literature Review

During the process of economic development, some cities have been intensively developed into industrial cities or hubs due to geographic and transportation advantages, as well as abundant resources [1]. During their industrialization process, substantial resources are invested to meet the needs of industrial cities. However, with changes in the global economic landscape, technological advancements, or shifts in national development strategies, these old industrial cities may fail to keep pace with the times, losing their status as industrial centers and fading from public view. Such cities are referred to as old industrial cities [2]. After industrial relocation, numerous factory buildings, infrastructure, and even industrial cultural assets were left, becoming industrial heritage [3]. How to effectively manage these industrial heritage sites is a topic that frequently discussed among government and academy. The adaptive reuse of the former industrial heritage can be beneficial. Dell'Anna [4] employed the economic method to quantify the economic benefits of adaptive reuse of industrial heritage in Turin, Italy, demonstrating a significant positive impact on nearby residential property values, with a capitalized benefit of over EUR 16 million within an 800 m radius of the transformed site. Günçe and Mısırlısoy [5] introduced a project in Turkey which converted the industrial building into the Kadir Has University, and its contribution to socio-cultural sustainability by preserving cultural identity, enhancing local engagement through educational and cultural activities, and revitalizing the surrounding community, thereby transforming industrial sites into vibrant socio-cultural hubs. The reuse object can be a building or an industrial hub, as well as a whole city. Konior and Pokojska [6] introduced the revitalization of entire postindustrial cities in Poland, emphasizing tourism as a key strategy, which can drive economic, social, and cultural revitalization, transforming these cities into vibrant, attractive destinations by revitalizing their industrial heritage. The process of dealing with industrial heritage is complex. However, the local government can play a crucial role through policies. Chen et al. [7] and Veldpaus and Wacogne [8] both pointed out that the policies' positive effect on the management, reuse and sustainable development of the postindustrial cities and industrial heritage. Perfetto et al. [9] introduced that tourism conversion is an efficient way to reuse the industrial heritage sustainably.

Erkman [10] introduced the concept of "Industrial Ecology" in the late 1980s, envisioning that industrial activities could be analyzed and managed in a way analogous to biological ecosystems. Ideally, this theory aims to integrate industrial processes and promote the recycling of resources, mirroring the functions of natural ecosystems [11]. Korhonen [12] explained that the ultimate goal of industrial ecology is to contribute to the sustainable development process. Industrial ecology offers numerous advantages. In the economic aspect, Saavedra et al. [13], through a bibliometric analysis of international literature, highlighted that industrial ecology can advance the circular economy in terms of concepts, technologies, and policies, thereby achieving improved sustainability outcomes. Herczeg et al. [14] conducted an extensive analysis of existing studies on supply chain systems, supplemented by interviews with key stakeholders, to demonstrate that industrial ecosystems can repurpose industry's by-products as raw materials for others, optimizing material and energy usage, minimizing waste disposal and resource loss, and yielding multidimensional benefits in environmental and social aspects. Xavier et al. [15] used a case to demonstrate that implementing the 3Rs principle can achieve sustainable waste management. They showed that closed-loop supply chains improve the efficiency of the circular economy and further explored the interconnectedness between industrial ecology, industrial metabolism, and sustainability. The circular economy is a fundamental theory and approach in industrial ecology, which contrasts sharply with the traditional linear industrial model of "take-make-use-dispose." The core principle of the circular economy is to maximize the efficiency of resource use by promoting reuse, repair, refurbishment, and recycling, thereby maintaining the highest possible value of products and materials throughout their lifecycle [16]. The handling of industrial waste in manufacturing is often seen as a burden, however, by structurally reforming existing business models into the circular economy and integrating waste into other supply chains, these by-products can be repurposed as valuable resources. This strategic shift underscores waste reuse as a crucial component of industrial ecology [17]. Beyond industrial waste, the factories themselves can also be repurposed and reused. Raising designers' awareness of recyclability during the design process can promote the advancement of circular economy strategies and increase the reuse rate of industrial production spaces [18]. Additionally, abandoned factories and other forms of industrial heritage are being repurposed and integrated into the industrial ecological cycle. While often perceived as less significant than other types of industrial waste or heritage, these buildings possess unique architectural styles, cultural characteristics, and expansive spaces that make them well-suited for redevelopment and alternative uses [19]. Berta et al. [20] used the case of the Shougang/Er-Tong mechanical factory in Beijing, China, to demonstrate how the regeneration of industrial wastelands can facilitate urban transformation and support the development of innovative urban policies, giving both economic and cultural benefits to the city.

Agenda-setting theory is a theory in the field of politics and communication, which was developed out of the election process in the USA in 1972 by Maxwell McCombs and Donald Shaw [21]. The main argument of the theory posits that although the mass media cannot directly affect how the public thinks, it can substantially shape what the public focuses on. In other words, the government can strategically influence public discourse by directing media coverage to set specific agendas [22]. Feezell [23] conducted an experimental study to test whether exposure to political information on Facebook influences participants' perceptions of certain policy issues, thus validating the agenda-setting effect. The findings indicated that the information audiences encounter on social media shapes their perception of its importance. Consequently, governments can leverage media coverage to rapidly elevate a topic or quickly inform the public about specific events. For instance, the government can strategically frame messages to disseminate cybersecurity knowledge to the public, thereby increasing societal awareness of cybersecurity protection [24]. Similarly, during pandemics, the government can use social media platforms to effectively communicate public health information, enhancing public awareness of preventive health behaviours [25, 26]. Moreover, effective agenda-setting can play a crucial role in advancing sustainable development initiatives. Jamil and Appiah-Adjei [27] employed a qualitative approach through in-depth interviews and inductive thematic analysis to examine media practices in Pakistan. Their findings revealed that local media could influence public behaviour by setting media agendas that emphasize eco-friendly issues, thereby promoting ecological development on the public agenda. Additionally, agenda-setting can be a strategic tool for governments to achieve specific objectives. Lee and Xu [28] analyzed the tweets of candidates during the 2016 U.S. presidential election and their influence on voter reactions, illustrating the significant impact of online agenda-setting in political campaign strategies. The government can also leverage agenda-setting to develop or revitalize specific industries. Deng et al. [29] utilized the Theory of Planned Behavior alongside three factors, including media coverage, macro destination image, and micro destination image, to examine tourists' intentions to revisit Macau in the post-pandemic period. Zhang and Jing [30] analyzed various media texts and public agenda samples from Ganzi, Sichuan Province, comparing tourism destination images across different agendas. Their findings indicate a strong correlation between public agendas and media agendas in the construction of tourism destination images, highlighting the significant role of media in shaping city image. This suggests that analyzing publicly available media content can provide insights into government promotional strategies, offering valuable references for the development of other cities.

While there is extensive research on the management of industrial heritage in old industrial cities, much of it focuses on the discussion of one successful case, which tends to be exceptional and rarely addresses general patterns or models. Besides, although agenda-setting theory has been widely used in studies on tourism and city image research, and the reuse of industrial heritage through tourism has also been explored, no research has applied agenda-setting theory to the treatment of industrial heritage. This study, therefore, integrates industrial ecology and agenda-setting theory, conducting a thematic analysis of news reports across different periods.

By examining the successful pathways of two old industrial cities in achieving the sustainable development of their industrial heritage, the study proposes a recovery pathway for old industrial cities grounded in industrial ecology.

3 Methodology

This study employs a large language model to perform a thematic clustering analysis of news texts across different time periods, focusing on the pathway models of industrial heritage reuse in the old industrial cities of Harbin and Zibo, China. The findings aim to provide valuable insights and guidance for other cities seeking to repurpose their industrial heritage.

3.1 Data Collection and Cleaning

Factiva, developed by Dow Jones, is a comprehensive news database that collects global news coverage, including Chinese media, and is widely utilized in the field of communication research. In this study, 263 relevant news articles were manually collected, and keywords were identified through word frequency analysis, which were then used to search on the Factiva database. Media companies were chosen based on criteria such as authority, youth audience reach, and inclusion in the database, resulting in the selection of People's Daily, Xinhua News Agency, ChuTian Metropolis Daily (Jimu News), Jiemian News, and Chengdu Economic Daily (Hongxing News) as data sources. The search timeframe began from the respective start dates of the trends in Zibo and Harbin and concluded on May 31, 2024. Data was retrieved by searching only within titles and lead paragraphs, excluding duplicates. The results were then refined by manually filtering out irrelevant ones, with the final data presented in Table 1.

According to the agenda-setting theory, the topics of media coverage were set at different stages. Thus, this study separated the stages by analyzing the publishing time of the data, resulting in 3 stages in each city, shown in Fig. 1. In Fig. 1, the x-axis represents the number of news reports, while the y-axis represents the timeframe. Based on the trend in the amount of news reports, Zibo's timeline is divided into

Table 1 The final data of this study

City	Timeframe	Raw data	Cleaned data
Zibo, China	1 Mar 2023–31 May 2024	196	114
Harbin, China	1 Dec 2023–31 May 2024	186	249

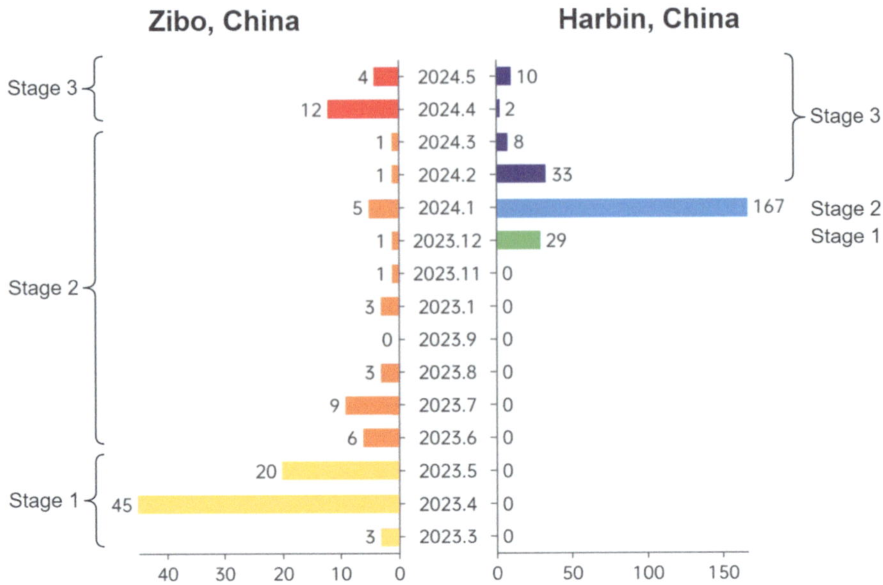

Fig. 1 The stages distribution in Zibo and Harbin, China

Stage 1 (March–May 2023), Stage 2 (June 2023-March 2024), and Stage 3 (April–May 2024), while Harbin's timeline is segmented into Stage 1 (December 2023), Stage 2 (January 2024), and Stage 3 (February-May 2024).

3.2 Data Processing

This study employed the Chinese-roberta-wwm-ext pre-trained large language model (LLM), developed by the Harbin Institute of Technology iFLYTEK Joint Laboratory, for text vectorization [31], such methods using LLM to process text data for clustering analysis or qualitative research are trends in social science research [32, 33]. The model is based on BERT and utilizes Whole Word Masking (WWM) technology, which is specifically designed for Chinese texts, and is trained on localized Chinese datasets. It demonstrated an accuracy of 97.7% in paragraph-level Chinese news text classification experiments and has been widely applied in various Chinese natural language processing tasks [34].

When processing data with a model based on BERT, the maximum allowable text length is 512 characters, while news articles typically exceed this limit, necessitating preprocessing. Common strategies include extracting characters only from the beginning, or the end, or sampling from both sides. Considering that Chinese news articles often consist of multiple paragraphs, each potentially addressing different themes, this study adopted a segmentation approach using Python to enhance

Table 2 The results of data processing

	Zibo, China	Averaged words	Harbin, China	Averaged words
First stage	1018 paragraphs	113 words	239 paragraphs	118 words
Second stage	487 paragraphs	117 words	1614 paragraphs	117 words
Third stage	338 paragraphs	121 words	561 paragraphs	118 words

processing feasibility and research accuracy. Paragraphs shorter than 50 characters were removed, and the resulting number of paragraphs is detailed in Table 2.

Then, the chinese-roberta-wwm-ext model was used to vectorize each paragraph, extracting the vector data from the topmost Transformer layer to obtain 768-dimensional vectorized data for each paragraph.

3.3 *Topics Clustering*

The K-Means algorithm may not perform well for the high dimensional vectorized data, while the Spectral Clustering is more suitable. Spectral clustering requires a given number of clusters, and varying numbers of clusters can significantly affect the clustering results. The silhouette coefficient typically assesses the clustering performance, where higher values indicate better clustering quality. This study identified the optimal number of clusters using a plot of cluster numbers versus silhouette coefficients and finalizes the cluster count based on the interpretability of the results. The plot and the numbers of clusters are shown in Figs. 2 and 3.

After clustering, keywords were extracted from all texts within each cluster to facilitate thematic analysis, which aimed to summarize the topics of each cluster. This study utilized Jieba, a Python library commonly used for Chinese text mining, to segment Chinese text into individual words, enabling further analysis. After segmentation, the Harbin Institute of Technology's Chinese stopwords list was used to remove stopwords. Subsequently, the TextRank algorithm was applied to score and rank the segmented words based on their importance, identifying representative terms for each cluster and generating word clouds. TextRank, grounded in Graph Theory, constructs a graph of word co-occurrences within a defined window size and calculates scores to rank keywords based on their significance [35].

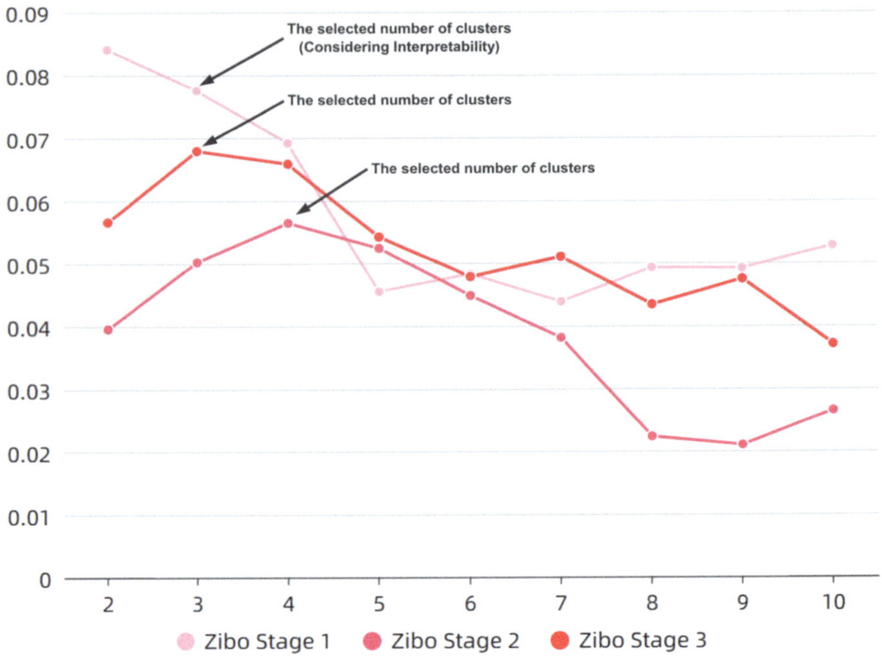

Fig. 2 The plot of the number of clusters versus the Silhouette coefficient in Zibo, China

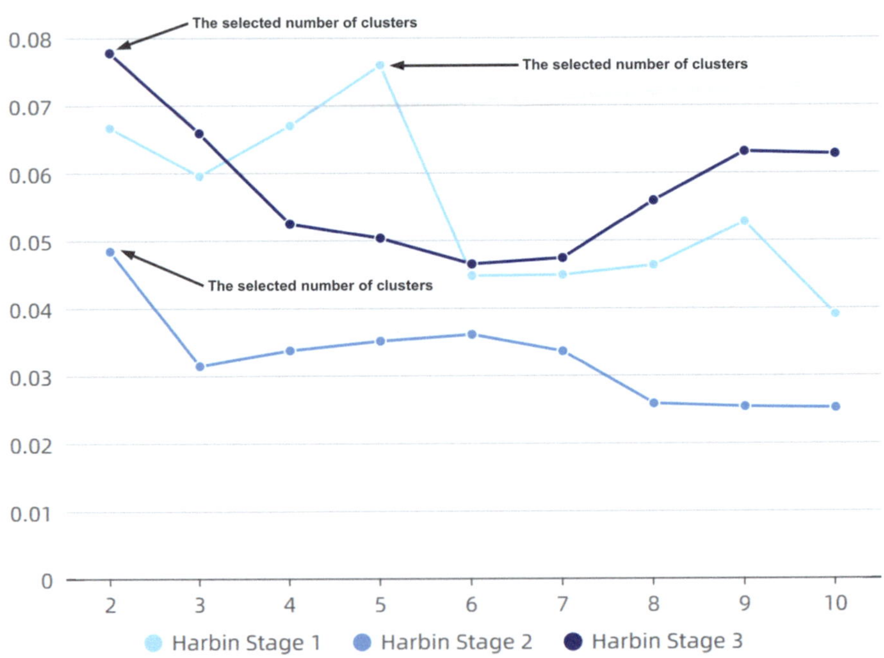

Fig. 3 The plot of the number of clusters versus the Silhouette coefficient in Harbin, China

4 Discussion

4.1 The Materials Preparation for Industrial Heritage Sustainable Development

"Industrial heritage" is an umbrella term that covers both tangible and intangible elements that are left by the process of industrialization. Tangible industrial heritages include the factories, infrastructure, labours' square, etc. Intangible industrial heritages compose the culture, communities, stories and identification that are related to the local industries. In Zibo and Harbin cases, both the tangible and intangible elements were used for reusing, which can be considered as following aspects:

Physical Heritage: In Harbin, many former factory buildings have been converted into commercial areas. For example, the Harbin Jilian Machinery Factory has been transformed into the Xicheng Hongchang Industrial Park, which is now used for commercial purposes while preserving industrial architectural features, such as a large "ant" sculpture assembled from old industrial components. Additionally, following the construction of the Chinese Eastern Railway, an increasing number of Russians settled in Harbin, establishing businesses and shops around Harbin Central Street, turning it into a significant testament to the city's industrialization and urbanization.

Culture and Political Heritage: In Zibo, labour culture and workers' movements have been a significant part of the city's history. From the late nineteenth century to the early twentieth century, mining workers in Zibo's key mining areas continuously fought for their rights. After the founding of the Chinese Communist Party, following Marxist education and outreach, a number of activists emerged from these labour movements, some of whom went on to become Communist Party members. In this context, Zibo's cultural and political heritage has shaped an image of its residents as honest and pragmatic workers.

Industries Heritage: Industrial heritage refers to distinctive local industries that have been preserved during the process of social development and were once integral to the city's economy. For example, Zibo's famous ceramic crafts are a notable case. The 1954 Ceramic Culture and Creative Park in Zibo was converted from the former state-owned Zibo Porcelain Factory, a major ceramic enterprise, and it retains the old factory buildings and equipment from the 1950s while integrating modern artistic design elements. Zibo's ceramic trade is also well-known. The Ceramic and Glazed Art Park, formerly a market for the ceramic trade, attracted a significant number of tourists during the tourism surge in 2023.

Living Behaviours: Harbin is home to China's first brewery, the Ulubulevsky Brewery, founded by a Russian businessman in 1900. The development of the beer industry in Harbin influenced local lifestyles, establishing a culture of beer drinking among the residents. Today, many beer-themed venues in Harbin, such as Beer

Cultural Street and the Harbin Beer Museum, have been converted from abandoned factory buildings and have become integral to the local lifestyle.

4.2 The Agenda-Setting

According to agenda-setting theory, the news reports in both cities are categorized into three distinct stages over time, with each stage exhibiting distinctive thematic characteristics.

4.2.1 Zibo, China

In the first stage (March–May 2023), media reports on Zibo focused on three main themes, predominantly revolving around the viral popularity of "Zibo BBQ" on social media, alongside brief introductions to the city's historical and cultural context. In the word cloud (Fig. 4), terms such as "Barbecue," "Qi State," "heritage sites," "industry," and "Chemical Industry" appeared frequently. Headlines such as "Zibo BBQ restaurants fill up with university students within a minute of opening," "University students flock to Zibo for BBQ," and "Zibo BBQ, the enduring legend" repeatedly trended on social media. Within a short period, Zibo, a former industrial city, gained significant exposure on social media, becoming a popular destination for BBQ enthusiasts. During this stage, in response to the surge in tourism, the Zibo government effectively capitalized on the trending topic to boost the city's visibility while highlighting its cultural and industrial heritage. The city's public transportation system also adapted to accommodate the substantial influx of tourists.

In the second stage (June 2023-March 2024), media coverage of Zibo shifted to four main themes: influencers, culture and humanities, crisis management, and business environment. This stage was marked by a strong emphasis on narrative strategy, with the government actively collaborating with social media influencers to promote the city's image through personal stories, particularly those that highlighted the industrious, hospitable, and modest character of local residents. For instance, Fig. 5 shows keywords such as "people," "life," "internet celebrities" and "creativity."

Simultaneously, cultural narratives became a key focus of media coverage, with reports suggesting that "to sustain Zibo's popularity, we must draw attention to our rich traditions of calligraphy and painting." In addition to its industrial heritage, Zibo's location in Shandong Province, where is home to Confucius and the cradle of Confucian culture, endows it with a deep cultural legacy. These elements were explored and promoted by the media during this period. The focus of media reports shifted towards cultural and humanistic aspects, presenting the city's industrial and regional culture to the public through intangible cultural narratives.

In the third stage (April–May 2024), media reports on Zibo covered three main themes: the government's commitments to improving tourism services, the diversification of cultural and tourism products, and the continued promotion of tourism.

Fig. 4 The word cloud of each topic in Zibo's stage 1

In the word cloud (Fig. 6), terms such as "demand," "cultural tourism," "BBQ," and "tourism" appeared frequently. Although the city's popularity had somewhat declined by this time, the government continued to reference the initial viral events to sustain and deepen public interest in the city.

4.2.2 Harbin, China

In the first stage (December 2023), a negative event briefly made Harbin a topic of public attention, but due to the prompt response from the government and tourist attractions, the city successfully transformed the incident into a public relations opportunity. Harbin capitalized on the recovery of tourism to attract large numbers of visitors. In the word cloud (Fig. 7), terms such as "experience," "ticket refund," "service," "ice and snow," "city management," "growth," and "tourism" appeared frequently, reflecting the city's handling of the crisis, improvements in tourist services, and increased tourism activity. During this stage, the government's agenda-setting efforts quickly redirected public discourse toward positive coverage, significantly boosting Harbin's visibility.

In the second stage (January 2024), the focus of media coverage shifted from the incident to Harbin's architecture, landmarks, and natural scenery, such as the Harbin

Fig. 5 The word cloud of each topic in Zibo's stage 2

Central Street, the Songhua River, and Ice and Snow World, shown in the Fig. 8. Reports also highlighted historically significant and regionally distinctive buildings, such as the Harbin Central Street and Saint Sophia Cathedral, both were related to the Russian Empire and the Chinese Eastern Railway, now serving as a prominent symbol of Harbin's cultural heritage [36]. In this stage, Harbin's media narrative differed significantly from that of Zibo, focusing on the city's industrial, cultural, and architectural identity.

In the third stage (February–May 2024), the media emphasized Harbin's efforts to "stimulate the growth of other industries," referring the Fig. 9. Coverage revisited Harbin's tourism boom while discussing how to leverage the post-peak tourism period to revitalize and develop other industries, ensuring the long-term sustainable use of its industrial heritage.

4.3 The Pathway Model

In the sustainable development of industrial cities like Harbin and Zibo, not only is strategic planning essential, but effective resource management is also required. The achievements in Harbin and Zibo are attributed to the integration of materials

Fig. 6 The word cloud of each topic in Zibo's stage 3

preparation and agenda setting. This study combines these two components in both cases to construct a strategic pathway model for the sustainable development of industrial heritage in old industrial cities, as illustrated in Fig. 10.

As illustrated in the figure, the pathways of Zibo and Harbin share the same objectives in the first and third stages, aiming to rapidly increase public visibility on social media and in the public discourse, while maintaining their prominence and influence to achieve sustainable development through distinct approaches. However, in the second stage, Zibo and Harbin differed in their strategies. Zibo focused on utilizing intangible elements of industrial heritage for agenda-setting, such as highlighting the pragmatic and honest characteristics of workers and residents shaped during the industrialization process, and narrating historical political activities related to labor movements to promote sustainable reuse. In contrast, Harbin emphasized tangible industrial heritage elements, such as notable locations and buildings from the industrialization era, including Central Street and Saint Sophia Cathedral, as well as beer culture activities and themed districts converted from former breweries.

The government's selection of different types of industrial heritage for sustainable development reflects specific preferences, which are influenced by the characteristics of the industrial heritage itself. For example, prior to this agenda-setting effort, Harbin was already a well-known city with a prominent industrial past. Its industrial heritage, along with its cultural and historical architecture, was already familiar to the

Fig. 7 The word cloud of each topic in Harbin's stage 1

public. Therefore, the government strategically utilized tangible elements that could effectively enhance public engagement and used these as a foundation to promote the reuse of other intangible heritage elements. Conversely, Zibo, a smaller city in Shandong that was not widely recognized, did not have distinctive architectural or urban features resulting from its industrialization. Hence, the direct use of tangible industrial heritage did not yield the desired results. Instead, the city leveraged culturally resonant narratives, such as the stories and images of local workers and communities, to facilitate the reuse of tangible industrial heritage.

Fig. 8 The word cloud of each topic in Harbin's stage 2

Fig. 9 The word cloud of each topic in Harbin's stage 3

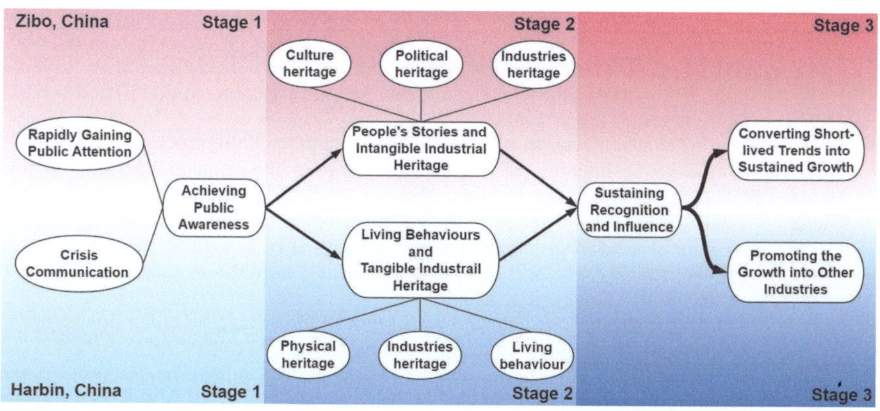

Fig. 10 The pathway model of sustainable development of industrial heritages in old industrial city

5 Conclusion

This study analyzed successful cases of industrial heritage reuse in the old industrial cities of Harbin and Zibo, China. By employing agenda-setting through media coverage and utilizing large language models for analysis, the study constructed a pathway model for the sustainable utilization of industrial heritage in these cities.

The findings indicate that the agenda-setting pathways in both cities exhibited similar patterns. Initially, both cities rapidly gained public attention, and eventually, they expanded successful reuse practices to other types of heritage and industries. However, during the intermediate stage, the types of industrial heritage selected differed due to each city's level of visibility and the characteristics of the heritage itself. Less well-known cities tend to prefer intangible industrial heritage for reuse, whereas more prominent cities are inclined to use tangible industrial heritage to achieve rapid success. Therefore, for the sustainable development of industrial heritage in post-industrial cities, governments need to establish effective public agendas, carefully select resources at each phase, and align strategies with their unique contexts to achieve optimal results.

References

1. Park, H.-S., Rene, E.R., Choi, S.-M., Chiu, A.S.F.: Strategies for sustainable development of industrial park in Ulsan, South Korea—from spontaneous evolution to systematic expansion of industrial symbiosis. J. Environ. Manage. **87**(1), 1–13 (2008)
2. Altenburg, T., Assmann, C.: Green Industrial Policy: Concept, Policies, Country Experiences. UN Environment; German Development Institute/Deutsches Institut für Entwicklungspolitk (DIE), German (2017)
3. Cizler, J.: Activation of former industrial areas: creative and civil sector as main actors in overcoming the institutional limitations in Serbia. Sociol. Space/Sociologija i Prostor **57**(1) (2019)
4. Dell'Anna, F.: What advantages do adaptive industrial heritage reuse processes provide? An econometric model for estimating the impact on the surrounding residential housing market. Heritage **5**(3), 1572–1592 (2022)
5. Günçe, K., Mısırlısoy, D.: Questioning the adaptive reuse of industrial heritage and its interventions in the context of sustainability. Sociol. Study **5**(9), 718–727 (2015)
6. Konior, A., Pokojska, W.: Management of postindustrial heritage in urban revitalization processes. Sustainability **12**(12), 5034 (2020)
7. Chen, J., Judd, B., Hawken, S.: Adaptive reuse of industrial heritage for cultural purposes in Beijing, Shanghai and Chongqing. Struct. Surv.. Surv. **34**(4/5), 331–350 (2016)
8. Veldpaus, L., Wacogne, R.: Industrial heritage and conservation planning, changing governance practices, examples from Europe. In Stegmeijer, E., Veldpaus, L. (Eds.) A Research Agenda for Heritage Planning. Edward Elgar Publishing (2021)
9. Perfetto, M.C., Vargas-Sánchez, A., Presenza, A.: Managing a complex adaptive ecosystem: towards a smart management of industrial heritage tourism. J. Tour. Sustain. Well-Being **4**(3), 243–264 (2016)
10. Erkman, S.: Industrial ecology: a new perspective on the future of the industrial system. Swiss Med. Wkly. **131**(37–38), 531–538 (2001)
11. Graedel, T.E.: On the concept of industrial ecology. Annu. Rev. Energy Env.. Rev. Energy Env. **21**(1), 69–98 (1996)

12. Korhonen, J.: Theory of industrial ecology. Prog. Ind. Ecol. Int. J. **1**(1/2/3), 61 (2004)
13. Saavedra, Y.M.B., Iritani, D.R., Pavan, A.L.R., Ometto, A.R.: Theoretical contribution of industrial ecology to circular economy. J. Clean. Prod. **170**, 1514–1522 (2018)
14. Herczeg, G., Akkerman, R., Hauschild, M.Z.: Supply chain collaboration in industrial symbiosis networks. J. Clean. Prod. **171**, 1058–1067 (2018)
15. Xavier, L.H., Giese, E.C., Ribeiro-Duthie, A.C., Lins, F.A.F.: Sustainability and the circular economy: a theoretical approach focused on e-waste urban mining. Resour. Policy. Policy **74**, 101467 (2021)
16. Reike, D., Vermeulen, W.J.V., Witjes, S.: The circular economy: new or refurbished as CE 3.0?—Exploring controversies in the conceptualization of the circular economy through a focus on history and resource value retention options. Resour. Conserv. Recycl. **135**, 246–264 (2018)
17. Perey, R., Benn, S., Agarwal, R., Edwards, M.: The place of waste: changing business value for the circular economy. Bus. Strateg. Environ.Strateg. Environ. **27**(5), 631–642 (2018)
18. Dokter, G., Thuvander, L., Rahe, U.: How circular is current design practice? Investigating perspectives across industrial design and architecture in the transition towards a circular economy. Sustain. Prod. Consump. **26**, 692–708 (2021)
19. Bottero, M., D'Alpaos, C., Oppio, A.: Ranking of adaptive reuse strategies for abandoned industrial heritage in vulnerable contexts: a multiple criteria decision aiding approach. Sustainability **11**(3), 785 (2019)
20. Berta, M., Bottero, M., Ferretti, V.: A mixed methods approach for the integration of urban design and economic evaluation: industrial heritage and urban regeneration in China. Environ. Plann. B: Urban Anal. City Sci. **45**(2), 208–232 (2018)
21. Wanta, W., Alkazemi, M.F.: Agenda-setting: history and research tradition. Int. Encyclop. Media Effects **2**, 12–25 (2017)
22. Wahl-Jorgensen, K., Hanitzsch, T. (Eds.): The Handbook of Journalism Studies. Routledge (2009)
23. Feezell, J.T.: Agenda setting through social media: the importance of incidental news exposure and social filtering in the digital era. Polit. Res. Q. **71**(2), 482–494 (2018)
24. De Bruijn, H., Janssen, M.: Building cybersecurity awareness: the need for evidence-based framing strategies. Gov. Inf. Q. **34**(1), 1–7 (2017)
25. Al-Dmour, H., Masa'deh, R., Salman, A., Abuhashesh, M., Al-Dmour, R.: Influence of social media platforms on public health protection against the COVID-19 pandemic via the mediating effects of public health awareness and behavioral changes: integrated model. J. Med. Internet Res. **22**(8), e19996 (2020)
26. Tursunbayeva, A., Franco, M., Pagliari, C.: Use of social media for e-Government in the public health sector: a systematic review of published studies. Gov. Inf. Q. **34**(2), 270–282 (2017)
27. Jamil, S., Appiah-Adjei, G.: The dire need of ecological sustainability in Pakistan: the agenda setting role of news media to address environmental problems. Media Asia 1–21 (2024)
28. Lee, J., Xu, W.: The more attacks, the more retweets: Trump's and Clinton's agenda setting on Twitter. Public Relat. Rev. **44**(2), 201–213 (2018)
29. Deng, C.D., Peng, K.-L., Shen, J.H.W.: Back to a post-pandemic city: the impact of media coverage on revisit intention of Macau. J. Qual. Assur. Hosp. Tour. **24**(1), 1–23 (2023)
30. Zhang, Y., Jing, Y.: Research on the effectiveness of network agenda setting of tourism destination image dissemination: take Ganzi as an example. Highl. Bus. Econ. Manage. **8**, 440–447 (2023)
31. Cui, Y., Che, W., Liu, T., Qin, B., Wang, S., Hu, G.: Revisiting pre-trained models for chinese natural language processing. Find. Assoc. Comput. Linguist. EMNLP **2020**, 657–668 (2020)
32. Grandeit, P., Haberkern, C., Lang, M., Albrecht, J., Lehmann, R.: Using BERT for qualitative content analysis in psychosocial online counseling. In: Proceedings of the Fourth Workshop on Natural Language Processing and Computational Social Science, pp 11–23 (2020)
33. Wang, G., Huang, Z., Li, D., Liao, J.: Analysis of ASEAN water resource policies in the context of belt and road initiative: a perspective on sustainable development goal 6 (SDG6). In: Wan Z (Ed.) Water Resources Management and Water Pollution Control, pp. 119–137. Springer Nature Switzerland (2024)

34. Cui, Y., Che, W., Liu, T., Qin, B., Yang, Z.: Pre-Training with whole word masking for Chinese BERT. IEEE/ACM Trans. Audio Speech Lang. Process. **29**, 3504–3514 (2021)
35. Mihalcea, R., Tarau, P.: TextRank: bringing order into texts. In: Proceedings of the 2004 Conference on Empirical Methods in Natural Language Processing, pp 404–411 (2024)
36. Geng, S., Chau, H.-W., Wang, T., Yan, S.: Influences of Russian culture on Harbin's urban planning from 1898 to 1931 and conservation recommendations. J. Chin. Archit. Urban. **3**(1), 1–25 (2021)

Mapping Research Trends on Sustainable Development Goals in China: A CiteSpace Visualization and Analysis

Guansu Wang, Deling Bai, Shiqi ji, and Xinru Liu

Abstract Sustainable Development Goals (SDGs), introduced by the United Nations, are becoming an important topic and framework of academic research globally. This study conducts a bibliometric analysis of Chinese scholars' research on Sustainable Development Goals using data from the Web of Science Core Collection and the CiteSpace software. The analysis covers the period from 2015 to 2024, focusing on publication trends, institutional collaborations, interdisciplinary research, and key research themes. Results show a significant increase in SDG-related research, with major contributions from institutions such as the Chinese Academy of Sciences. Key research areas include climate change, food security, renewable energy, circular economy, and the role of education and corporate social responsibility in sustainable development. The study identifies emerging trends and provides insights into the contributions of Chinese scholars to global sustainability efforts. By visualizing these trends, this research aims to inform future studies and policy developments in the field of sustainable development.

Keywords Sustainable development goals · Bibliometric analysis · CiteSpace · Climate change · Renewable energy · Circular economy · China · Web of Science

1 Introduction

In recent years, the rapid expansion of the global economy and unsustainable resource exploitation has led to numerous global challenges, such as climate change, resource scarcity, environmental pollution, public health issues, poverty, and social inequality

G. Wang (✉) · S. ji · X. Liu
School of Humanities, Zhuhai College of Science and Technology, Zhuhai, China
e-mail: s2128778@siswa.um.edu.my

G. Wang
Asia-Europe Institute, Universiti Malaya, Kuala Lumpur, Malaysia

D. Bai
Faculty of Creative Arts, Universiti Malaya, Kuala Lumpur, Malaysia

Y. Zhu (ed.), *Proceedings of the 4th International Symposium in Environmental Science and Industrial Ecology*, https://doi.org/10.1007/978-981-96-1578-0_11

[18, 19]. These issues affect not only individual countries or regions but also pose a threat to global sustainable development. In response, the United Nations adopted the 2030 Agenda for Sustainable Development at its 70th General Assembly in New York in September 2015, which established 17 goals and 169 specific targets, collectively known as the Sustainable Development Goals (SDGs). The introduction of the SDGs provides a framework for addressing global challenges and sets a common goal for international cooperation. The concept of "sustainable development" is not only concerned with current resource use and economic growth but also with the future survival and development of the human species.

In this context, Sustainable Development (SD) has become one of the most discussed topics in contemporary academia. Given that the targets of the Sustainable Development Goals (SDGs) are interrelated, achieving one goal often impacts others. For instance, the goal of eradicating poverty is closely connected with objectives related to education, gender equality, and health. Consequently, scholars frequently approach this topic through interdisciplinary collaboration. A review of the literature reveals that the international academic community primarily investigates this issue from the following perspectives. Firstly, many scholars focus on the implementation and governance of the SDGs, emphasizing the importance of effective governance structures in achieving sustainable development goals. These studies often analyze the effectiveness of policy formulation, execution, and monitoring at national and regional levels. For instance, some research highlights that the achievement of SDGs depends on effective governance arrangements, but the understanding of governance models and aspects required for different SDG targets remains unclear [13]. Research from this perspective often employs case studies and comparative analysis to assess experiences and lessons learned from different countries in SDG implementation. Additionally, other scholars examine the interaction between economic growth and environmental protection, exploring how to balance these two aspects. This research focuses on the interactions between SDGs, particularly the impact of economic development on the environment and how sustainable economic policies can promote environmental goals. These studies typically use quantitative analysis methods, employing models and indicators to evaluate the effects of various policies on the economy and environment [3].

In addition, scholars in the social sciences emphasize the relationship between SDGs and social equity and inclusion, particularly how to ensure that no one is left behind. This perspective examines the interests and impacts of different social groups in the implementation of SDGs and investigates how policy design can safeguard the rights of vulnerable populations. Research methods in this area often include qualitative approaches, such as interviews and focus groups, to gain insights into social dynamics and individual experiences [13]. Another research direction focuses on the role of education in promoting the achievement of SDGs. Numerous studies indicate that education is a critical element in achieving sustainable development goals, as increasing public awareness and knowledge can effectively advance sustainable practices. This research frequently uses literature reviews and case studies to evaluate the contribution of educational programs to the SDGs [5].

As the world's second-largest economy and the largest developing country, China faces both opportunities and challenges in sustainable development. On the one hand, China has achieved significant progress during its rapid economic growth, greatly improving the standard of living for its population. On the other hand, the environmental issues and resource pressures resulting from economic development have become increasingly prominent. Problems such as air pollution, water scarcity, and land desertification are worsening, severely impacting people's quality of life and health. Consequently, the Chinese government has incorporated sustainable development goals as key indicators in its 13th and 14th Five-Year Plans, promoting coordinated development across economic, social, and environmental dimensions. Specific measures include: poverty alleviation, environmental protection and renewable energy. China's various initiatives related to the SDGs play a crucial role in advancing and implementing these goals on a global scale.

This study aims to conduct a bibliometric analysis of literature published by Chinese scholars in the Web of Science Core Collection using CiteSpace. Through this analysis, the study seeks to evaluate the development trends of SDG-related research from a Chinese perspective, identify key research hotspots, and provide guidance for future research.

2 Methodology

2.1 Data Collection

In this study, the Web of Science (WoS) Core Collection citation database was used as the primary source for data collection, processing, and analysis. WoS is a widely utilized literature retrieval and bibliometric analysis database in academic research, recognized as one of the most reputable citation databases, ensuring the inclusion of high-quality and impactful research [22].

The data was collected using a targeted search query to focus on contributions from Chinese scholars concerning the SDGs. The search criteria employed were AD = (China) AND TS = ("sustainable development goals" OR "SDGs" OR "millennium development goals"), filtering for publications including articles, review articles, proceeding papers, and book chapters. The specific time range for the search spanned from 2015-09-25 to 2024-09-19. This period was selected to encompass the entire scope of research following the official adoption of the SDGs by the United Nations on September 25, 2015, up to the most recent data available, providing a comprehensive view of the ongoing research activities and developments. Including "millennium development goals" in the search helps in capturing the foundational research that preceded the SDGs, thus providing a full spectrum analysis of the evolution of sustainability research among Chinese scholars. The 3978 identified literature records, along with citation information, bibliographic details, abstracts, keywords, and other relevant data, were exported as a .txt file format for bibliometric analysis. This format allows for comprehensive processing and analysis of the collected data.

2.2 Bibliometric Analysis

Bibliometrics uses mathematical and statistical methods to quantitatively analyze the production, growth, maturation, and consumption of scientific publications. As a result, it has been widely used as an important tool to evaluate and analyze the development status of research fields, including researcher output [20, 26], institutional collaboration [10], and economic development [2] and so on. This study employs CiteSpace for data analysis. CiteSpace, an open-source software package developed at Drexel University and based on the Java platform [9], aims to dynamically track emerging trends in knowledge domains, with a focus on identifying key turning points in the development of knowledge fields.

3 Result

3.1 Visualization of Publication Trends

As shown in Fig. 1, the number of publications by Chinese scholars on Sustainable Development Goals (SDGs) has generally been on the rise from 2015 to 2024. In the early stages of the global sustainable development agenda (2015–2018), China's research on SDGs was in its infancy. However, with the advancement of national policies, the number of publications increased significantly, jumping from 139 papers in 2019 to 447 in 2022. This indicates that SDG-related topics have garnered broader research attention and policy support. In the last two years, the number of publications surged further, reaching 1073 papers in 2023 and 1022 in 2024. This growth may be attributed to the increasing domestic and international emphasis on sustainable development issues, as well as the rise in research funding [15].

3.2 Visualization of Research Institutions

Figure 2 and Table 1 display the collaboration network and influence of major Chinese research institutions in the field of Sustainable Development Goals (SDGs). As shown in Fig. 2, the concentrated nodes indicate dense collaboration among various institutions, particularly between leading universities and research institutes such as Beijing Normal University, Tsinghua University, and Peking University. The gradient colors (from blue to red) demonstrate an increasing number of collaborations over

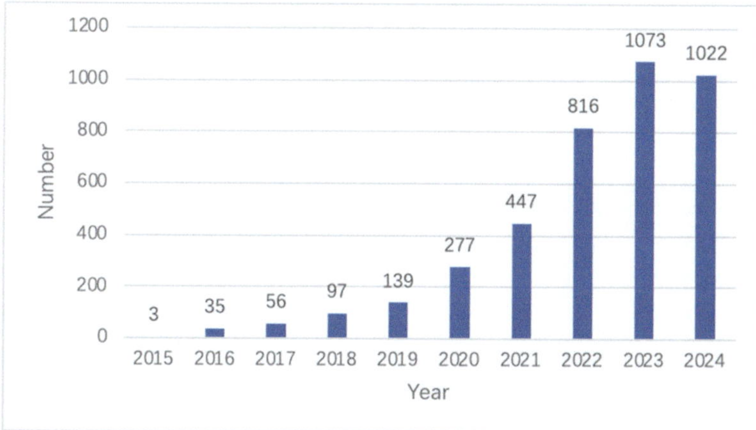

Fig. 1 Number of publications and trends in SDG research by Chinese scholars (2015–2024)

time, especially between 2022 and 2024, reflecting the growing research activity in the SDG field. The numerous nodes and links highlight extensive cooperation and exchange both within China and internationally in SDG-related research.

According to Table 1, the Chinese Academy of Sciences has the highest centrality in the network (Centrality = 0.05), making it a key node. The network structure is relatively concentrated, with a large number of nodes surrounding the central institution, indicating its pivotal role as a bridge in the overall network. This suggests that the Chinese Academy of Sciences has widespread collaborations and influence in

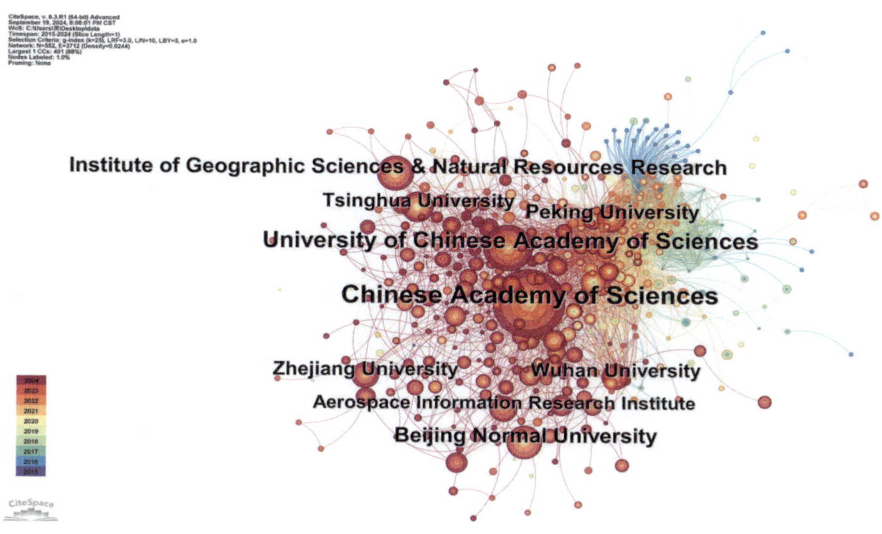

Fig. 2 Co-occurrence map of major research institutions

Table 1 Top 10 research institutions focusing on SDGs

Institutions	Count	Centrality
Chinese Academy of Sciences	649	0.05
University of Chinese Academy of Sciences	305	0.01
Beijing Normal University	179	0.01
Institute of Geographic Sciences and Natural Resources Research	171	0
Peking University	137	0.01
Tsinghua University	132	0.01
Zhejiang University	122	0.02
Wuhan University	117	0.03
Aerospace Information Research Institute	107	0.02
University of London	91	0.03

SDG-related research. Additionally, Wuhan University and the University of London also show high centrality (0.03), signifying their prominent roles in international collaboration and influence. Other institutions, such as Zhejiang University and the Beijing Academy of Aeronautics and Astronautics, also demonstrate strong research capabilities and collaboration networks. The collaboration with international institutions, particularly through joint research projects and exchange activities, enhances the global perspective and practical application of research in the SDG field.

3.3 Co-occurrence Visualization of Disciplines

Figure 3 highlights the interactions and connections between various disciplines in the research on Sustainable Development Goals (SDGs). Environmental Science, Green and Sustainable Science and Technology, and Environmental Studies are the three core nodes, showing high publication volume and centrality, indicating their dominant role in SDG research. Environmental Science, being the largest node with the highest centrality (0.19), underscores the central importance of environmental issues in sustainable development research.

Additionally, closely related disciplines such as Public Health, Energy and Fuels, and Remote Sensing are prominent in the map, reflecting the interdisciplinary nature of modern research, particularly in addressing complex global challenges like climate change, energy crises, and ecological conservation. The time-based color gradient (from blue to red) in the map reveals the rapid rise of emerging fields like Green Technology and Sustainable Science in recent years.

These dense connections among disciplines not only indicate the growing academic attention to SDGs and the increasing trend of cross-disciplinary collaboration, but also highlight the critical role of scientific research in shaping social and environmental policies [27].

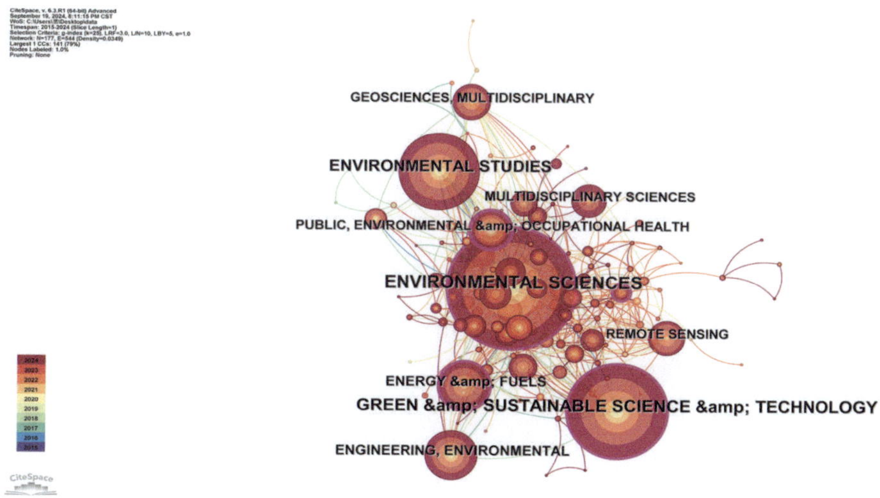

Fig. 3 Co-occurrence map of major research disciplines

3.4 *Keyword Co-occurrence Network Visualization Analysis*

As shown in Fig. 4, the co-occurrence network contains 622 nodes and 4393 links, with a network density of 0.0227. Each circular node represents a corresponding keyword, and the size of the node is proportional to the frequency of the keyword's appearance. The larger the node, the more likely it is to be a hotspot in Chinese scholars' research on SDGs. It is evident that there are relatively few scattered nodes, indicating that the connections between various themes are relatively close. This suggests that Chinese scholars' research on SDGs tends to focus on a few concentrated topics, reflecting a high level of interconnection between different research areas. The dense network demonstrates that the focal points in China's SDG-related research are highly aligned and coherent across different disciplines.

According to the keyword frequency analysis (Table 2), the term "sustainable development goals" appears most frequently, with 673 occurrences and a centrality of 0.03. "Sustainable development" appears 442 times with a centrality of 0.41. Following these are terms such as "impact" and "China."

Notably, aside from the top four general keywords, more specific research-related keywords show significant prominence. "CO_2 emissions" appears most frequently, with 356 occurrences and a centrality of 0.03, followed by "economic growth" with 325 occurrences and a centrality of 0.02. Other high-frequency keywords include "climate change," "renewable energy," and "consumption." These results highlight the focused attention on environmental and economic aspects of SDG research in China, particularly concerning climate change and sustainable energy solutions.

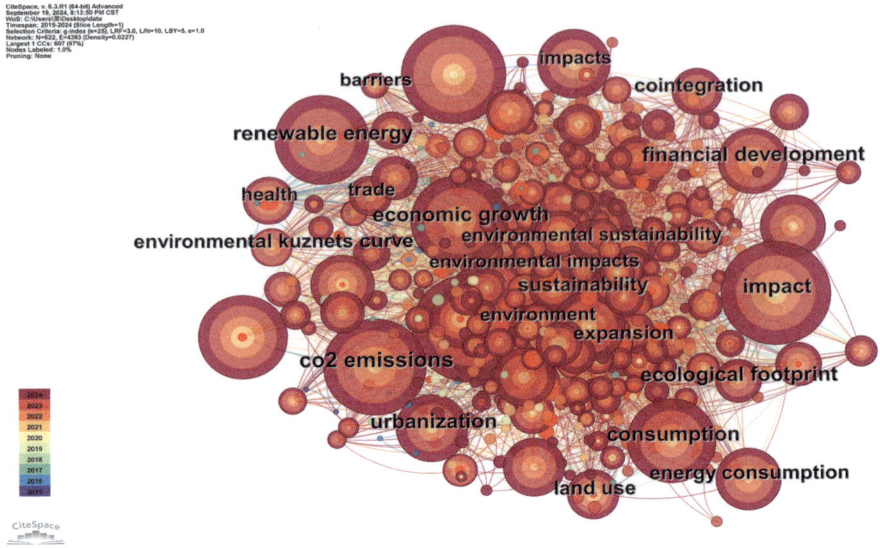

Fig. 4 Keyword co-occurrence network of Chinese scholars' SDG research

Table 2 Clustering labels of Chinese scholars' SDG research

Keywords	Count	Centrality
Sustainable development goals	673	0.03
Sustainable development	442	0
Impact	420	0.05
China	367	0.02
CO_2 emissions	356	0.03
Economic growth	325	0.02
Climate change	293	0.02
Renewable energy	287	0.02
Consumption	278	0.04

3.5 Keyword Clustering Analysis

Based on the keyword co-occurrence analysis, clustering was performed to generate the keyword clustering map of Chinese scholars' research on SDGs (Fig. 5). Six clusters were formed in the map, with a Q-value of 0.4278 (> 0.3), indicating that the clustering is effective. The average silhouette value is 0.7042 (> 0.5), demonstrating high consistency among cluster members, making the clustering results reasonable. As seen in Table 3, the average silhouette values for all clusters are above 0.5, indicating a high concentration of research themes within each category.

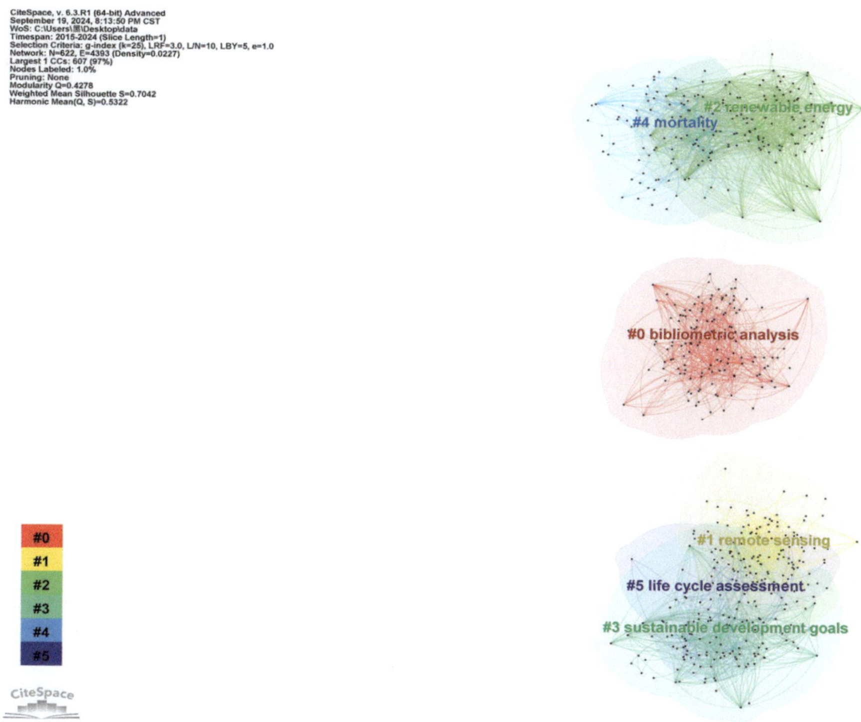

Fig. 5 Keyword clustering map of Chinese scholars' SDG research

The research hotspots of Chinese scholars in SDG-related studies mainly focus on the following three areas: Sustainable Development Assessments of Climate Change and Food Security; Technological and Financial Innovation in Renewable Energy and Circular Economy; Sustainable Development in Education, Corporate Responsibility, and Green Entrepreneurship.

3.6 Keyword Burst Detection Analysis

Figure 6 illustrates the surge in keyword citations in Sustainable Development Goals (SDG)-related research between 2015 and 2024, highlighting the topics that have gained significant attention in scientific research. The data shows that keywords like "health," "Sustainable Development Goals (SDGs)," and "climate change" exhibit the highest burst indexes, reflecting the global emphasis on health, environmental changes, and sustainability policies as current and future research hotspots.

From a temporal perspective, most keyword bursts began in 2016, likely aligning with the implementation of the United Nations' SDGs. Notably, the keyword "Sustainable Development Goals" has shown a sustained high citation intensity since

Table 3 The final data of this study

Label (LLR)	ClusterID	Silhouette	Label (LSI)
Bibliometric analysis	0	0.599	Sustainable development goals; climate change; food security; food waste management; greenhouse gases I sustainable development; bibliometric analysis; text mining; science mapping; floods
Remote sensing	1	0.643	Sustainable development goals; ecosystem services; loess plateau; carbon sequestration; land degradation I sustainable development; light types classification; human population size; mainland Southeast Asia; urban sustainability
Renewable energy	2	0.835	Renewable energy; carbon emissions; ecological footprint; technological innovation; chemical oxygen demand I CO_2 emissions; financial development; bri countries; fossil fuels electricity capacity; renewable power capacity
Sustainable development goals	3	0.738	Sustainable development goals; climate change; food security; greenhouse gases; food waste management I sustainable development; populous cities; pollution; landscape metrics; sustainable development index
Mortality	4	0.724	Sustainable development goals; affective commitment; teacher wellbeing; corporate social responsibility; educational institute culture I sustainable development; environmental education; sustainable entrepreneurship; green entrepreneurship; health systems evaluation
Life cycle assessment	5	0.76	Sustainable development goals; environmental impacts; agricultural residues; sustainability contributions; qardl I life cycle assessment; circular bioeconomy; sustainable waste management; fuel co-firing; plus model

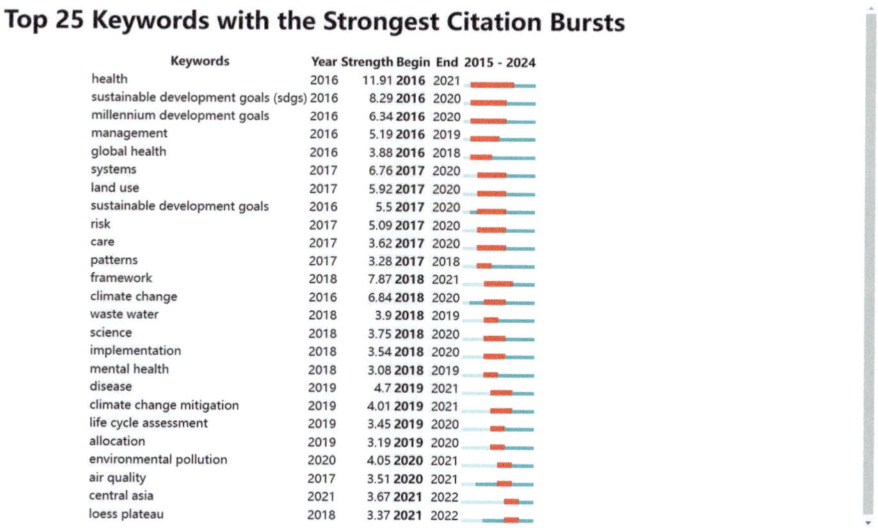

Fig. 6 Keyword burst detection in Chinese scholars' SDG research (2015–2024)

2016, indicating growing global interest in scientific research aimed at achieving these goals.

In addition, research areas such as "mental health," "water resource management," and "environmental pollution" have also shown increasing trends, likely reflecting heightened public and policymaker interest in the intersection of health, environmental protection, and resource management. As time progresses, more specific keywords such as "climate change mitigation" and "life cycle assessment" have started to gain attention, indicating a shift towards more targeted implementation and evaluation studies in these areas.

4 Discussion

4.1 Assessment of Sustainable Development in Climate Change and Food Security

As illustrated by Figs. 2 and 6, this topic is one of the most prominent areas of research among Chinese scholars in the field of SDGs. It encompasses the intersection of climate change mitigation, food security, and ecosystem service protection, with a focus on addressing global environmental challenges through sustainable development. This field can be interpreted from three perspectives.

Firstly, the adverse impacts of climate change, such as extreme weather events, greenhouse gas emissions, and ecological imbalances, are affecting agricultural production and food security. Analysis in this area requires evaluating various climate adaptation measures, such as reducing carbon emissions, adopting climate-resilient crops, and optimizing water resource management [12, 29].

Secondly, food security is a core component of sustainable development. Climate change, land degradation, and the sustainability of food production and distribution directly impact global food supply. Research focuses on improving food security through methods such as ecological agriculture, waste management, and reducing food waste [7, 8].

Lastly, ecosystem services, including carbon sequestration, clean water sources, and soil nutrient cycling, are crucial for supporting human life and economic activities. Protecting and restoring these services is essential, especially in the context of land degradation and environmental pollution [21, 23].

China's focus on climate change, food security, and ecosystem services in SDG research reflects an active response to global environmental challenges. By reducing carbon emissions, improving the sustainability of food production and distribution, and protecting ecosystem services, China demonstrates its commitment and actions toward addressing environmental issues. These efforts are significant for advancing global sustainable development.

4.2 Technological and Financial Innovations in Renewable Energy and Circular Economy

Keyword co-occurrence and clustering analysis reveal that Chinese scholars place significant emphasis on technological and financial innovations in renewable energy and the circular economy. This focus highlights the importance of the interaction between renewable energy technologies and circular bio-economy in China, as well as the crucial role of technological innovation and financial investment in achieving sustainable development goals. As fossil fuels become increasingly depleted and their environmental impacts become more pronounced, renewable energy sources such as solar, wind, and biomass energy are seen as key solutions to carbon emission issues. Research in this area concentrates on accelerating the innovation and deployment of renewable energy technologies to replace fossil fuel power generation and reduce greenhouse gas emissions [16, 31].

Furthermore, the circular bio-economy involves achieving both economic and environmental sustainability through resource regeneration and recycling. Technologies such as agricultural residue management, waste management, and co-firing are integral components of the circular bio-economy, helping to reduce resource waste and environmental pollution [6, 24]. Lastly, technological innovation and financial development are pivotal. Technological innovation is at the heart of the energy transition, while financial development is essential for supporting green technology and promoting corporate sustainability. Policy support and funding are critical factors for the success of technological innovation and the bio-economy [1, 25].

Chinese scholars' focus on technological and financial innovations in renewable energy and the circular economy reflects the country's emphasis on technological advancement and financial support in addressing fossil fuel depletion and environmental challenges. Accelerating the innovation and deployment of renewable energy technologies and integrating practices from the circular bio-economy demonstrate China's proactive efforts in advancing sustainable development goals.

4.3 The Role of Education, Social Responsibility, and Green Entrepreneurship in Sustainable Development Goals

Keyword clustering and burst detection analyses indicate that Chinese scholars focus significantly on the synergistic roles of education, corporate social responsibility (CSR), and green entrepreneurship in advancing Sustainable Development Goals (SDGs), emphasizing the positive impact of social and cultural factors in this process. Environmental and sustainability education plays a crucial role in shifting attitudes toward natural resources both at the individual and societal levels. Through educational institutions and cultural dissemination, fostering environmental awareness and social responsibility in the next generation is a key step toward achieving sustainable development [4, 11].

CSR involves corporate contributions to environmental and social sectors, particularly in addressing climate change, reducing carbon emissions, and promoting community sustainability. The active promotion of CSR contributes to the achievement of global SDGs [14, 30]. Green and sustainable entrepreneurship represents the entrepreneurial spirit that drives environmental protection through innovative business models. The development of green technologies and eco-friendly products not only helps reduce pollution but also creates economic opportunities, aiding in the transition to a green economy [17, 28].

The focus of Chinese scholars on the interplay between education, CSR, and green entrepreneurship underscores the significance of social and cultural factors. Environmental education enhances public awareness, CSR advances environmental and community development, and green entrepreneurship fosters an eco-friendly economy. Collectively, these factors contribute to China's proactive progress toward achieving sustainable development goals.

5 Conclusion

This study provides a comprehensive bibliometric analysis of Chinese scholars' contributions to Sustainable Development Goals (SDGs) research from 2015 to 2024. The results reveal a steady growth in publication output, particularly in areas like climate change mitigation, food security, renewable energy, and circular economy.

Chinese institutions, especially the Chinese Academy of Sciences, play a pivotal role in this field, fostering both domestic and international collaborations. The research highlights the growing interdisciplinary nature of SDG studies, with strong connections between environmental science, public health, energy, and technological innovation.

Additionally, the study emphasizes the importance of education, corporate social responsibility (CSR), and green entrepreneurship in advancing SDGs. By promoting awareness, responsibility, and innovation, these social and cultural factors contribute significantly to sustainable development efforts. The findings underscore the proactive role of Chinese scholars in addressing global sustainability challenges, offering valuable insights for future research and policy formulation in the pursuit of global sustainability. Further research should focus on strengthening interdisciplinary collaboration and exploring more targeted strategies for achieving SDGs across different regions and sectors.

References

1. Adewale Alola, A., Ozturk, I., Bekun, F.V.: Is clean energy prosperity and technological innovation rapidly mitigating sustainable energy-development deficit in selected sub-Saharan Africa? A myth or reality. Energy Policy **158**, 112520 (2021)
2. Afjal, M.: Bridging the financial divide: a bibliometric analysis on the role of digital financial services within FinTech in enhancing financial inclusion and economic development. Humanit. Soc. Sci. Commun. **10**(1), 1–27 (2023)
3. Alcamo, J., Thompson, J., Alexander, A., Antoniades, A., Delabre, I., Dolley, J., Marshall, F., Menton, M., Middleton, J., Scharlemann, J.P.W.: Analysing interactions among the sustainable development goals: findings and emerging issues from local and global studies. Sustain. Sci. **15**(6), 1561–1572 (2020)
4. Almasri, N.A., Smythe, T., Hadders-Algra, M., Olusanya, B.O.: Global research on developmental disabilities collaborators (GRDDC): prioritising rehabilitation in early childhood for inclusive education: a call to action. Disabil. Rehabil. **45**(19), 3155–3159 (2023)
5. Armitage, C.S., Lorenz, M., Mikki, S.: Mapping scholarly publications related to the sustainable development goals: do independent bibliometric approaches get the same results? Quant. Sci. Stud. **1**(3), 1092–1108 (2020)
6. Barros, M.V., Salvador, R., De Francisco, A.C., Piekarski, C.M.: Mapping of research lines on circular economy practices in agriculture: from waste to energy. Renew. Sustain. Energy Rev. **131**, 109958 (2020)
7. Ben Hassen, T., El Bilali, H.: Impacts of the Russia-Ukraine war on global food security: towards more sustainable and resilient food systems? Foods **11**(15), 2301 (2022)
8. Bizikova, L., Jungcurt, S., McDougal, K., Tyler, S.: How can agricultural interventions enhance contribution to food security and SDG 2.1? Glob. Food Secur. **26**, 100450 (2020)
9. Chen, C.: CiteSpace II: detecting and visualizing emerging trends and transient patterns in scientific literature. J. Am. Soc. Inform. Sci. Technol. **57**(3), 359–377 (2006)
10. Farooq, R.: A review of knowledge management research in the past three decades: a bibliometric analysis. VINE J. Inf. Knowl. Manag. Syst. **54**(2), 339–378 (2024)
11. Fernandez Rivas, D., Boffito, D.C., Faria-Albanese, J., Glassey, J., Afraz, N., Akse, H., Boodhoo, K.V.K., Bos, R., Cantin, J., (Emily) Chiang, Y.W., Commenge, J.-M., Dubois, J.-L., Galli, F., De Mussy, J.P.G., Harmsen, J., Kalra, S., Keil, F.J., Morales-Menendez, R., Navarro-Brull, F.J., Noël, T., Ogden, K., Patience, G.S., Reay, D., Santos, R.M., Smith-Schoettker, A.,

Stankiewicz, A.I., Van Den Berg, H., Van Gerven, T., Van Gestel, J., Van der Stelt, M., Van De Ven, M., Weber, R.S.: Process intensification education contributes to sustainable development goals. Part 1. Educ. Chem. Eng. **32**, 1–14 (2020)
12. Fujimori, S., Hasegawa, T., Takahashi, K., Dai, H., Liu, J.-Y., Ohashi, H., Xie, Y., Zhang, Y., Matsui, T., Hijioka, Y.: Measuring the sustainable development implications of climate change mitigation. Environ. Res. Lett. **15**(8), 085004 (2020)
13. Glass, L.-M., Newig, J.: Governance for achieving the sustainable development goals: how important are participation, policy coherence, reflexivity, adaptation and democratic institutions? Earth Syst. Gov. **2**, 100031 (2019)
14. Gunawan, J., Permatasari, P., Tilt, C.: Sustainable development goal disclosures: do they support responsible consumption and production? J. Clean. Prod. **246**, 118989 (2020)
15. Hailiang, Z., Chau, K.Y., Waqas, M.: Does green finance and renewable energy promote tourism for sustainable development: empirical evidence from China. Renew. Energy **207**, 660–671 (2023)
16. He, J., Yang, Y., Liao, Z., Xu, A., Fang, K.: Linking SDG 7 to assess the renewable energy footprint of nations by 2030. Appl. Energy **317**, 119167 (2022)
17. Herman, E.: The interplay between digital entrepreneurship and sustainable development in the context of the EU digital economy: a multivariate analysis. Mathematics **10**(10), 1682 (2022)
18. Huang, T.-H., Chang, C.-H.: Improving response diversity through commonsense-aware empathetic response generation. In: Chang, Y.-C., Huang, Y.-C. (eds.) Proceedings of the 34th Conference on Computational Linguistics and Speech Processing (ROCLING 2022), pp. 299–306. The Association for Computational Linguistics and Chinese Language Processing (ACLCLP) (2022)
19. Khojasteh, S.C., Argikar, U.A., Cheruzel, L., Cho, S., Crouch, R.D., Dhaware, D., Heck, C.J.S., Johnson, K.M., Kalgutkar, A.S., King, L., Liu, J., Ma, B., Maw, H., Miller, G.P., Seneviratne, H.K., Takahashi, R.H., Wang, S., Wei, C., Jackson, K.D.: Biotransformation research advances—2022 year in review. Drug Metab. Rev. **55**(4), 301–342 (2023)
20. Mishra, M., Desul, S., Santos, C.A.G., Mishra, S.K., Kamal, A.H.M., Goswami, S., Kalumba, A.M., Biswal, R., Dos Da Silva, R.M., Santos, C.A.C., Baral, K.: A bibliometric analysis of sustainable development goals (SDGs): a review of progress, challenges, and opportunities. Environ. Dev. Sustain. **26**, 11101–11143 (2024)
21. Peng, K., Jiang, W., Ling, Z., Hou, P., Deng, Y.: Evaluating the potential impacts of land use changes on ecosystem service value under multiple scenarios in support of SDG reporting: a case study of the Wuhan urban agglomeration. J. Clean. Prod. **307**, 127321 (2021)
22. Pranckutė, R.: Web of Science (WoS) and Scopus: the titans of bibliographic information in today's academic world. Publications **9**(1), 12 (2021)
23. Qiu, H., Hu, B., Zhang, Z.: Impacts of land use change on ecosystem service value based on SDGs report—taking Guangxi as an example. Ecol. Ind. **133**, 108366 (2021)
24. Ronzon, T., Sanjuán, A.I.: Friends or foes? A compatibility assessment of bioeconomy-related sustainable development goals for European policy coherence. J. Clean. Prod. **254**, 119832 (2020)
25. Rout, S.K., Gupta, M., Sahoo, M.: The role of technological innovation and diffusion, energy consumption and financial development in affecting ecological footprint in BRICS: an empirical analysis. Environ. Sci. Pollut. Res. **29**(17), 25318–25335 (2022)
26. Vlase, I., Lähdesmäki, T.: A bibliometric analysis of cultural heritage research in the humanities: the Web of Science as a tool of knowledge management. Humanit. Soc. Sci. Commun. **10**(1), 1–14 (2023)
27. Wittmann, V., Meissner, D.: The nexus of global challenges and global studies: a transdisciplinary global sustainability science curriculum. Discov. Sustain. **5**(1), 43 (2024)
28. Włodarczyk, B., Firoiu, D., Ionescu, G.H., Ghiocel, F., Szturo, M., Markowski, L.: Assessing the sustainable development and renewable energy sources relationship in EU countries. Energies **14**(8), 2323 (2021)
29. Yang, R., Li, X., Mao, D., Wang, Z., Tian, Y., Dong, Y.: Examining fractional vegetation cover dynamics in response to climate from 1982 to 2015 in the Amur River Basin for SDG 13. Sustainability **12**(14), 5866 (2020)

30. Yarram, S.R., Adapa, S.: Board gender diversity and corporate social responsibility: is there a case for critical mass? J. Clean. Prod. **278**, 123319 (2021)
31. Zhao, J., Sinha, A., Inuwa, N., Wang, Y., Murshed, M., Abbasi, K.R.: Does structural transformation in economy impact inequality in renewable energy productivity? Implications for sustainable development. Renew. Energy **189**, 853–864 (2022)

Blueprint for the 'Dual Carbon' Strategy and Exploration of Mutual Aid Economy Under Green Transformation

Shuming Chen, Xianjin Li, Shangyi Jin, and Xiaohui Zou

Abstract Amid global climate change, nations have set emission targets. China's "dual carbon" strategy aims for carbon peaking and neutrality, showcasing proactive climate governance and economic green transformation. This paper proposes scientific methods for achieving these goals, advocating a multifaceted approach beyond individual actions or carbon taxes/trading. It emphasizes China's commitment to the "3060" targets, green development, and international cooperation. The paper clarifies carbon neutrality misconceptions, outlines six critical balances for the strategy, and proposes a five-step method for fundamental principles. An implementation roadmap spanning short, medium, and long-term phases is developed. Furthermore, the potential of a mutual-aid economy as a future model is explored, suggesting it will evolve in the AI and carbon neutrality era. The study concludes that a mutual-aid economy represents the future of societal and technological development, offering a novel path to resolve conflicts like employment and carbon consumption rights through green carbon rights and digital currency exchange.

Keywords Cross-boundary perspective · Dual carbon strategy · Implementation road-map · Climate change · Green transformation · Mutual aid economy

1 Introduction

In the face of the escalating severity of global climate change [1], nations worldwide have set ambitious emission reduction targets as a collective response to this pressing global challenge. As the largest developing country, China has proactively embarked

S. Chen · X. Li · X. Zou (✉)
Tianyanzhi Technology Industry Co., Ltd., Beijing, China
e-mail: 949309225@qq.com

S. Jin · X. Zou
Tiancheng Zhi (Beijing) Big Data Technology Co., Ltd., Beijing, China

X. Zou
Interdisciplinary Knowledge Modeling Research Group, Peking University, Beijing, China

Y. Zhu (ed.), *Proceedings of the 4th International Symposium in Environmental Science and Industrial Ecology*, https://doi.org/10.1007/978-981-96-1578-0_12

on the dual carbon strategy, striving to achieve carbon peaking and ultimately carbon neutrality [2]. This strategic move not only underscores China's commitment to global climate governance but also heralds a pivotal shift towards fostering a green transformation in its economic and social development trajectories.

The essence of this strategy lies in proposing rigorous scientific methodologies tailored to meet the dual carbon targets and delving deep into the theoretical underpinnings that support them. Carbon neutrality, specifically, stands as a cornerstone milestone in China's comprehensive climate action plan, necessitating a multifaceted approach that transcends mere individual low-carbon behaviors, carbon taxes, or carbon trading mechanisms. In response to the intensifying global climate crisis, China has consistently demonstrated its dedication by committing to the ambitious 3060 dual carbon target, actively promoting green development initiatives, and fostering international cooperation to tackle climate change head-on [3–5].

Moreover, this paper underscores the paramount importance of innovative strategies within the realm of the mutual-aid economy as a catalyst for accelerating progress towards these goals. It emphasizes the necessity for comprehensive and systematic planning, coupled with the enhancement of clean energy consumption and efficiency, as integral components of this transformative journey. By exploring the potential of the mutual-aid economy as a future economic paradigm, this paper posits that it will emerge as one of the primary economic forms in the era of AI and carbon neutrality, contributing significantly to the global effort towards sustainability.

2 Background

Addressing Common Misconceptions Surrounding Carbon Neutrality from a Cross-boundary Perspective.

The pursuit of carbon neutrality in the face of global climate change necessitates a nuanced understanding that transcends simplistic solutions. Firstly, the notion that universal carbon reduction alone is a panacea for achieving carbon neutrality is misguided. While individual actions, such as adopting low-carbon lifestyles, undoubtedly contribute, they cannot stand as the sole foundation for reaching this milestone. For instance, isolated efforts by individuals living in energy-intensive lifestyles, like cycling instead of driving a luxury car, are but drops in the ocean compared to the collective emissions generated by industrial processes and large-scale energy consumption. True carbon neutrality necessitates comprehensive, society-wide mandates and regulations that govern all sectors of the economy.

Secondly, carbon taxes, while a tool in the climate policy toolkit, are not an effective standalone solution for regulating carbon neutrality. Carbon taxes, inherently results-driven, can create disparities and have limited impact without a well-established framework for distributing carbon emissions equitably across production, services, and consumption. Establishing a comprehensive carbon emission distribution system that encompasses the entire energy supply chain—from production to consumption—is imperative before implementing carbon taxes effectively.

Thirdly, carbon trading should not be misconstrued as a means to carbon neutrality but rather seen as a byproduct of a well-designed climate policy framework. Without a robust system in place to manage carbon emissions at the individual and industrial levels, carbon trading can exacerbate imbalances. The essence of a fair carbon trading system lies in recognizing that every individual on Earth has an equal right to the carbon emissions necessary for basic survival, necessitating a bottom-up approach that begins with regulating individual carbon consumption and extends to logistics and production processes.

Furthermore, it is crucial to dispel the misconception that carbon neutrality equates to restricting energy demand. Energy is the lifeblood of economic growth, and limiting its use is not the objective. Rather, carbon neutrality aims to decouple economic progress from environmental degradation by promoting clean energy sources, enhancing energy efficiency, and fostering a green economy that supports both environmental protection and sustainable development.

Drawing from a myriad of sources, including the IPCC, UNFCCC agreements, OECD reports [6, 7], and various government policies, this discourse underscores the importance of overcoming prevalent misconceptions about carbon neutrality. In the context of China's dual-carbon strategy [8, 9], the reconstruction of its green carbon sequestration system necessitates a shift from simplistic notions like "green equals carbon sequestration" [10–12] to a more nuanced understanding that only incremental greenness leads to meaningful carbon sequestration.

The analysis of China's energy supply–demand balance highlights the ultimate responsibility of individuals, families, and organizations for their energy consumption, tied closely to their demand drivers. Achieving carbon neutrality paves the way for a controllable, equitable, and just social operational system. Moreover, it emphasizes the need to address the inherent disconnect in China's current energy supply–demand structure and the gradual transition from investment-attracting policies like land taxation to carbon neutrality-driven measures [13–15].

In conclusion, promoting measures for China's overall energy supply–demand balance involves redefining energy source roles, formulating incremental control policies, and establishing industrial support systems rooted in the carbon neutrality strategy. Key principles include maximizing clean and stored energy usage in demand systems and ensuring stable, controlled energy for essential social and national needs. Based on these principles and the existing power grid structure, comprehensive optimization projects aim to minimize energy dispatching losses, guide local governments in industrial transfers, and optimize energy efficiency across advantageous industries. Lastly, while emerging industries like the metaverse may garner attention, they must align with carbon neutrality objectives, underscoring the need for a critical assessment of their environmental impact and potential for fostering sustainable practices [16–18].

3 Materials and Methods

3.1 General Information

From a cross-boundary perspective, achieving carbon neutrality necessitates a multi-faceted strategy that transcends sole reliance on universal carbon reduction. For China's dual carbon strategy to succeed, a delicate balance must be struck between energy supply–demand dynamics, economic growth, and environmental protection. This involves harmonizing policy formulation and implementation, technological innovation, domestic and international cooperation, alongside short- and long-term objectives. The methodology encompasses clarifying goals, formulating strategies, executing actions, monitoring, evaluating, and continuously refining these efforts, forming the cornerstone of a logical framework for carbon neutrality.

① **Annual energy demand of human society** (Table 1):

$$D_{total} = D_{rigid} + D_{decentralized} + D_{non-rigid} + D_{distributed} \tag{1}$$

Explanation: D_{total}, D_{rigid}, $D_{decentralized}$, $D_{non\text{-}rigid}$ and $D_{distributed}$ represent the total annual energy demand for fixed-space–time scale rigid demand, fixed-space–time decentralized rigid demand, fixed-space–time non-rigid demand, and distributed space–time non-rigid demand, respectively (based on statistical or forecast data).

② **Annual energy supply of human society** (Table 2):

$$S_{total} = S_{stable,fossil} + S_{stable,clean} + S_{unstable,clean} + S_{distributed,clean} + S_{emergency,fossil} \tag{2}$$

Explanation: S_{total}, $S_{stable,fossil}$, $S_{stable,clean}$, $S_{unstable,clean}$, $S_{distributed,clean}$ and $S_{emergency,fossil}$ represent the annual supply of stable-scale fossil fuels, stable-scale clean energy, unstable-scale clean energy, unstable distributed clean energy, and emergency fossil fuels, respectively (based on statistical or forecast data).

Table 1 Annual energy demand of human society table

Demand type	Annual energy demand
D_{total}	Total annual energy demand
D_{rigid}	Rigid demand
$D_{decentralized}$	Decentralized rigid demand
$D_{non\text{-}rigid}$	Non-rigid demand
$D_{distributed}$	Distributed demand

Table 2 Annual energy supply of human society table

Supply type	Annual energy supply
S_total	Total annual energy supply
S_stable,fossil	Stable fossil fuels
S_stable,clean	Stable clean energy
S_unstable,clean	Unstable clean energy
S_distributed,clean	Distributed clean energy
S_emergency,fossil	Emergency fossil fuels

3.2 The Six Critical Balances

Carbon Emission and Absorption: Aiming for equilibrium between emissions from human activities and natural sources versus carbon sequestration by green vegetation. Strategies include forest enhancement, vegetation optimization, and reduced high-carbon activities.

Clean Energy and Overall Demand: Balancing stable clean energy (e.g., nuclear, hydro, solar) with societal needs. This necessitates clean energy expansion, consumption pattern optimization, and efficiency improvements.

Distributed Clean Energy and Manufacturing: Matching large-scale distributed clean energy with manufacturing demands. Deployment of clean energy facilities in industrial regions is key.

Operational Energy and Conversion: Balancing operational energy needs with efficient energy conversion technologies. Smart energy management and storage solutions are essential.

Urban Building Energy and Hydrogen: Satisfying urban building energy demands with clean hydrogen. Promoting hydrogen usage and developing supporting infrastructure is vital.

Uncontrollable Emissions and Artificial Absorption: Addressing uncontrollable emissions through artificial carbon capture and storage, complemented by afforestation (Fig. 1).

These six balances embody a holistic approach, ensuring harmony between economic development and carbon reduction, short-term measures and long-term goals, national strategies and regional actions, technological innovation and traditional industries, government guidance and market mechanisms, and domestic and international cooperation.

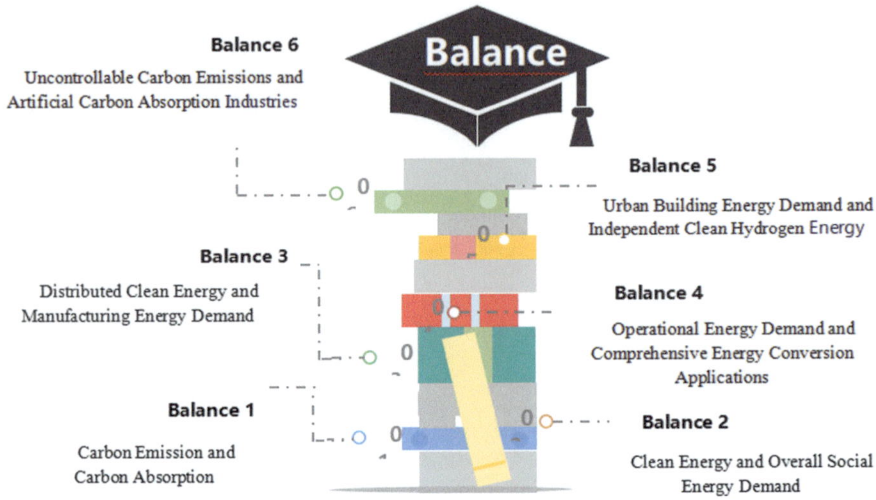

Fig. 1 The six balances key to implementing China's dual carbon strategy

3.3 Methodological Steps

Identify Carbon Emission Behaviors: Quantify natural and human-induced emissions to prioritize reduction efforts (Fig. 2).

Assess Carbon Absorption Capacity: Evaluate natural and artificial sequestration potential to inform enhancement strategies.

Analyze Energy Supply–Demand Structure: Assess the balance between energy demand and supply, including source stability and proportion.

Formulate Strategies: Develop tailored strategies for emission reduction, absorption enhancement, and supply–demand optimization.

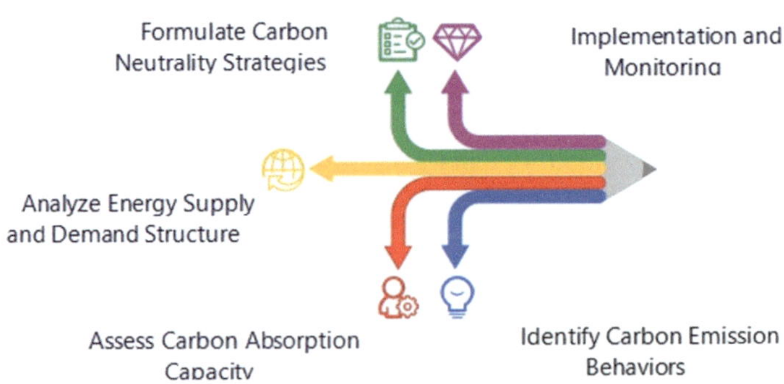

Fig. 2 Methodological steps involved in obtaining this first principle

Implementation and Monitoring: Execute strategies while regularly monitoring their impact on emissions, absorption, and energy balance, ensuring continuous improvement.

4 Results: Dual Carbon Strategy Implementation Road-Map

4.1 Short-Term (2023–2025)

Enhance policy frameworks by refining regulations on carbon trading markets and boosting non-fossil energy share in total energy mix.

Foster technological breakthroughs by investing in green R&D, spurring innovation and adoption of eco-friendly technologies.

Drive industrial transformation by phasing out polluting, energy-intensive sectors and fostering green, low-carbon industries.

4.2 Medium-Term (2026–2035)

Construct a clean, low-carbon energy system, augmenting clean energy proportions and curbing fossil fuel usage.

Expedite the development of a low-carbon transportation network, championing electric vehicles and optimizing transport infrastructure.

Upgrade urban and rural green infrastructure, emphasizing eco-friendly buildings, transportation, and other sustainable developments.

4.3 Long-Term (2036–2060)

Achieve nationwide carbon neutrality through persistent efforts and comprehensive strategies. Fortify carbon sinks by safeguarding and rejuvenating forests, grasslands, and other ecosystems, enhancing their sequestration capabilities (Fig. 3).

Actively engage in global climate governance, contributing significantly to the global fight against climate change.

Near-Term Planning Priorities: Initiate pilot projects within China's Carbon Civilization Planning Zones, categorizing regions into seven distinct "carbon behavior zones" based on consumption patterns and sequestration capacities.

High Carbon Consumption Zones: Intensive coal-based areas like mining regions with on-site power generation.

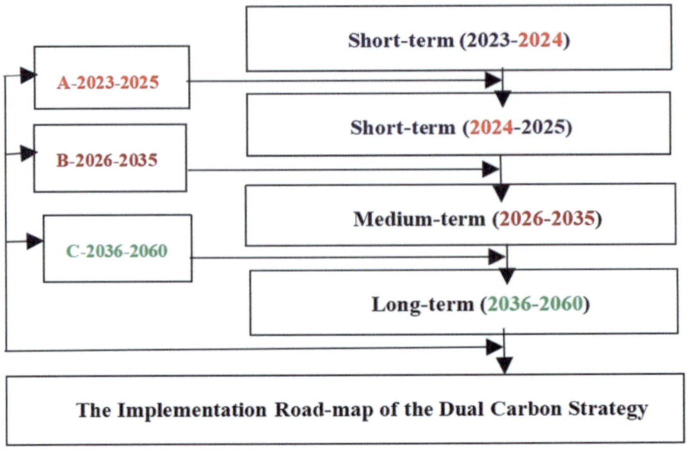

Fig. 3 The year 2024 is the most crucial year of the first phase in the road-map

Low Carbon Consumption Zones: Urban centers and high-speed rail corridors where controlled low-carbon lifestyles prevail.

Carbon Neutral Zones: New rural areas striking a balance between carbon emissions and absorption.

Zero Carbon Consumption Zones: Deserts, gobies, and transportation corridors leveraging abundant renewable resources for energy self-sufficiency.

Low Carbon Sequestration Zones: Uninhabited or agricultural lands with natural sequestration potential.

High Carbon Sequestration Zones: Forests and grasslands naturally adept at carbon storage.

Specialized Carbon Storage Zones: Designated areas for large-scale, technology-driven artificial carbon capture and storage.

The Pivotal Year of 2024 and the Future of Carbon Neutrality

The year 2024 marks a crucial juncture in the initial phase of carbon peaking and neutrality efforts, bridging the gap between 2023 and 2025. This period is significant not only chronologically but also intellectually, as it necessitates the rectification of prevalent misconceptions surrounding previous domestic and international strategies (as highlighted in Part II's review). Moreover, the fundamental principles derived from our methodology (Part III) must be promulgated widely in 2024, ushering in a new era of human–machine collaboration.

At the core of carbon neutrality lies a balance between two primary emission behaviors—natural direct emissions from living organisms and industrial activities, and two absorption behaviors—natural sequestration processes and artificial carbon capture initiatives. Additionally, a delicate equilibrium must be maintained between annual energy demand, encompassing essential and non-essential fixed and distributed space–time requirements, and supply, comprising various energy sources from stable to unstable and emergency options.

To genuinely achieve carbon neutrality and foster the green economy, a comprehensive approach rooted in these first principles is imperative. As artificial intelligence intertwines with carbon neutrality, the global economy is poised to evolve into three distinct forms: the new knowledge economy, the green carbon economy, and the mutual assistance economy. These forms underpin the development of China's domestic circular economy, mirroring the dynamics of global human security, scientific computing, and social science knowledge systems.

The virtual realm is also undergoing a transformation, with the internet and IoT emerging as foundational networks. These economies will foster specialized networks—the knowledge internet, carbon internet, and assistance internet—with the knowledge internet serving as a vital backbone. Consequently, traditional internet economies like e-commerce, sharing, and influencer-driven models will gradually give way to the mutual assistance economy, offering a new frontier for public participation in the global digital landscape [19–21].

Furthermore, the traditional development pillars of the digital, green, technological, and consumer economies will converge and reshape into these three novel economic forms, redefining the landscape of the carbon neutrality and information intelligence era [22, 23].

5 Discussion the New Mutual Aid Economy

5.1 Basic Logical Structure of the New Mutual Aid Economy

The national central bank or relevant financial institutions issue digital currency and accept donations of RMB or assets from social charitable organizations to form a national digital currency fund pool. Based on green carbon rights, every citizen is allocated digital currency annually, with special groups receiving a higher proportion. Mutual aid services participated by citizens become the largest source of credit information, reflecting the self-interest of providers and the welfare needs of recipients.

5.2 The New Mutual Aid Economy as a Real Economy

The new mutual aid economy is a low-energy-consuming form of real economy. Its prosperity and growth do not lead to a corresponding increase in wealth consumption, such as energy, raw materials, and food, thus avoiding pressure on China's strategic resource supply and demand balance. As long as basic living needs are met, the development of this economy will not substantially impact inflation. Theoretically, it can continue to develop without long-term restrictions and should become a core component of each country's domestic circular economy and a basic form for promoting social common prosperity.

5.3 First Principle Thinking of Mutual Aid Economy in Human Society

Human direct consumption includes material, service, and spiritual consumption, roughly in a 40%:40%:20% ratio. Effective and high-quality direct material consumption will not exceed half of an individual's total.

The wealth distribution, consumption ratio, and happiness deficiency in society reveal that most private wealth is concentrated in a small, wealthy group, whose happiness deficiency stems from a lack of spiritual consumption. The middle and grassroots groups' happiness deficiencies arise from a lack of service consumption and comparisons with wealthier groups.

In the carbon-neutral era, equal carbon consumption rights are core to human universal values. A special coupling exists between individual direct consumption, basic carbon consumption ownership, individual happiness, and carbon neutrality. The fundamental purpose of carbon neutrality is to build an environment suitable for human survival and happiness.

The upcoming digital and intelligent era will bring revolutionary changes to the human productive force system, combined with the need for carbon neutrality. Almost all productive activities will move towards orderly intelligent production, with human participation less than 5%. Ensuring employment and survival rights, driving consumption towards carbon neutrality, and the sustainable growth of service and spiritual consumption create conditions for a new form of human social activity.

Individuals do not need to participate directly in productive activities for employment. Through orderly intelligent production, sufficient direct consumption materials can be provided. Based on interests and hobbies, individuals provide mutual aid services, forming their own employment and obtaining remuneration according to transparent rules. This continuously growing mutual aid brings lasting happiness and addresses the fundamental structural contradictions of universal employment and carbon consumption rights in the digital and intelligent era.

In summary, mutual aid economy in human society is the inevitable development path at the current stage.

6 Conclusion

In conclusion, the Blueprint for the 'Dual Carbon' Strategy and Exploration of Mutual Aid Economy under Green Transformation is a complex and systematic endeavor that requires deep participation and close collaboration from multiple sectors and industries, as well as the concerted efforts and sustained investment of the entire society. By formulating a detailed implementation road-map, we can more effectively drive the achievement of the 'Dual Carbon' strategic goals and attain significant results in the process.

In the face of the dual challenges posed by rapid technological development and global carbon neutrality, the establishment of a mutual aid economic system emerges as an innovative solution. This system not only aligns with the inevitable trend of highly developed technological societies but also provides the public with new avenues for achieving employment, creating green and healthy lifestyles, and supporting the implementation of carbon neutrality through the issuance of digital currencies and the creation of a socialized job system. Simultaneously, it promotes the comprehensive enhancement of individual and family well-being and the formation of a lifelong training system.

As a low-energy-consuming physical economy, the prosperity and development of the new mutual aid economy will not exert pressure on China's strategic resource supply and demand balance or trigger inflation. Therefore, it should be regarded as a core component of China's domestic circular economy and the fundamental economic form for promoting common prosperity. Looking forward, we will continue to monitor the progress of the 'Dual Carbon' strategy and mutual aid economy, contributing our wisdom and strength to driving green transformation and sustainable development [24].

References

1. Lu, X., Jiang, Y.: Research on issues and countermeasures of global climate change governance. Heilongjiang Environ. Bull. **37**(07), 111–113 (2024)
2. Shen, G., Yan, H., Liu, X.: Construction of energy and carbon digital operation and management service platform under the dual carbon strategy. Value Eng. **43**(22), 114–116 (2024)
3. Xu, Z.: Digital technology enables and boosts the realization of dual carbon goals. Inf. Syst. Eng. **08**, 9–11 (2024)
4. Establishing ecological civilization to promote green development. Resour. Conserv. Environ. Prot. (02), 150 (2020)
5. Li, G.: Strengthening international cooperation is the inevitable choice for effectively addressing climate change challenges. China Environ. Manage. **15**(04), 5–7 (2023)
6. Guo, D., Wang, X., Zou, S.: Research on government's two-way tax preference strategies in the carbon inclusive system. Mod. Manag. **44**(04), 180–190 (2024)
7. The sixth assessment report of the intergovernmental panel on climate change released. Science **75**(03), 10 (2023)
8. Dai, Z.: On the status and effectiveness of relevant legal documents under the United Nations Framework Convention on Climate Change: also on the enlightenment to China's climate diplomacy negotiation. Int. Law Stud. **01**, 94–110 (2017)

9. Zhai, D.: A study on the interaction between Sino-US bilateral climate relations and the process of the United Nations Framework Convention on Climate Change. Pac. J. **30**(03), 1–12 (2022)
10. Ye, X.: The role of the Organization for Economic Cooperation and Development (OECD) in formulating lifelong education policies. University **02**, 130–135 (2023)
11. Wang, W., Ren, H.: Can international carbon emission trading promote China's green product exports? Prices Mon. (07), 57–67 (2024)
12. Dong, R.: International Energy Agency: a new era of global clean energy technology manufacturing is dawning. China Electr. Power Enterp. Manage. **01**, 94–96 (2023)
13. Chen, J., et al.: Deepening the reform of oil and gas system and promoting low-carbon development of the industry with the "dual carbon" goals: a review of China's oil and gas policies in 2020. Int. Petrol. Econ. **029**(002), 62–67 (2021)
14. Li, J., et al.: Review of the development status of green hydrogen production and application technologies under the "dual carbon" goals. Distrib. Energy **6**(4), 9 (2021)
15. Le, X., et al.: Review of agricultural soil carbon sink research. Agric. Technol. **40**(22), 3 (2020)
16. Wu, J., Yang, H.: Review of research methods for calculating agricultural net carbon sink. Agric. Econ. **10**, 3 (2020)
17. Han, M., Jin, Y.: Principles and Applications of Green Engineering. Tsinghua University Press (2005)
18. Yan, Z., et al.: Research on evaluation index system and method of green construction. China Saf. Sci. Technol. **8**(7), 5 (2012)
19. Tapscott, D.: The Economics of the Data Age: Rethinking Opportunity and Risk in the Networked Intelligence Era. China Machine Press (2016)
20. Wang, H., et al.: Developing green and low-carbon economy in Heilongjiang Province under the new normal by drawing on international experience. Intellect. Econ. **3**, 5 (2019)
21. Xu, Y., Du, L.: The development of carbon fund under the new normal of green economy. Lanzhou Acad. J. **12**, 5 (2015)
22. Zhang, Y., Hu, J.: Theoretical inheritance and practical progress of carbon finance: a review of the academic conference on carbon finance and green economic development. Contemp. Econ. Res. **2013**(2), 88–90
23. Xiao, B., et al.: Green carbon energy circular economy technology. In: China Biomass Energy/Coal-to-Oil Technology Exchange and Investment Seminar (2005)
24. LNCS homepage: http://www.springer.com/lncs. Accessed 21 Nov 2016

Strategic Solutions to Provide a Sustainable Environment and Automotive Use in Modern Underground Cities

Salah H. R. Ali, Gehan A. I. Elsayed, and Amr S. H. Ali

Abstract The movement of people, purchases, and recreational trips using automobiles has become more intensive and requires new concepts to operate in a safe environment, especially in modern underground cities construction projects. This work explores the pivotal role of engineering technologies in enhancing the quality of natural daylight within underground cities. The study investigates advanced lighting utilizing, architectural design strategies with sustainable transportation, and innovative materials to mitigate the inherent constraints of subterranean living. The research assesses the effectiveness of technological interventions to provide comfortable living and safe, sustainable transportation. It therefore addresses the design of light wells, utilize of smart materials, and the incorporation of cutting-edge lighting systems to optimize the distribution of natural light. Additionally, the study evaluates the psychological and physiological impacts of enhanced daylight exposure on residents, emphasizing the importance of human-centric design in subterranean environments. Key findings highlight the significance of daylight simulations in the planning phase project, state-of-the-art engineering utilizing to predict sunlight patterns to achieve great benefit in providing conceptual solutions to provide a suitable environment for automotive operation. The research also emphasizes the importance of incorporating modern technology systems as atriums, light wells, and reflective surfaces to maximize natural light penetration. Furthermore, the research investigates the energy implications of relying on artificial lighting in conjunction with natural daylight. It explores sustainable energy strategic solutions, such as solar technology,

S. H. R. Ali
Pyramids Higher Institute of Engineering and Technology (PHI), 6th of October, Egypt

Precision Engineering, National Institute of Standard (NIS), Haram, Giza, Egypt

Misr University for Science and Technology (MUST), 6th of October, Giza, Egypt

Ahram Canadian University (ACU), 6th of October, Giza, Egypt

G. A. I. Elsayed
College of Engineering, Jouf University, Jouf, Saudi Arabia

A. S. H. Ali (✉)
College of Southern Europe, Malta, Malta
e-mail: Amr.Salah.Ali94@gmail.com

Y. Zhu (ed.), *Proceedings of the 4th International Symposium in Environmental Science and Industrial Ecology*, https://doi.org/10.1007/978-981-96-1578-0_13

according to implementation considerations and using automotive to minimize the environmental impact of energy consumption in modern underground cities projects.

Keywords Sustainable energy · Automotive use · Strategic solutions · Safe environment · Underground city projects

1 Introduction

The underground city construction project is considered one of the largest engineering projects in the world, because it requires high digital technologies and a huge amount of capital. In addition to, those projects must be entrepreneur work. Therefore, managing such projects requires full knowledge and cooperation with many different technological specialties. The concept of underground city projects is expanding as technology and environmental concerns impact the future of urban life, especially with the use of automobiles. These cities range in both complexity and size, from tiny underground residences to large networks of tunnels and chambers that provide the same functions as surface-level towns. Underground cities have served a variety of purposes throughout history and across civilizations. Several projects and new concepts also provide insight into how automotive can be integrated into modern underground urban environments. The technology of integrating automotive use into underground cities therefore faces a unique challenge and opportunity [1–5]. A major challenge faced in subterranean environments is the limited access to natural daylight. Natural light is essential for increasing the quality of life and sustainability in underground places, as it provides several advantages that assist both human well-being and the environment. The distribution of natural light in underground places requires architects to develop novel solutions that allow sunlight to penetrate. Light wells, reflective surfaces, or novel openings that allow daylight into deeper regions can be used as strategies to improve the environment's aesthetics and functionality [5–9]. Using Solatube or other daylighting utilizing to deliver natural light into underground cities can be a realistic and innovative way to do it. Where, Solatube is a brand of tubular daylighting devices (TDDs). By incorporating Solatube for daylighting systems, underground cities can enhance the quality of life for residents, promote sustainability, and reduce dependence on artificial lighting.

Thus, the aim of the research is to discuss the importance of the Solatube system and provide technological solutions to achieve natural light and improve the quality of life and sustainability, especially with the operation of automotive in a safe environment in underground spaces. The used methodology is 1st: review the existing literature on underground cities in the field of the study topic, 2nd: explore case studies of underground structures that successfully use natural light, 3rd: concepts of integrating automotive use in a safe environment at modern underground cities, 4th: analyze the challenges faced and lessons learned from using natural light and providing a suitable and sustainable environment for automotive use in underground cities, and then draw conclusions.

2 Historical Context of Underground Cities

Actually, the construction of underground cities has been a fascinating aspect of human history, often driven by the need for defense and shelter during times of war or other threats. We will explore the historical context of underground cities, focusing on the ancient Hittites and Derinkuyu in Cappadocia, Turkey. Hittite capital cities, such as Hattusa, were known for their impressive fortifications, but the Hittites also constructed underground chambers for various purposes. These underground structures served as storage rooms, living quarters, and possibly as shelters during times of invasion or conflict. The underground city of Derinkuyu is believed to date back to the Hittite period, around the eighth to seventh centuries BCE, although some sections might have been expanded and modified by later civilizations, including early Christians. The city was designed with various features, including ventilation shafts, wells, kitchens, and storage rooms, indicating a comprehensive understanding of urban planning and engineering [10, 11].

2.1 The Concept of Underground City Projects

To provide protection during times of conflict, several underground cities were built with defense in mind. The natural insulation of the ground also supplied protection from environmental conditions. Underground spaces are often naturally insulated, requiring less energy for heating and cooling, making them energy-efficient. Recently, some modern cities have incorporated underground spaces for transportation systems (subways), utility tunnels, and parking facilities to optimize land use and reduce congestion. As concerns about environmental sustainability, some urban planners explore the idea of creating habitable underground spaces to reduce the ecological footprint of cities [12].

2.2 Challenges and Benefits of Building Underground Cities

Constructing underground cities requires careful engineering to ensure stability, proper ventilation, and safety. There are some challenges and benefits for underground cities, which are as follows [3]. Where the Challenges of constructing underground cities are:

I. Limited accessibility to natural light: Underground spaces, by definition, lack direct access to natural light. This poses a significant challenge in creating spaces that feel open, inviting, and well-connected to the external environment.
II. Technical challenges: Designing underground structures to harness and distribute natural light involves overcoming technical challenges. This includes

developing efficient light wells, light tubes, or reflective surfaces that can channel sunlight into deeper spaces.

III. Spatial limitations: Underground spaces often have limited surface area exposed to the outside environment. This makes it challenging to bring in sufficient natural light to illuminate the entire space evenly.

IV. Environmental considerations: Depending on the location and depth of the underground structure, environmental factors such as soil conditions, water table levels, and geological considerations may impact the feasibility of incorporating natural light [5].

2.3 Important Requirements When Constructing Underground Cities

I. Energy efficiency: Utilizing natural light reduces the reliance on artificial lighting, leading to energy savings. This is not only environmentally friendly but also contributes to lower operational costs for the structure [3–7].

II. Health and well-being: Exposure to natural light has been linked to improved mood, productivity, and overall well-being. Incorporating natural light in underground spaces can contribute to a more comfortable and enjoyable environment for occupants.

III. Aesthetic appeal: Natural light enhances the aesthetic appeal of spaces. It brings out the textures and colors of architectural elements, creating a visually pleasing atmosphere. This is particularly important in spaces like museums, galleries, or residential areas.

IV. Biophilic design: Integrating natural elements, including light, into underground spaces aligns with the principles of biophilic design. This approach seeks to connect people with nature, even in-built environments, promoting a sense of harmony and balance.

2.4 The Role of Natural Light in Achieving Sustainability in Underground Cities

Natural light plays a crucial role in improving the quality of life and sustainability in underground spaces, offering various benefits that contribute to human well-being and environmental considerations. Natural light enhances visual comfort by providing a more uniform and pleasant illumination compared to artificial lighting. This helps reduce eye strain, headaches, and fatigue among occupants, thereby improving productivity and comfort levels. Natural light reduces the dependency on artificial lighting, resulting in energy savings and decreased carbon footprint. This aligns with sustainability goals by promoting energy-efficient design and reducing the environmental impact associated with electricity consumption. Natural daylight is often accompanied by natural ventilation, promoting air exchange and reducing the need for mechanical ventilation systems [6].

3 Utilizing and Design Strategies for Natural Daylighting

Designing natural day lighting for underground cities involves implementing utilizing strategies that mimic the effects of natural sunlight to create a comfortable and visually appealing environment. The following techniques are some of the available strategies for achieving natural day lighting in underground cities.

3.1 Solar Fiber Optic Systems

The fiber optic cables can be used to capture and channel sunlight from above ground into underground spaces. A collector system on the surface directs sunlight into the fiber optic cables, which then distribute the light throughout the underground spaces [1, 13].

3.2 Light Tubes or Solatubes

Light tubes known as Solatubes or sun pipes, are highly reflective tubes that capture sunlight at the surface and channel it into the underground areas. They can be installed with lenses or diffusers to spread natural light evenly throughout the space. Using Solatube or similar daylighting systems can be a practical and innovative way to bring natural daylight into underground cities. Solatube is a brand of tubular daylighting device that captures sunlight at the roof level and then channels it through a highly reflective tube into interior spaces as showing in Fig. 1. Here are some considerations for using Solatube in underground city design such as [8, 14]: lighting efficiency, light distribution, flexible design, technology advancements, maintenance, emergency lighting, and integration with other sustainable technology.

Underground cities successfully depends on natural daylight are not as common, as the nature of being underground limits direct access to sunlight. However, some underground spaces are designed with innovative architectural features to bring in natural light. As example, the lowline underground park is an ambitious and innovative project that seeks to be a sustainable one. Unlike traditional parks, the lowline utilizes cutting-edge solar technology to bring natural sunlight underground, creating a lush and green oasis in what was once a forgotten space. This subterranean park aims to offer a unique and immersive experience, showcasing a variety of plant life and providing a serene escape from the hustle and bustle of city life. With its focus on sustainability and creative use of space, the lowline represents a groundbreaking

Fig. 1 Using advanced utilizing in natural lighting the underground cities [14]

initiative in urban design. The project captures the spirit of innovation that defines New York City, pushing the boundaries of traditional urban planning and reimagining the possibilities that lie beneath the surface.

4 Smart Underground Automotive Transportation System (SUATS)

To develop an intelligent, natural and energy-efficient transportation system in underground cities, it is necessary to integrate many systems with intelligent and advanced utilizing [15–21]. Especially since natural lighting is always linked to converting natural light into energy to achieve the desired sustainability. Therefore, it was possible to provide conceptual solutions that focus on the sustainability of automobile operation within this context, as shown in Table 1.

5 Integrating Automotive Use in Modern Underground Cities

To ensure the successful continuation of life in modern underground cities, it is necessary to provide an integrated and safe environment for the use of automotive. Various projects and concepts offer insights into how cars can be incorporated into underground urban environments as follows.

Table 1 The conceptual solutions for the underground intelligent transportation system

Means	Concept	Technology	Advantages	Indication/figure
Electric and autonomous vehicles (EVs and AVs)	Utilize electric autonomous vehicles (AVs) designed specifically for underground environments to ensure efficient, safe, and sustainable transportation	• Electric Powertrains: EVs with high-capacity batteries or wireless charging capabilities to eliminate emissions • Autonomous Navigation: Advanced sensors (LiDAR, radar, cameras) and AI algorithms for precise navigation and collision avoidance • Vehicle-to-Infrastructure (V2I) Communication: Enables vehicles to interact with the underground infrastructure for optimized routing and traffic management	• Zero emissions improve air quality in enclosed spaces • Autonomous technology reduces human error and enhances safety	 Ref. [22]
Smart charging infrastructure	Develop a network of smart charging stations that support fast and efficient charging of electric vehicles	• Wireless Charging Pads: Embedded in the road or parking areas for seamless charging while vehicles are stationary or in motion (dynamic charging) • Solar-Powered Charging Stations: Located at strategic points within the underground network, utilizing solar energy collected on the surface • Smart Grid Integration: Manages energy distribution and ensures efficient use of available power, incorporating renewable energy sources and storage systems	• Convenient and efficient charging, reducing downtime for vehicles • Utilizes renewable energy, enhancing sustainability	 Ref. [23]

(continued)

Table 1 (continued)

Means	Concept	Technology	Advantages	Indication/figure
Advanced ventilation and air purification systems	Ensure clean air quality in underground tunnels and vehicle pathways through state-of-the-art ventilation and purification systems	• Air Quality Sensors: Continuously monitor pollutants and other air quality parameters • HEPA and Activated Carbon Filters: Remove particulate matter and harmful gases from the air • UV-C Sterilization: Disinfects air to eliminate pathogens • Automated Ventilation Controls: Adjust airflow based on real-time air quality data and vehicle density	• Maintains a healthy environment by continuously purifying the air • Adaptable to varying conditions, ensuring optimal air quality at all times	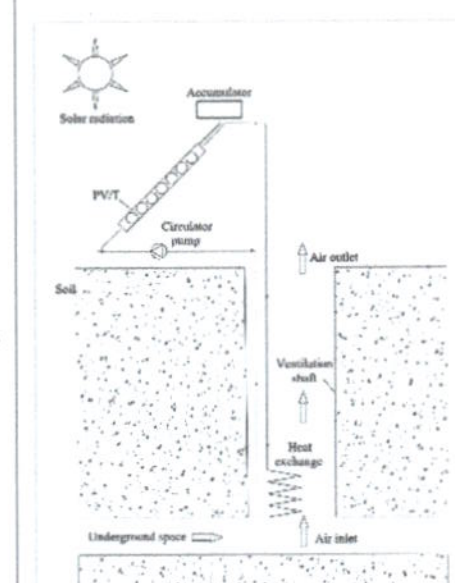 Ref. [24]
Intelligent traffic management system	Implement a comprehensive traffic management system to optimize vehicle flow and reduce energy consumption	• Real-Time Traffic Monitoring: Use cameras and sensors to monitor traffic conditions • AI-Based Traffic Control: Algorithms analyze data to optimize traffic lights, manage congestion, and direct vehicles through less congested routes • Connected Vehicle Ecosystem: Vehicles communicate with each other (V2V) and infrastructure (V2I) to share information and coordinate movement	• Reduces traffic congestion and improves flow efficiency • Minimizes energy consumption by reducing stop-and-go driving	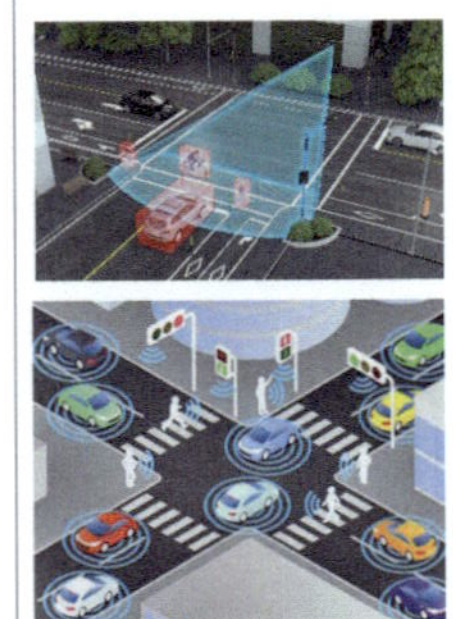 Ref. [25]

(continued)

Table 1 (continued)

Means	Concept	Technology	Advantages	Indication/figure
Green tunnel design and materials	Design tunnels using sustainable materials and incorporate elements that enhance the natural feel of the environment	• Biophilic Design: Integrate green walls and ceilings with plants that can thrive underground with appropriate lighting and irrigation • Sustainable Materials: Use recycled and low-impact materials for construction • Energy-Efficient Lighting: LED lighting systems that mimic natural daylight and are controlled by smart sensors to adjust based on time and occupancy	• Creates a pleasant and natural environment, reducing the psychological impact of being underground • Sustainable construction methods reduce the overall environmental footprint	 Ref. [26]
Emergency response and safety systems	Ensure the safety of passengers and vehicles with advanced emergency response systems	• Emergency Communication Systems: Enable real-time communication with control centers and emergency services • Automated Fire Suppression: Sensors detect fires and activate suppression systems instantly • Evacuation Protocols: Clearly marked and illuminated escape routes, along with automated guidance systems	• Enhances safety and provides rapid response in case of emergencies • Reduces risks associated with underground environments	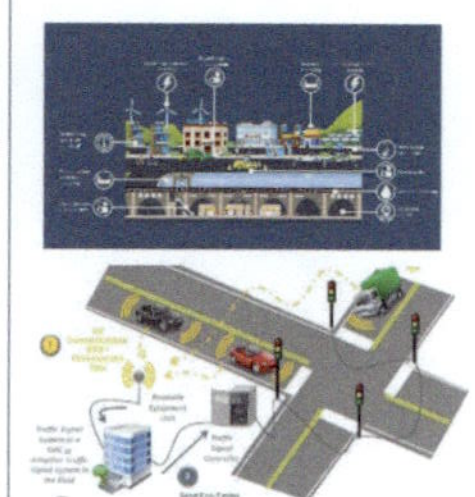 Ref. [27]

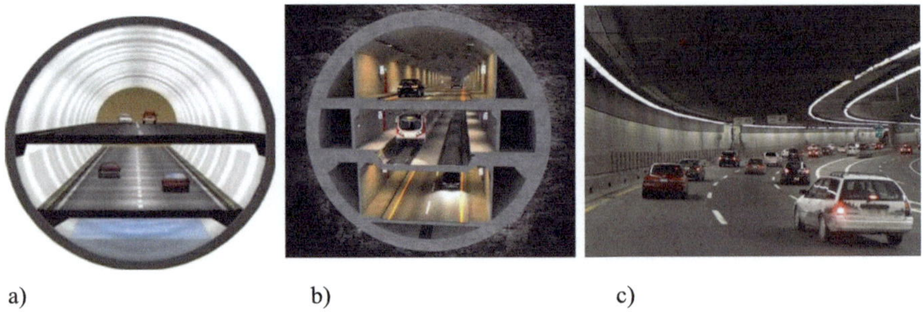

Fig. 2 Some of underground tunnels around the world. a Kuala Lumpur smart tunnel [26]. b The world's first triple-deck tunnel opened in Istanbul in 2020 [28]. c One of Boston's original subway tunnels opens to the public, USA [29]

5.1 Underground Road Tunnels

SMART Tunnel, Kuala Lumpur, Malaysia: The Stormwater Management and Road Tunnel (SMART) in Kuala Lumpur is a dual-purpose tunnel that functions both as a road tunnel and a flood relief system. The tunnel helps mitigate traffic congestion by providing an alternative route through the city center while also serving as a crucial flood control mechanism during heavy rains, as shown in Fig. 2a. Figure 2b presents the Eurasia Tunnel, Istanbul, Turkey. The Eurasia Tunnel is a major transport link under the Bosphorus Strait, completed in 2016, and has won two international awards (International Tunneling Association (ITA), Tunnel Project in 2015 and Engineering News-Record (ENR), Best Global Project, Bridge/Tunnel in 2016). This tunnel has distinguished itself by connecting the European and Asian sides. It accommodates private vehicles and public sustainable, safe transportation, easing traffic congestion on surface roads. While Fig. 2c shows one of Boston's original subway tunnels opens to the public.

5.2 Underground Transport Hubs

Shinjuku Station, Tokyo, Japan: Shinjuku station incorporates underground parking facilities and access points for vehicles in a safe environment, seamlessly integrating with the extensive rail and subway network serving millions of commuters daily, see Fig. 3.

Fig. 3 Underground parking facilities and access points for vehicles at Tokyo, Japan [30]

Fig. 4 Underground spaces for retail, dining, and entertainment at Atlanta, Georgia, USA [31]

5.3 Smart Infrastructure Integration

Underground Infrastructure in Smart Cities: Various smart city initiatives explore integrating automotive use into underground environments using advanced utilizing such as automated parking systems, intelligent traffic management, and real-time data analytics.

5.4 Urban Expansion Projects

Underground Atlanta Streetscape Project, Georgia, USA: The ongoing revitalization project aims to enhance underground spaces for retail, dining, and entertainment while addressing vehicular circulation and parking needs, as shown in Fig. 4.

5.5 Environmental Considerations

Underground Parking in Green Cities: Some cities prioritize sustainability by incorporating underground parking facilities into green infrastructure projects, minimizing the environmental impact of car use in urban areas.

6 Analyses and Discussions

Examples demonstrated the potential for integrating automotive use into underground modern cities in safe sustainable environment. Each project faces unique logistical, environmental, and social challenges. Comprehensive case studies analyzing the effectiveness, efficiency, and sustainability of these initiatives would provide valuable insights for future urban planning and development efforts. Achieving natural daylight in underground cities poses a unique set of challenges; there are potential solutions in this field as the following:

1. Light Shaft Technology: Develop advanced light shaft technology to channel sunlight from the surface to underground spaces. This involves creating reflective surfaces and using optical devices to capture and direct sunlight efficiently.
2. Fiber Optic Systems: Implement fiber optic systems to transmit sunlight from the surface to underground areas. This involves using bundles of optical fibers to capture and distribute natural light throughout the underground spaces.
3. Encouraging the Use of Electric Vehicles: Residents should be encouraged to use electric automobiles will have the greatest impact in reducing sources of environmental pollution in underground cities.
4. Bioluminescent Technology: Explore bioluminescent organisms or synthetic materials that can emit light. This could be integrated into the architectural design of underground spaces, providing a natural-looking and sustainable source of light. Bioluminescent technology represents an innovative and sustainable approach to addressing the lighting needs and a safe transportation of underground cities.

7 Implementation Considerations

In fact, implementation depends on many considerations, the most important of which are the following five:

1. Scalability and Flexibility: Design systems to be easily scalable and adaptable to changes in demand and technological advancements.
2. Energy Efficiency: Prioritize the use of renewable energy sources and high-efficiency utilizing to minimize the environmental impact.
3. Integration and Interoperability: Ensure seamless integration of all subsystems and promote interoperability to optimize performance and user experience.
4. Reducing Sources of Pollution: Predominance in the use of electric cars.
5. User Experience: Focus on creating a user-friendly environment that promotes comfort, safety, and convenience for passengers.

By integrating these considerations as smart utilizing, underground cities can develop a transportation system that is natural, safe, and highly energy-efficient, providing a sustainable solution for future urban mobility challenges.

8 Conclusions

The possibility of integrating the use of automotive into modern underground cities is very important issue. Therefore, the following conclusions can be reached:

- The project managers and planners should collaborate with automotive engineers to address technical, logistical, environmental and social challenges to ensure efficient and safe sustainable use in future urban development of underground cities projects.
- The strategic use of engineering technology utilizing to harness natural daylight in underground cities is not only a practical technological solution to architectural challenges but also a forward-thinking approach that prioritizes the well-being, sustainability and resilience of urban communities, as well as improving the health of people living in underground cities. As technology continues to advance, integrating these innovations will play a crucial role in shaping the future of underground city design and development.
- Creating natural daylight in underground cities requires a multidisciplinary strategy that includes architecture, urban planning, engineering technology, and sustainability. The key findings and contributions focus on optimizing designs, incorporating innovative utilizing, prioritizing human well-being and safety, and promoting energy efficiency to create underground environments that are not only functional but also conducive to a high quality and luxury of life according to implementation considerations.
- It has become to incorporate innovative design strategies to channel natural light and ventilation into underground spaces by using atriums, light wells, and reflective surfaces, reduce dependence on artificial lighting and ventilation systems with safe sustainable transportation.

Eventually, incorporating utilizing to harness natural daylight into the design and development of underground city projects is a critical and innovative approach to achieve great benefit in providing conceptual solutions and provide a suitable environment for automotive operation, with recommending the use of electric vehicles. Since this strategic solutions addresses many of the challenges associated with underground cities living, it has become necessary to include it in modern designs due to its many benefits to the lives of residents and the environment.

References

1. De Cusatis, C.: Handbook of Fiber Optic Data Communication: A Practical Guide to Optical Networking. Academic Press, London (2014). ISBN-10: 0123742161
2. National Research Council: Underground Engineering for Sustainable Urban Development (2013). https://doi.org/10.17226/14670
3. Zhao, X., Chen, Y., Wu, Y., Guo, D., Chen, Z.: Enhancing urban landscapes through underground space utilization: public perceptions. Sustainability **16**(11), 4501 (2024). https://doi.org/10.3390/su16114501

4. Elmarakby, E., Elkadi, H.: Comprehending particulate matter dynamics in transit-oriented developments: traffic as a generator and design as a captivator. Sci. Total Environ. **931**, 172528 (2024). https://doi.org/10.1016/j.scitotenv.2024.172528
5. Yuan, Y.: Challenges in underground construction. ICE Proc. Munic. Eng. **161**(1), 25–34 (2008). https://doi.org/10.1680/muen.2008.161.1.25
6. Chrysothemis, P.: Underground space and urban sustainability: an integrated approach to the city of the future. In: Changing Cities IV Spatial Design, Landscape and Socioeconomic Dimensions (2019)
7. Admiraal, H., Cornaro, A.: Future cities, resilient cities—the role of underground space in achieving urban resilience. Undergr. Space **5**(3), 223–228 (2020). https://doi.org/10.1016/j.undsp.2019.02.001
8. Butko, D.J.: The sound of daylight: the visual and auditory nature of designing with natural light. In: Light in Engineering, Architecture and the Environment (2011). https://doi.org/10.2495/light110081
9. Admiraal, H., Cornaro, A.: Why underground space should be included in urban planning policy—and how this will enhance an urban underground future. Tunn. Undergr. Space Technol. **55**, 214–220 (2016). https://doi.org/10.1016/j.tust.2015.11.013
10. Information on: https://i.pinimg.com/originals/28/39/1c/28391c1e5f02b8a0856549c79ad2b84a.jpg. Accessed 21 May 2024
11. Information on: https://www.bing.com/images/search?q=Ancient+Underground&FORM=IRMHRS. Accessed 24 May 2024
12. Information on: https://www.bing.com/images/search?q=underground+shopping+street+in+guangzhou+underground+mall+1&FORM=AWIR. Accessed 25 June 2024
13. Fiber optics for daylighting. BuildingGreen (2016). Information on: https://www.buildinggreen.com/blog/fiber-optics-daylighting. Accessed 25 June 2024
14. Information on: https://th.bing.com/th/id/OIP.iQN-OGURPAOM0Y4CPpD6ewHaFj?rs=1&pid=ImgDetMain. Accessed 21 July 2024
15. Mukhopadhyay, A., et al.: Prioritized allocation of emergency responders based on a continuous-time incident prediction model. In: International Conference on Autonomous Agents and Multiagent Systems. International Foundation for Autonomous Agents and Multiagent Systems (2017)
16. Bettelini, M., Rigert, S.: Ventilation and safety of long and deep tunnels—state of the art and new perspectives. Paper presented at the ITA-AITES World Tunnel Congress WTC2016, San Francisco, CA, 22–28 Apr 2016. ISBN: 978-1-5108-2262-7
17. Kumar, H., Singh, P.K.: Analyzing data aggregation in wireless sensor network. In: Proceedings, 01–03, pp. 4024–4029 (2017)
18. Zhang, J., et al.: Flow prediction in patio-temporal networks based on multitask deep learning. IEEE Trans. Knowl. Data Eng. (2019)
19. Wen, Y.M., et al.: Integrated design for underground space environment control of subway stations with atriums using piston ventilation. Indoor Built Environ. (2020)
20. Hossam, A.G.: Fast Charging and Resilient Transportation Infrastructures in Smart Cities (2022)
21. Alper Kerem, A.: Fast charging station utilizing for electric vehicles. Gazi Üniv. Fen Bilimleri Derg. Part C Tasarım Teknol. **8**(3), 644–661 (2020). https://doi.org/10.29109/gujsc.713085
22. Information on: https://scet.berkeley.edu/wp-content/uploads/autonomous-cars.jpg. Accessed 21 July 2024
23. Information on: https://ocn.org.au/wp-content/uploads/2022/04/ev-charging-square-400x400.jpg. Accessed 21 July 2024
24. Information on: https://p.turbosquid.com/ts-thumb/bY/TFmIZ3/Lq/garage_final_0001/jpg/1643390678/1920x1080/fit_q87/93e41e6b57e55a3aa5eca403f8b8b5f524dd99cf/garage_final_0001.jpg. Accessed 21 July 2024
25. Information on: https://www.electronicwings.com/storage/ProjectSection/Projects/9939/smart-traffic-control-system-using-image-processing/icon/smart-city-solution.png. Accessed 21 July 2024

26. Information on: https://emalte.com.au/top-6-projects/kuala-lumpur-smart-tunnel/
27. Information on: https://www.adeunis.com/wp-content/uploads/2019/06/xSmart-City-web2.jpg.pagespeed.ic.jLAj8UtloE.webp. Accessed 24 July 2024
28. Information on: https://driveeurope.co.uk/2015/03/02/istanbul-tripledeck-tunnel/ (2020)
29. Information on: https://www.bing.com/images/search?q=Boston%e2%80%99s+tunnels&form=HDRSC3&first=1. Accessed 24 July 2024
30. Information on: https://www.alamy.com/stock-photo-shinjuku-station-tokyo-japan-19714691.html, Shinjuku station, Tokyo, Japan. Accessed 24 July 2024
31. Information on: https://packupthecar.blogspot.com/2011/04/atlanta-georgia-april-2011-attractions.html. Accessed 23 July 2024

Evaluation of Environmental Impact for Electric Automobiles

Salah H. R. Ali, Mostafa M. A-Razek, and Amr S. H. Ali

Abstract Electric automobile manufacturing technology projects are now playing a new vital role in some countries around the world, which will of course affect the behavior of people and economy in the future. This work is concerned with an in-depth evaluation of the environmental factors to understand and explain the advantages and disadvantages using statistical comparative methods. In addition to anticipating how its future impact on society. The examination begins by looking at the environmental impact, namely differences in emissions, carbon footprints, and life-cycle sustainability from production to disposal. Economic and engineering performance aspects are also evaluated Moreover, the development of battery management technology using AI, charging infrastructure, and power drive design are thoroughly explored. This paper aims to provide a balanced comparison to evaluate and understand the current landscape and predict solutions for the development of electric automobile technology to improve the quality of life of passengers and people. Thus, to helps stakeholders including researchers and policy makers.

Keywords Electric automobile · Environmental impact · AI in electric power battery management

S. H. R. Ali · M. M. A-Razek
Pyramids Higher Institute of Engineering and Technology (PHI), 6th of October, Giza, Egypt

S. H. R. Ali
Precision Engineering, National Institute of Standard (NIS), Haram, Giza, Egypt

Misr University for Science and Technology (MUST), 6th of October, Giza, Egypt

Ahram Canadian University (ACU), 6th of October, Giza, Egypt

A. S. H. Ali (✉)
College of Southern Europe, Malta, Malta
e-mail: Amr.Salah.Ali94@gmail.com

Y. Zhu (ed.), *Proceedings of the 4th International Symposium in Environmental Science and Industrial Ecology*, https://doi.org/10.1007/978-981-96-1578-0_14

1 Introduction

Global regulatory trends require reduced exhaust emissions, especially from automobiles. Electric automobiles (EAs) represent a crucial transformation in the automotive industry, driven by the pressing need to solve environmental issues, and reduce reliance on fossil fuel to improve energy security. Over the last decade, technical breakthroughs and supporting regulations have expedited the adoption of EAs, establishing them as a cornerstone of sustainable mobility. The beginnings of electric automobiles may be traced back to the early nineteenth century, when innovators began experimenting with electric propulsion technologies. In the 1820s, Robert Anderson created the first primitive electric carriage powered by non-rechargeable primary cells [1]. However, it wasn't until the late twentieth and early twenty-first centuries that EAs became viable alternatives to internal combustion engine automobiles (ICEAs), mainly to advancements in battery technology and power electronics [2]. The development of lithium-ion batteries, in particular, has been critical, providing higher energy density, a longer lifespan, and faster charging capabilities than previous battery types [3]. In addition to technology advancements, government regulations and incentives have played an important role in encouraging EA adoption. Tax subsidies, rebates, and investments in charging infrastructure have reduced restrictions on entry for both consumers and producers [4]. For example, the Zero Emission Vehicle (ZEV) regulation first developed by the California Air Resources Board (CARB), which was implemented in the 1990s, played an important role in encouraging automakers to develop and sell zero-emission automobiles [5].

Advancements in battery technology have been the backbone of the modern EA movement. Lithium-ion batteries, which have become the industry standard, provide significant benefits over other battery technologies. They have a higher energy density, allowing for longer vehicle range, and can be recharged more quickly. Furthermore, the cost of lithium-ion batteries has been falling, making EAs more accessible to the typical user [6]. Solid-state batteries, which promise even better energy densities and more safety, are now being developed and might further revolutionize the EA industry in the future years [7]. Electric motors, a vital part of EAs, have also seen substantial advances. Permanent magnet synchronous motors (PMSMs) are popular because of their excellent efficiency and power density [8]. Tesla popularized induction motors, which are strong and cost-effective, but somewhat less efficient than PMSMs [9]. The development of improved power electronics has enabled greater control and efficiency of these motors, hence improving EA performance [10].

Many nations have enacted a variety of policies to stimulate the adoption of EAs, including tax credits, rebates, and subsidies for both users and manufacturers. For example, in the US, the federal government provides tax credits for the purchase of EAs, and numerous states offer extra incentives [4]. In Europe, nations such as Norway have achieved high EA adoption rates using a mix of incentives such as tax breaks, free parking, and access to bus lanes [11].

The environmental benefits of electric automobiles are enormous. Unlike conventional automobiles, which generate greenhouse gases and pollutants while driving, EAs create no exhaust emissions. This transition not only contributes to limiting climate change, but it also improves urban air quality, lowering the health risks associated with pollution [12]. Furthermore, as the electrical grid is increasingly supplied by renewable energy sources, EAs' lifecycle carbon footprint is predicted to fall even further [13]. Countries with a high proportion of renewable energy in their electrical mix, such as Iceland and Norway, have the potential to achieve almost zero lifecycle emissions for EAs [14]. However, in areas where coal and other fossil fuels dominate the electrical mix, EAs' environmental advantages may be severely diminished. As a result, shifting to a cleaner energy infrastructure is critical for maximizing the potential of EAs as a sustainable transportation option [15].

Despite these benefits, wide adoption of electric cars faces several difficulties. The availability of charging infrastructure, battery recycling and disposal, and the initial purchase cost remain significant barriers. Furthermore, battery manufacture, namely the mining of raw materials such as lithium, cobalt, and nickel, creates environmental and ethical problems that must be addressed in order for EAs to be sustainable [16]. Charging infrastructure continues to be a major barrier to the prevalent adoption of EAs. While fast-charging networks are increasing, they are still not ubiquitous, and potential EA purchasers continue to be concerned about range anxiety [17]. Furthermore, the initial purchase cost of EAs is often greater than that of similar ICEAs, however this difference is diminishing as battery costs fall [6]. Battery recycling and disposal are both essential challenges. The growing number of EAs on the road will ultimately generate substantial amounts of battery trash, which must be managed responsibly. Developing effective recycling procedures and guaranteeing ethical raw material procurement are critical for reducing the environmental impact of electric vehicle batteries [18].

While the massive engineering projects financed by automobile manufacturers and their feeder industries around the world will remain the locomotive of the global economy [19]. Because the economy based on digital manufacturing and entrepreneurship is now at the heart of the lives of professionals and people [20]. The accuracy of geometrical dimension and tolerances (GD&T) using modern technology will remain the key to ensuring that automobiles are manufactured of high-quality, whether conventional or electric [19, 21, 22].

On the other hand, the phenomenon of global warming that affected us all this summer is a strange phenomenon that should indicate the need to reduce exhaust emissions. Therefore, it is clear that there is still a general trend and strong demand for further understanding and interpretation of the performance of electric vehicles and their economic and environmental impact, especially to clarify the vision for decision makers as well as to increase the understanding of researchers, target customers and society, especially as their production density increases and their widespread use begins.

Thus, the work provides an overview of the engineering performance, economic impact beside environmental impact of electric vehicles. Section 2 examines the environmental impact of electric vehicles in more details. Section 3 refers to the performance metrics of electric vehicles and innovations, including recent technology using artificial intelligence in electric vehicle battery development. Conclusions are presented at the end of the paper as in Sect. 4.

2 Comprehensive Environmental Impacts

The environmental benefits of electric automobiles (EAs) over conventional internal combustion engine (ICE) automobiles have drawn a lot of attention. They pledge to lessen the negative consequences of climate change, enhance air quality, and cut greenhouse gas emissions. But because EAs have an environmental impact throughout their whole lifecycle from production to disposal a detailed analysis is required to determine their genuine sustainability. The zero tailpipe emissions of EAs are among its most well-known environmental advantages. During operation, EVs do not release any pollutants, such as carbon dioxide (CO_2), nitrogen oxides (NOx), or particulate matter (PM), like ICEAs do. Air quality has improved, particularly in metropolitan areas, as a result of this notable decrease in local air pollutants [12]. In addition, EAs' regenerative braking systems increase energy efficiency by repurposing energy that would otherwise be wasted [9]. EAs have no tailpipe emissions, but how much of an environmental impact they have is determined by the carbon intensity of the power needed to charge them. The advantages of EAs may be lessened in areas where the electrical system is mostly dependent on fossil fuels. But the carbon footprint of EVs reduces significantly when renewable energy sources like hydropower, solar, and wind power become more and more integrated into the system [13]. Studies have revealed that EAs often have fewer lifespan emissions than ICE automobiles, even when charging from a grid that is dominated by coal. The benefit is amplified when using greener energy blends. For example, EAs lifecycle emissions are substantially lower than ICEs car lifespan emissions in Norway, where hydropower is the primary energy source [13].

The energy-intensive and heavily polluting aspect of the production process for EAs is the manufacturing of the batteries. The processing and extraction of raw minerals such as nickel, cobalt, and lithium add to the environmental impact. But over time, it's anticipated that improvements in battery science and production efficiency will lower these emissions [23]. The extraction of raw materials is where EAs' environmental effect starts. Mining operations for metals needed in batteries, like as cobalt and lithium, have the potential to seriously damage the environment and pollute it. Furthermore, producing batteries requires a significant amount of energy, which frequently results in greater emissions than when making ICE automobiles [24]. Improvements to recycling procedures and the development of more environmentally friendly battery technology are under progress. According to Ref. [7], solid-state batteries are a promising innovation that has the potential to lessen the environmental effect of battery manufacture. They have better energy densities and utilize less toxic elements.

According to International Organization of Motor Vehicle Manufacturers (OICA) provides annual statistics on the global vehicle population, US DOE and EPA: provides data on the average fuel efficiency of vehicles in the US, which can be indicative of global trends and Intergovernmental Panel on Climate Change (IPCC) provides guidelines on greenhouse gas inventory and carbon intensity factors. Table 1 indicates the total CO_2 emissions form ICE cars at last decade.

According to International Energy Agency (IEA): Global EV Outlook reports, US Department of Energy (DOE): Fuel Economy data for electric vehicles and International Energy Agency (IEA): Global data on electricity generation and CO_2 emissions. Table 2 indicates the total CO_2 emissions form EAs electricity generation at last decade.

There are potential and problems at the end-of-life period of electric vehicles. No new raw material extraction is required because batteries include valuable components that may be recycled. But present rates of recycling lithium-ion batteries are rather low, and improper management of the processes can have negative environmental effects [18]. It is critical to create recycling techniques that are both effective and sustainable. The reduction of the environmental effect of EAs at the end of their lifespan will be largely dependent on laws and policies that support battery recycling and the safe disposal of components.

According to the previous standards the best procedure to deal with waste batteries are Battery Identification and Sorting: To guarantee proper recycling procedures, batteries should be properly identified and sorted by type (e.g., lead-acid, lithium-ion, nickel–cadmium), Ensuring the safe transportation of batteries, particularly those categorized as hazardous goods, requires adherence to rules, Respecting local, national, and international environmental laws and norms is known as environmental compliance and Ensuring that personnel who handle batteries receive training on appropriate handling, processing, and storage techniques is known as worker safety

Table 1 Total CO_2 emissions form ICE cars at last decade

Year	Number of ICE cars (billion)	Average fuel efficiency (L/100 km)	Average millage (km/year)	Total CO_2 emissions (million ton)
2014	1.2	8	15,000	3326.4
2015	1.25			3465
2016	1.3			3603.6
2017	1.35			3742.2
2018	1.4			3880.8
2019	1.45			4019.4
2020	1.5			4158
2021	1.55			4296.6
2022	1.6			4435.2
2023	1.65			4573.8
Total				39,500

Table 2 Total CO_2 emissions form EAs electricity generation at last decade

Year	Number of EAs (million)	Average energy consumption (kWh/100 km)	Average millage (km/year)	Total CO_2 emissions (thousand metric tons CO_2)
2014	0.5	20	15,000	750
2015	1.0			1500
2016	2.0			3000
2017	3.5			5250
2018	5.0			7500
2019	7.5			11,250
2020	10.0			15,000
2021	15.0			22,500
2022	20.0			30,000
2023	25.0			37,500
Total				134,250

training. Even after taking into consideration the emissions from the production of electricity, EAs still emit fewer greenhouse gases than equivalent gasoline-powered automobiles, according to a 2015 research by the Union of Concerned Scientists. There will be noticeable health advantages from this reduction in emissions, such as a decline in the prevalence of lung illness, asthma, and other respiratory disorders. There are immediate health advantages from the decrease in air pollution brought about by the use of EAs. Because ICEAs emissions are connected to respiratory and cardiovascular ailments, improved air quality lowers their prevalence. Widespread EA adoption might avert thousands of preventable lives each year in the United States alone, according to a study published in the journal Environmental Research Letters [25]. EAs contribute to less noise pollution because they are typically quieter than ICE cars. In metropolitan areas, where traffic noise is a major issue, this reduction in noise might enhance the quality of life. Nonetheless, to preserve road safety, it's critical to make sure EAs are still audible to cyclists and pedestrians [26].

Reducing the environmental effect of EA when electric automobiles are combined with renewable energy sources, their environmental advantages are enhanced. EAs will be more sustainable if policies support the development of smart grid technology and renewable energy infrastructure. The efficiency and stability of renewable energy systems can be further enhanced by vehicle-to-grid (V2G) technology, which enable electric vehicles to return energy to the grid, ongoing research and development in battery technology. Solid-state batteries have the potential to drastically lower the environmental impact of battery manufacture and disposal due to their safer chemistries and greater energy densities and Effective recycling techniques may recover valuable resources and lessen the demand for extracting new raw materials. It will be essential to implement policies that encourage recycling and to create more effective recycling technology.

3 Performance and Innovations of Electric Automobiles

The promise of electric automobile to lower greenhouse gas emissions and rely less on fossil fuels has drawn a lot of interest as a sustainable alternative to ICEs automobiles. EAs provide maximum torque instantly at 0 revolutions per minute (RPM), enhancing initial performance, while ICEAs reach maximum torque at higher RPMs, which can reduce overall efficiency [27]. For example, high-end EAs can go up to 300 miles between charges, whereas gasoline-powered vehicles can go up to 400 miles between tanks. Nevertheless, these factors might eventually close the gap with gasoline-powered automobiles [28].

EAs exhibit lower maintenance requirements than ICEAs due to their simpler mechanical structure, which includes fewer moving parts and eliminates the need for oil changes. Additionally, EAs experience less mechanical wear on components such as brakes. Conversely, ICEAs necessitate regular maintenance for complex systems, including engines, transmissions, and exhausts, leading to increased long-term costs. Although the initial cost of EAs is higher, their reduced maintenance demands can result in cost advantages over time. Table 3 shows the main different between the EAs and ICEAs in performance metrics.

However, ICEAs often have an advantage in long-distance highway driving, where their extended fuel range and rapid refueling capabilities can outperform the current limitations of EAs charging infrastructure. When choosing between an EA and traditional ICEAs, it's important to weigh several key factors. Consider the cost, including the initial purchase price, maintenance, and fuel savings, as EAs typically have higher upfront costs but lower long-term expenses. Assess the driving range and availability

Table 3 The performance evaluation of EAs versus ICEAs

Parameters	Automobile type	
	EAs [4]	ICEAs [5]
Range	150–400 miles per charge depending on model (author's note: changes at increasing rates from time to time)	300–400 miles per tank depending on model
Charging/refueling time	Typically 30 min for fast charging to 80%	About 5 min for a full tank
Energy efficiency	3.0–4.0 miles per kWh	20–30 miles per gallon
Performance	Instant torque, smooth acceleration	Torque build-up, generally less smooth acceleration
Operational cost	Lower cost per mile, but higher initial purchase price	Higher cost per mile, lower initial purchase price
Maintenance	Fewer moving parts, generally lower maintenance costs	More moving parts, higher maintenance requirements
Battery/engine life	Battery degradation over time, typically 8–10 years	Engine lifespan is variable, often 10–15 years

of charging infrastructure, particularly if you frequently take long trips. Finally, think about your performance needs; EAs offer instant torque and excel in urban driving, while ICEAs may be more advantageous for long-distance highway travel due to their extended range and quick refueling.

The performance of electric vehicles is mostly driven by battery technology. Energy density, cost reduction, and battery chemistry have all advanced significantly in recent years. While lithium-ion batteries continue to be the industry standard, recent advancements in solid-state battery technology provide increased safety, quicker charging times, and better energy density. The time needed to fully charge an electric vehicle's battery is called its charging time. The level 1, level 2, or DC fast charging of the charger and the battery capacity of the car affect how long a charge takes. According to US DOE, charging systems can be classified into three categories:

Level 1 (120 V): Adds about 4–5 miles of range per hour.
Level 2 (240 V): Adds about 25–30 miles of range per hour.
DC Fast Charging: Adds about 60–100 miles of range in 20–30 min.

Several innovative technologies and techniques are needed to fast-charge EAs. By supplying a direct current (DC) to the EA battery instead of using the onboard charger of the car, high-power DC fast charging systems enable significantly quicker charging rates. Electromagnetic fields are used in wireless inductive charging to transmit energy between a receiver on the car and a charging pad on the ground. This approach may accelerate the charging procedure and does away with the requirement for physical connections. EAs battery technologies have advanced dramatically over the years. Lead-acid batteries ($Pb–PbO_2$) are the first rechargeable batteries, but their low specific energy and energy density are their main drawbacks. Because of worries about the memory effect and environmental impact, nickel–metal hydride (Ni–MH) batteries have mostly replaced nickel–cadmium (Ni–Cd) batteries, which were popular in the 1990s due to their higher energy density. The market for EAs is now dominated by lithium-ion batteries, which function well even at low temperatures but are prone to self-discharge in really cold weather. The high energy density and efficiency of advanced technologies such as sodium–sulfur (Na–S) and sodium–nickel chloride (Na–NiCl) batteries in high-temperature applications are being investigated. In the meanwhile, cutting-edge battery technologies including magnesium-ion batteries, lithium–air, lithium–metal, and lithium iron phosphate are under research to enhance energy density, durability, and charging times, promising further improvements in EA performance [29]. Every battery type has advantages and disadvantages according to various criteria. Solid-state batteries, an emerging technology, promise major increases in energy density, longevity, and safety. Currently, lithium-ion batteries dominate the market due to their high energy density and relatively excellent balance of other aspects. For these more recent technologies, cost containment and infrastructure adaption continue to be major obstacles. This fascinating and quickly developing topic of sustainability is being driven by worries about resource availability and environmental effect, as well as research into improved recycling technologies and alternative materials [30].

Continuous advancements in electric drivetrain technology have a major impact on the shift to EAs. Electric drivetrains have distinct benefits including rapid torque and regenerative braking, and they are easier to use and more economical than ICEAs drivetrains, which depend on intricate mechanical components. The EA drivetrain's electric motors are its core component, and recent developments have greatly increased both its output and efficiency. This is because of their great performance at different speeds, high efficiency, and power density. The DC brushed, DC brushless, Induction (Asynchronous), Synchronous, and Switched Reluctance motors are among the motors that are commonly utilized by electric car makers. The selection of an appropriate motor depends on key parameters such as the power-to-weight ratio, torque-speed characteristics, and overall efficiency. A DC brush motor is a type of electric motor that runs by passing current through brushes to the motor windings in order to create rotational motion. A DC brush motor's main parts are the rotor (armature), stator, brushes, and commutator. These parts work together to produce a magnetic field that propels the rotor. This particular motor type is well-known for its easy speed regulation, high starting torque, and straightforward control, which makes it perfect for a variety of uses including power tools, toys, and small domestic appliances. Due to brush wear and tear, DC brush motors require routine maintenance to avoid higher operating costs and periodic replacement of the brushes. These motors are often less efficient than their brushless counterparts in terms of efficiency because friction and heat produced by the brushes cause energy losses. DC brush motors can also spark and create electrical noise at the brushes, which lowers dependability and may cause interference with neighboring delicate electronic equipment. These elements are crucial to take into account when assessing the overall effectiveness and appropriateness of DC brush motors for certain applications. An electric motor that runs on DC without the use of brushes is known as a DC brushless motor since it uses electronic controllers to manage its operation. The motor is made up of a stator with windings and a rotor with permanent magnets. To produce rotational motion, the electronic controller precisely sequences the supply of current to the windings in the stator. Higher efficiency, less maintenance because there are no brushes, quieter operation, and a longer lifespan is just a few benefits of this design. DC brushless motors are especially well-suited for applications like computer cooling fans, drones, and electric automobile because of these advantages. Despite the many benefits that DC brushless motors provide, their cost is often higher than that of brushed motors because of the intricacy of their electronic controllers and the extra parts needed to make them function. Sophisticated electronics and control systems are also required because to the increasing complexity, which can make design and integration into different applications more difficult. Furthermore, the performance and dependability of DC brushless motors can be affected by environmental factors like humidity and temperature, which they are more susceptible to. When choosing DC brushless motors for particular applications, these variables especially need to be carefully taken into account, especially in severe settings.

An electric motor that runs on alternating current (AC) and produces torque via the electromagnetic induction principle is called an induction motor, often referred to as an asynchronous motor. The motor is made up of a rotor that rotates because it is induced by a revolving magnetic field created by the windings in the stator. Because of their reputation for durability, ease of maintenance, and straightforward design, induction motors are widely used in industrial settings as well as in fans, pumps, and other machinery. When working with large loads, induction motors may require the usage of extra starting mechanisms due to their generally lower starting torque. They can use more energy since they are often less efficient than synchronous motors in terms of efficiency, especially at lower speeds. Furthermore, induction motors are less appropriate for applications requiring precise speed regulation due to their limited ability to adjust speed without the need for extra equipment. When choosing an induction motor for a particular application, several aspects should be taken into account.

Synchronous motors are AC motors that run at a steady pace in sync with the supply current's frequency. This means that the rotor and stator revolve in perfect unison, producing a revolving magnetic field. The motor consists of an electromagnet (which needs an external power source for excitation) or a permanent magnet (which has windings in its stator). Synchronous motors are perfect for applications like power factor correction, big industrial drives, and synchronous condensers because of their excellent efficiency, accurate speed control, and capacity to work at leading power factors. The additional parts required for rotor excitation in synchronous motors can complicate the design and raise the total cost of the system, making them more complicated. They also cannot start when under load, thus an additional motor or independent starting mechanism is needed to get the rotor up to synchronous speed. In addition, the need for excitation systems and more complex control mechanisms makes synchronous motors often more costly than induction motors. When determining whether synchronous motors are appropriate for a certain application, several variables should be taken into account.

A switched reluctance motor (SRM) is an electric automobile that runs on the reluctance torque concept, in which the rotor turns to lower the reluctance of the magnetic circuit. The motor is mechanically strong since it has a basic stator and rotor with prominent poles and doesn't need any permanent magnets or rotor windings. SRMs are especially well-suited for use in electric automobiles, industrial motors, and robots because of their great efficiency, simple design, and exceptional performance at high speeds. Significant torque ripple is a common occurrence for switched reluctance motors (SRMs), which can cause noise and vibrations while operating and could affect performance in precision applications. Furthermore, the management of phase switching by SRMs requires complicated control algorithms and electrical drive systems, adding to the entire system's complexity. SRMs work better at high speeds than other motor types, although they could be less effective at low speeds and beginning torque. When determining if SRMs are appropriate for a certain application, these variables need to be taken into account [31]. In EAs, BLDC motors and PMSMs are favored because of their increased efficiency, improved torque control, and smoother operation. Additionally, induction motors are employed, especially where affordability and durability are important factors. The vehicle's budgetary limits and unique performance needs are taken into consideration while choosing a motor type.

The automotive industry is undergoing a change thanks to artificial intelligence (AI), especially in the realm of electric passenger automobiles. The application of AI technology to improve the efficiency, security, and user experience of EPCs is examined in this article. The influence of recent developments in AI on EAs is the main topic of discussion. These developments include machine learning, computer vision, and autonomous driving systems. The automobile industry has seen a substantial transformation with the incorporation of AI technology in electric passenger vehicles, mostly due to the need for safer, more intelligent, and more efficient transportation options. AI improves the capabilities of EPCs in a number of areas, including advanced driver assistance systems (ADAS) and autonomous driving, as well as battery management optimization.

3.1 AI Battery Management System

In order to guarantee battery performance and safety in electric automobiles, the Battery Management System (BMS) is essential. It keeps an eye on important variables including temperature, voltage, current, and state of charge (SoC) to guard against problems like deep discharge, overcharging, and overheating. These variables can be tracked in real time by current BMS systems thanks to sophisticated sensors and algorithms. In addition, the BMS defends the battery by putting safety measures in place to prevent overvoltage, under voltage, overcurrent, and thermal runaway. In the event that dangerous conditions are identified, the BMS has the ability to separate the battery from the load or charger. Additionally, the BMS evaluates the battery's State of Health (SoH), directing decisions about maintenance and replacement by forecasting deterioration and remaining usable life through analysis of past data and present performance. It controls the battery pack's cell balancing to maintain consistent voltage across all of the cells and avoid any of them from being overcharged or over drained, which might otherwise compromise performance and safety [32].

Constant current charging (CCC) and constant voltage charging (CVC) are the two main approaches used in traditional battery charging procedures. While it may seem simple, CCC entails charging the battery continuously, which increases the danger of overcharging and can reduce battery life if the current is not appropriately controlled. On the other hand, CVC lowers the danger of overcharging by applying a set voltage to the battery and allowing the current to progressively drop when the battery approaches full charge. While sophisticated algorithms have been created to improve charging efficiency and overcome some of these techniques' drawbacks, they can also add complexity, increase the danger of battery deterioration, and cause other long-term problems. These advancements are meant to strike a balance between longer-term battery health maintenance and enhanced performance. Through the analysis of variables such vehicle state of charge, journey distance, future pricing signals, and charging station availability, artificial intelligence (AI) smart charging systems use sophisticated algorithms to optimize charging times. Batteries are kept longer by using models of reinforcement learning and deep learning to improve charging efficiency, cut cycle times, and save driving expenses. Real-time monitoring and control are also made possible by these AI-driven methods, which facilitate better energy management and improve the integration of electric automobiles into smart energy networks [33].

Algorithms for machine learning are used to evaluate large-scale information and find patterns that inform decision-making, especially when it comes to electric car charging plans. These algorithms improve efficiency and convenience for drivers by predicting the best times to recharge based on factors like energy pricing, grid load, and user behavior. Furthermore, they have the ability to adjust to shifting circumstances, continually improving their efficacy and accuracy to offer suggestions that are more accurate over time. EAs charging behavior is predicted using machine learning models, which use a variety of techniques to improve efficiency and decision-making. Regression approaches (e.g., XGBoost, Linear Regression) and supervised learning models are used to predict departure times and energy usage. Decision trees (DT), Random Forest (RF), and Support Vector Regression (SVR) can help to make a variety of forecasts about charging needs. Gaussian Mixture Models (GMM) is an example of an unsupervised learning model that clusters charging activities to find unique patterns and improve charging infrastructure. Recurrent neural networks (RNN), long short-term memory (LSTM) networks, and convolutional neural networks (CNN) are examples of deep learning models that are used for advanced charging load forecasting and charging profile classification. These models handle complex and sequential data to improve operational efficiency and predictive accuracy [34].

3.2 *Advanced Driver Assistance Systems (ADAS)*

Smart EAs are not complete without Advanced Driving Assistance Systems (ADAS). These devices use a variety of sensors, cameras, radar, and artificial intelligence (AI) algorithms to keep an eye on the environment around the car, identify any threats, and help the driver make judgments. The goal of ADAS such automated emergency braking, adaptive cruise control, and lane-keeping assistance is to increase road safety by lowering the risk of collisions.

Sensor fusion's is significance for smart electric vehicles. The process of creating a thorough picture of the vehicle's environment through sensor fusion entails integrating data from many sensors. The system may overcome the limits of individual sensors and give more precise and dependable information by merging inputs from cameras, radar, lidar, and other sensors. In smart EAs, this improved perception is essential for efficient control and decision-making. The function of vehicular ad hoc networks (VANETs) for vehicle-to-vehicle (V2V) communication is also used by EV. Vehicle-to-vehicle (V2V) communication makes it possible for cars to communicate with other cars in the area about their position, speed, and direction. This promotes coordinated driving and increases road safety. In order to minimize fuel consumption and maximize road capacity, many cars ride closely together in a coordinated way, a phenomenon known as platooning, which is made possible by this technology [35].

State of charge (SOC), dynamic vehicle, battery, and electric motor models are integrated into an ADAS-based system to study the effect on energy usage. Because of the impacts of regenerative braking, as evidenced by the negative values seen in the power curve graphs, the results showed that, when the ADAS model is active, a reduction in the distance between cars favorably improves the battery's charge state. A notable reduction in battery charge status is noted as vehicle speed approaches the conclusion of the driving cycle, which is succeeded by a rise brought on by the forceful braking action and regenerative power [36].

4 Conclusions

It is clear that the issue of evaluating the economic and environmental performance and impact of electric vehicles has become very important and urgent at the present time, especially with the increase in production intensity and the beginning of widespread use. Although the electric car industry has become a reality and inevitable, through a clear comparison between fossil fuel-powered automobiles (ICEAs) and electric automobiles (EAs), it is clear that each has unique advantages and disadvantages that need more and more studies. Hence, the objectives of this paper were achieved in helping stakeholders from policy makers, researchers and consumers to see the current scene clearly and provide proposed solutions for the development of automotive technology. Other, further conclusions were reached as follows:

- EAs are especially well-suited for driving in cities, especially in underground modern cities, since they have zero emissions and less maintenance expenses. EAs also have the advantage of greater torque and Max speeds on intercity highways. However, due to their longer range and faster recharging times, ICEAs often perform better for long-distance travel.
- Hopes are still pinned on scientific research to improve battery health and charge management system, whether for constant current charging (CCC) or constant voltage charging (CVC) method, as well as the importance of enhancing charging efficiency and reducing charging frequency time using AI.
- There is also another very important dimension, which requires a huge amount of serious research and honest disclosure of its results, to determine the types of environmental pollutants emitted by electric vehicles and how they affect the health of the driver, passengers and pedestrians in the community. So, EAs need further research and development to predict become more competitive in a wider range of driving circumstances as the gap between them and ICEAs closes due to advancements in battery technology and charging infrastructure. This is particularly evident with the use of AI in battery management system and advanced driver assistance system.

Eventually, currently, in light of the concerns of electric automobiles users in global communities towards critical issues of both safety and enhancing charging systems. Therefore, experts and interested users are waiting to see how the Group of Seven major industrialized countries (G7) will fund research in the automotive industry. Will most of them lean towards developing fuel injection car systems of various types or towards developing electric car systems, especially during the next two decades? Hence, implementing strong security protocols (based primarily on safety and not just economics) must be a top priority to make the most of this technology and ensure a safe future for societies in a sustainable manner.

References

1. Skoog, S.: Power Sources for Hybrid Electric Vehicles. CODEN: LUTEDX/TEIE
2. Chan, C.C.: The state of the art of electric and hybrid vehicles. Proc. IEEE **90**(2), 247–275 (2002)
3. Whittingham, M.S.: Lithium batteries and cathode materials. Chem. Rev. **104**(10), 4271–4302 (2004)
4. IEA: Global EV Outlook 2021. IEA, Paris (2021)
5. CARB: Zero-Emission Vehicle Program. Information on: https://ww2.arb.ca.gov/our-work/programs/zero-emission-vehicle-program/about. Accessed 20 Nov 2023
6. Nykvist, B., Nilsson, M.: Rapidly falling costs of battery packs for electric vehicles. Nat. Clim. Change **5**(4), 329–332 (2015)
7. Janek, J., Zeier, W.G.: A solid future for battery development. Nat. Energy **1**(9), 1–4 (2016)
8. Jahns, T.M., Soong, W.L.: Pulsating torque minimization techniques for permanent magnet AC motor drives—a review. IEEE Trans. Ind. Electron. **43**(2), 321–330 (1996)
9. Chan, C.C.: The state of the art of electric, hybrid, and fuel cell vehicles. Proc. IEEE **95**(4), 704–718 (2007)
10. Su, Z., Wang, Y., Xu, Q., Fei, M., Tian, Y.C., Zhang, N.: A secure charging scheme for electric vehicles with smart communities in energy blockchain. IEEE Internet Things J. **6**(3), 4601–4613 (2018)
11. Figenbaum, E., Kolbenstvedt, M.: Learning from Norwegian battery electric and plug-in hybrid vehicle users (2016)
12. Holland, G., Tiggemann, M.: A systematic review of the impact of the use of social networking sites on body image and disordered eating outcomes. Body Image **17**, 100–110 (2016)
13. Hawkins, T.R., Singh, B., Majeau-Bettez, G., Strømman, A.H.: Comparative environmental life cycle assessment of conventional and electric vehicles. J. Ind. Ecol. **17**(1), 53–64 (2013)
14. Helmers, E., Marx, P.: Electric cars: technical characteristics and environmental impacts. Environ. Sci. Eur. **24**, 1–15 (2012)
15. Denholm, P., Sioshansi, R.: The value of compressed air energy storage with wind in transmission-constrained electric power systems. Energy Policy **37**(8), 3149–3158 (2009)
16. Grosjean, C., Miranda, P.H., Perrin, M., Poggi, P.: Assessment of world lithium resources and consequences of their geographic distribution on the expected development of the electric vehicle industry. Renew. Sustain. Energy Rev. **16**(3), 1735–1744 (2012)
17. Lutsey, N., Nicholas, M.: Update on electric vehicle costs in the United States through 2030. Int. Counc. Clean Transp. **12**, 1–12 (2019)
18. Harper, G., Sommerville, R., Kendrick, E., Driscoll, L., Slater, P., Stolkin, R., Anderson, P.: Recycling lithium-ion batteries from electric vehicles. Nature **575**(7781), 75–86 (2019)
19. Ali, S.H.R.: Automotive Engine Metrology, 1st edn. Taylor Francis, Florence, Kentucky (KY); Pan Stanford, Singapore. Printed in Canada (2017). ISBN 978-982-4669-52-8 (Hardcover), ISBN 978-1-315-36484-1 (eBook)

20. Ali, A.S.H.R., Ali, S.H.R.: Digital industry by entrepreneurship: the way to a flourish economy. J. Mod. Ind. Manuf. (JMIM) **1**(1), 1–7 (2022)
21. Ali, A., Ali, A.M.S., Amin, Y.W.R., Ali, S.H.R.: The Important Role of GD&T System in Mechanical Drawing, Design and Manufacturing for Students of Engineering Institutes. Paper: 2024-01-2052. SAE International, USA (2024)
22. Ali, S.H.R., Mohamd, O.M.: Dimensional and geometrical form accuracy of circular pockets manufactured for aluminum, copper and steel materials on CNC milling machine using CMM. Int. J. Eng. Res. Afr. **17**, 64–73 (2015). http://www.scientific.net/JERA.17.64
23. Ellingsen, L.A.W., Majeau-Bettez, G., Singh, B., Srivastava, A.K., Valøen, L.O., Strømman, A.H.: Life cycle assessment of a lithium-ion battery vehicle pack. J. Ind. Ecol. **18**(1), 113–124 (2014)
24. Gaines, L.: Lithium-ion battery recycling processes: research towards a sustainable course. Sustain. Mater. Technol. **17**, e00068 (2018)
25. Thompson, A.: Combating Tailpipe Emissions: An Analysis of State Electric Vehicle Incentives (2019)
26. Sandberg, U., Ejsmont, J.: Tyre/Road Noise. Reference Book (2002)
27. Gritsenko, A., Zadorozhnaya, E., Shepelev, V., Gimaltdinov, I.: Parameters of internal combustion engine efficiency while introducing additives in the oil. Tribol. Ind. **41**(4), 592 (2019)
28. Poornesh, K., Nivya, K.P., Sireesha, K.: A comparative study on electric vehicle and internal combustion engine vehicles. In: 2020 International Conference on Smart Electronics and Communication (ICOSEC), pp. 1179–1183. IEEE (2020)
29. Sanguesa, J.A., Torres-Sanz, V., Garrido, P., Martinez, F.J., Marquez-Barja, J.M.: A review on electric vehicles: technologies and challenges. Smart Cities **4**(1), 372–404 (2021)
30. Sun, X., Li, Z., Wang, X., Li, C.: Technology development of electric vehicles: a review. Energies **13**(1), 90 (2019)
31. Jape, S.R., Thosar, A.: Comparison of electric motors for electric vehicle application. Int. J. Res. Eng. Technol. **6**(09), 12–17 (2017)
32. Badran, M.A., Toha, S.F.: Employment of artificial intelligence (AI) techniques in battery management system (BMS) for electric vehicles (EV): issues and challenges. Pertanika J. Sci. Technol. **32**(2) (2024)
33. Aravind, R., Surabhi, S.N.R.D.: Smart charging: AI solutions for efficient battery power management in automotive applications. Educ. Adm. Theory Pract. **30**(5), 14257–21467 (2024)
34. Shahriar, S., Al-Ali, A.R., Osman, A.H., Dhou, S., Nijim, M.: Machine learning approaches for EV charging behavior: a review. IEEE Access **8**, 168980–168993 (2020)
35. Haque, T.S., Rahman, M.H., Islam, M.R., Razzak, M.A., Badal, F.R., Ahamed, M.H., Ryan, M.: A review on driving control issues for smart electric vehicles. IEEE Access **9**, 135440–135472 (2021)
36. Ataman, T., Biberci, M.A., Celik, M.B.: Simulation of advanced driving assistance systems for a dynamic vehicle model. Eng. Technol. Appl. Sci. Res. **14**(5), 16553–16558 (2024)

Kapok Pod Fibre as a Sustainable Biofuel Resource: Prediction of Cellulose Hydrolysis Using a Heuristic Algorithm Optimized Random Forest

Yitong Niu, Chee Keong Lee, and Cheu Peng Leh

Abstract Energy consumption of biofuels has increased over the past few years relative to fossil fuels, and biomass-based ethanol has attracted interest due to its versatility. The production of bioethanol is a complex process in which the hydrolysis of cellulose plays a decisive role in the final ethanol yield. Moreover, the effectiveness of biomass pretreatment is crucial to determine the enzymatic accessibility and hydrolysis capacity of biomass. Current common studies on biomass pretreatment usually set pretreatment parameters as independent variables without understanding the effect on biomass chemical composition. In this study, a Random Forest-based prediction model was developed to predict the enzymatic hydrolysis of cotton fibers for glucose yield based on the chemical composition (cellulose, hemicellulose, and lignin) of pretreated cotton fibers. Enzymatic hydrolysis of poplar wood samples pretreated with water, acid, and alkali was performed, and the fermentable glucose content obtained was determined, constituting a dataset of 33. To find suitable model parameters, the random forest model was tuned using random search, genetic algorithm, particle swarm algorithm, and simulated annealing algorithm. It was found that the model tuned using a simulated annealing algorithm gave the best results with R^2 of 0.92 and RMSE of 2.94 in the test set. The model tuned using a random search algorithm gave the second-best results, while the model tuned using a genetic algorithm and particle swarm algorithm showed underfitting. Therefore, the random forest model optimized based on the simulated degradation algorithm is more suitable for predicting the hydrolyzed glucose content of kapok fibers.

Keywords Machine learning · Bioethanol · Cellulose hydrolysis · Random forest · Heuristic algorithm

---

Y. Niu · C. K. Lee · C. P. Leh (✉)
School of Industrial Technology, Universiti Sains Malaysia, Penang, Malaysia
e-mail: cpleh@usm.my

Y. Zhu (ed.), *Proceedings of the 4th International Symposium in Environmental Science and Industrial Ecology*, https://doi.org/10.1007/978-981-96-1578-0_15

1 Introduction

In recent years, the global energy focus has shifted to renewable and sustainable energy sources to reduce dependence on fossil fuels and combat climate change. Fossil fuels have long been the primary source of energy but have contributed to significant greenhouse gas emissions and environmental degradation [1]. As the world faces the consequences of climate change, there is growing interest in cleaner and more sustainable alternative energy sources. Biofuels, particularly bioethanol, have emerged as a promising alternative with the potential to reduce carbon emissions. Bioethanol is made from a variety of biomaterials and can be used for transportation, industrial processes, and power generation [2]. This renewable fuel, if produced sustainably, could contribute significantly to a low-carbon economy [3]. As the demand for biofuels continues to grow, research has focused on improving the conversion of biomass to ethanol, especially given the abundance and renewability of plant material and agricultural residues. However, the conversion of biomass to ethanol is complex, and innovative approaches are needed to address the challenges associated with breaking down lignocellulosic materials [4].

Bioethanol is made from plant materials and is gaining attention as a renewable energy source due to its potential to reduce carbon emissions. It can be blended with gasoline or used alone in engines [5, 6]. Bioethanol is attractive because it is derived from renewable organic materials and is essentially carbon neutral, meaning that the carbon released during combustion is offset by the carbon absorbed by the plant during growth [7]. The primary sources of bioethanol production fall into three categories: first-generation bioethanol from sugar and starch crops such as corn and sugarcane, second-generation bioethanol from lignocellulosic biomass such as agricultural residues, and third-generation bioethanol from algae [8]. Concerns about the use of food crops for fuel have shifted interest to second-generation bioethanol made from non-food materials. These sources are more sustainable but are more challenging to convert to ethanol due to their complex structure. Pretreatment methods are essential to break down encumbrance biomass (lignin or hemicellulose), followed by enzymatic hydrolysis to convert cellulose into fermentable sugars. Kapok fiber is rich in cellulose and is a promising feedstock for non-traditional bioethanol production [9]. This study investigates how pretreatment and enzymatic hydrolysis of kapok can be optimized to increase bioethanol yield and, thus, biofuel production efficiency.

The process of converting biomass to bioethanol consists of several steps: pretreatment, enzymatic hydrolysis, fermentation, and distillation. Of these, pretreatment and hydrolysis are essential because they determine the efficiency of releasing sugars for ethanol production [10]. Pretreatment breaks down the complex structure of the lignocellulosic material to improve enzyme accessibility. If pretreatment is inadequate, the efficiency of the conversion process is reduced. Enzymatic hydrolysis uses enzymes such as cellulase to break down biomass into fermentable sugars, primarily glucose. The success of this step depends on the effectiveness of the pretreatment. After hydrolysis, fermentation uses microorganisms to convert these sugars into ethanol and finally purifies the ethanol by distillation so that it can be used as a biofuel. By ensuring efficient pretreatment and hydrolysis, bioethanol production from biomass can be optimized.

Despite significant progress in the field of bioethanol production, most studies have focused on optimizing biomass pretreatment conditions without sufficiently considering the interaction between pretreatment methods and changes in biomass chemistry. Traditional optimization methods, such as the design of experiments or response surface methods, usually consider pretreatment parameters (e.g., temperature, pH, time) as independent variables and enzymatic glucose production as the only response variable [11]. However, this approach ignores the complex interactions between pretreatment and biomass chemistry that play a vital role in the efficiency of enzymatic hydrolysis. Biomass consists of different proportions of cellulose, hemicellulose, and lignin, and the effectiveness of pretreatment strategies may vary greatly depending on the relative abundance of these components [12]. Changes in biomass structure during pretreatment (e.g., removal of hemicellulose or lignin) can directly affect the accessibility of cellulose to enzymes during hydrolysis [13]. Without a thorough understanding of how these compositional changes affect enzymatic hydrolysis, it is not easy to effectively optimize the bioethanol production process. Gaps in understanding the impact of biomass composition on pretreatment outcomes necessitate the use of advanced predictive models to capture the relationship between the chemical composition of biomass and its hydrolysis performance. Machine learning models (e.g., Random Forest) offer a promising solution by allowing researchers to predict enzymatic glucose production based on the chemical composition of pretreated biomass. These models can help determine which pretreatment conditions are best suited for different types of biomasses, leading to a more customized and efficient bioethanol production process [14]. Furthermore, while previous studies have attempted to optimize biomass conversion using a variety of modeling approaches, most have not taken full advantage of the potential of heuristic algorithms to optimize machine learning models to improve prediction accuracy.

The main objective of this study was to develop and optimize a Random Forest-based predictive model for estimating the glucose yield from enzymatic hydrolysis of kapok fibers based on their chemical composition (cellulose, hemicellulose, and lignin). By utilizing data from an experimental dataset of pretreated cottonwood samples, this study aimed to create a robust predictive model that accurately predicts glucose yields after different pretreatment methods (aqueous, acidic, and alkaline). The study also compares the effectiveness of various heuristic optimization algorithms (stochastic search, genetic algorithm, simulated annealing, and particle swarm optimization) for fine-tuning the random forest model. Through these efforts, this study aims to address the current gaps in understanding the relationship between biomass composition and hydrolysis outcomes and provide a reliable tool for predicting bioethanol production efficiency.

2 Method

2.1 Random Forest (RF)

RF is an integrated learning method that improves model stability and accuracy by constructing multiple decision trees and averaging their predictions. The technique was introduced by [15] and has gained popularity due to its robustness and versatility in handling various prediction problems. Random forests are very effective for both classification and regression tasks, especially in high-dimensional environments where the number of variables exceeds the number of observations, and they are more capable of handling small sample sizes, high-dimensional feature spaces, and complex data structures [16–18].

2.2 Random Search (RS)

RS search performs optimization by randomly selecting combinations of parameters in the parameter space. The principle is to select parameter combinations randomly in the parameter space instead of exhaustively enumerating all possible combinations. Compared to grid search, random search is more efficient in high-dimensional parameter spaces and is applicable to continuous, discrete and mixed parameter spaces [19].

2.3 Genetic Algorithms (GA)

Genetic algorithms (GA) were proposed by Prof. John H. Holland and his students at the University of Michigan in 1969. Holland found that the evolution of populations was conducive to enhanced learning when he was engaged in related research. Later, he explained the theoretical knowledge and basic steps of genetic algorithms in his works. Scholars such as Goldberg enriched and analyzed in detail the basis of this, which led to the present-day genetic algorithms.

Organisms in the process of survival, "natural selection, the survival of the fittest," in the natural elimination mechanism, further upgrading, species will also progress. Similar to biological evolution, genetic algorithms do not need to understand the problem itself; they start from an initial population, through the selection of the fitness value function, the survival of the fittest, through continuous selection and reproduction, and ultimately, the individuals with good fitness values are retained. The solution corresponding to the individual with the best fitness value is recognized as the optimal solution to the problem. The genetic algorithm is based on population search, which is done by selecting a population to search in the search space and selecting and replacing a better population. The genetic algorithm is based on such a model to carry out a heuristic search, which is the true sense of the meta-heuristic search algorithm. In recent years, genetic algorithms have been widely used in solving complex optimization problems and have been applied in various fields such as industrial engineering, artificial intelligence, bioengineering, and automatic control, which makes the algorithm get the attention of scholars.

2.4 Particle Swarm Algorithm (PSO)

Genetic algorithms (GA) were proposed by Prof. John H. Holland and his students at the University of Michigan in 1969. Holland found that the evolution of populations was conducive to enhanced learning when he was engaged in related research. Later, he explained the theoretical knowledge and basic steps of genetic algorithms in his works. Scholars such as Goldberg enriched and analyzed in detail the basis of this, which led to the present-day genetic algorithms.

The particle swarm algorithm (PSO) idea comes from the theory of artificial life and evolutionary computation [20], it is a stochastic global optimization algorithm based on group intelligence; the advantage lies in the faster convergence speed, the algorithm is simple, easy to be programmed, and implemented, and it has been widely used [21]. However, the conventional PSO algorithm quickly falls into the local optimum, and premature convergence occurs, a significant defect of the PSO algorithm. In order to overcome this defect, scholars have put forward many improvement ideas in recent years; one type of method is to improve the algorithm itself; the second type of method is to integrate with other algorithms to find the optimal and use the advantages of other algorithms to help it to jump out of the local optimum, which is one of the ways of integrating with the chaos algorithm. By integrating the chaos optimization algorithm into the PSO algorithm, the traversal of chaotic sequences can generate many neighboring points of the local optimal solution, creating more feasible solution spaces for the particles, which can help the inert particles to fly away from the local optimal point, and drive the whole particle swarm to traverse to all solution spaces, which overcomes the defect of the particles falling into the local optimal point to a certain extent [22].

Let the particle in the n-dimensional search space (solution space); there are m particles composed of particle swarm, where the position of the fth particle in the n-dimensional search space is represented as the n-dimensional vector $X_1 = (x_{11}, x_{12}, \ldots, x_{1n})$, the position of each particle represents an optimal solution in the search space parameters. After selecting the objective function, the fitness value of the ith particle can be calculated by bringing the position vector x_1 into the objective function, and according to the size of the fitness value, the particle can be judged to be superior or inferior. In addition, there is a velocity that describes the iterative displacement, where the flight velocity of the ith particle is denoted as $v_1 = (v_{11}, v_{12}, \ldots, v_{1n}).p_1 = (p_{11}, p_{12}, \ldots, p_{1n})$ is the optimal position experienced by the ith particle in its current search history, and the position fitness value is Pbest, called the individual extreme value. The other opposite, $G_1 = (g_{11}, g_{12}, \ldots, g_{1n})$, is the optimal position experienced in the current search history of the swarm, which has a position fitness value of Gbest, called the global extreme value. The particle updates its own velocity and position variables according to the following equation:

$$v_{IJ}^{k+1} = \omega \cdot v_{IJ}^{k} + c_1 \cdot r_1\left(p_{IJ}^{k} - x_{IJ}^{k}\right) + c_2 \cdot r_2\left(g_{IJ}^{k} - x_{IJ}^{k}\right) \tag{1}$$

$$x_{IJ}^{k+1} = v_{IJ}^{k} + x_{IJ}^{k} \tag{2}$$

The particle flight speed adjustment rule is:

$$\begin{cases} v_{IJ} = v_{\max}, & v_{IJ} > v_{\max} \\ v_{I} = -v_{\max}, & v_{Ij} \le v_{\max} \end{cases} \tag{3}$$

where $i = 1, 2, 3, \ldots, n$, is the number of particles; $j = 1, 2, 3, \ldots, n$, is the number of independent variables; k is the current number of iterations; ω is the inertia weighting coefficients; c_1, c_2 are a set of acceleration constants; and r_1, r_2 is a set of random numbers, whose values are in the range of (0, 1).

2.5 *Simulated Annealing Algorithm (SAA)*

SAA algorithm is a meta-heuristic algorithm based on physics. Metropolis and other scholars proposed the original simulated annealing algorithm in the early 1950s. At temperature t, a new state j, is generated under the current state i. The energies are E_i and E_j, respectively. If $E_i > E_j$, then the current state is transformed to j; otherwise, the probability that the current state becomes j can be expressed by the following equation:

$$p = \exp\left[\frac{-(E_j - E_i)}{kt}\right] \tag{4}$$

where k is the Boltzmann constant, it is verified that after repeating the process many times, the system will tend to the equilibrium state. In this equilibrium situation, the energy of the system is easily low. The probability distribution of each state will be close to the regular distribution, and this new method of accepting the state is known as the Metropolis criterion. This received little attention and response from scholars at that time. The process of slowly decreasing temperature can be simulated by the Metropolis criterion, which is used as a search strategy that allows the SAA algorithm to escape from the local optimum and eventually reach the optimal or near-optimal solution state. The disadvantages of SAA are that it is too slow in its convergence, has a long execution time, and the performance of the simulated annealing algorithm is more sensitive to the parameters involved.

3 Results and Discussion

After the experiments, I measured cellulose, hemicellulose, lignin, and glucose and constructed the dataset, out of 33 data, 80% (27 data) were used to train the model, and 20% (6 data) were used to test the model.

The glucose yield from enzymatic hydrolysis after cellulose hydrolysis was predicted using a random forest regression model, and parameter optimization was performed by genetic algorithm (GA), particle swarm optimization (PSO), simulated annealing algorithm (SAA), and random search (RS). The parameters of the adjusted random forest model are shown by Table 1.

The prediction of the optimized model on the test set is shown in Fig. 1.

It can be seen that the training results of the optimized RF prediction models, among which the models optimized by RS and SAA, are still good. In contrast, the models optimized by GA and PSO both show underfitting, indicating that the models perform poorly on the training data and fail to adequately capture the complexity of the data. This is mainly due to the limited size of the dataset and the lack of model complexity. Although the random forest model has a solid nonlinear fitting ability, the model complexity may not be sufficient to capture all patterns in the data when the data size is small.

Since the GA-RF and PSO-RF models were poorly trained, only the RS-RF and SAA-RF prediction models were tested. Figure 2 shows the results.

As shown in Fig. 1, the performance of RS-RF ($R^2 = 0.92$, RMSE = 2.03) is slightly better than that of SAA-RF ($R^2 = 0.90$, RMSE = 2.27) on the training data, which seems to be that the random search has an advantage in parameter optimization. However, in the performance of the test set, as shown in Fig. 2, the performance of SAA-RF ($R^2 = 0.92$, RMSE = 2.94) is instead significantly better than that of RS-RF ($R^2 = 0.83$, RMSE = 3.34). This may be due to the small sample size of the model and collecting more experimental data to improve the training and testing effect of the model and enhance the generalization ability of the model. Overall, the SAA-RF

Table 1 Parameter adjustment based on meta-heuristic algorithm

Parameter	Value			
	RS	GA	PSO	SAA
Node split evaluation criteria	MSE	MAE	MAE	MAE
Minimum number of samples for internal node splitting	2	21	2	3
Minimum number of samples for leaf nodes	1	12	50	1
Minimum weight of samples in leaf nodes	0	0.19	0.5	0
Maximum depth of tree	10	20	2	11
Maximum number of leaf nodes	50	134	105	27
Threshold for node partitioning impurity	0	0.27	0.23	0
Number of decision trees	100	51	50	53
Sampling with replacement	True	False	True	True

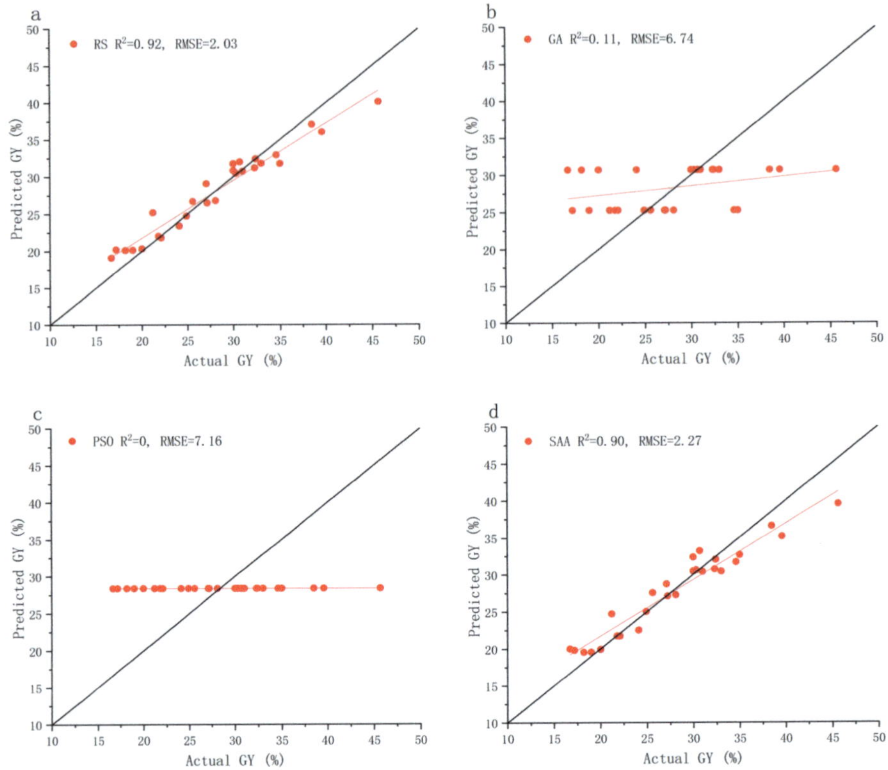

Fig. 1 Test set performance of the optimized random RF prediction model. **a** RS-RF, **b** GA-RF, **c** PSO-RF, **d** SAA-RF

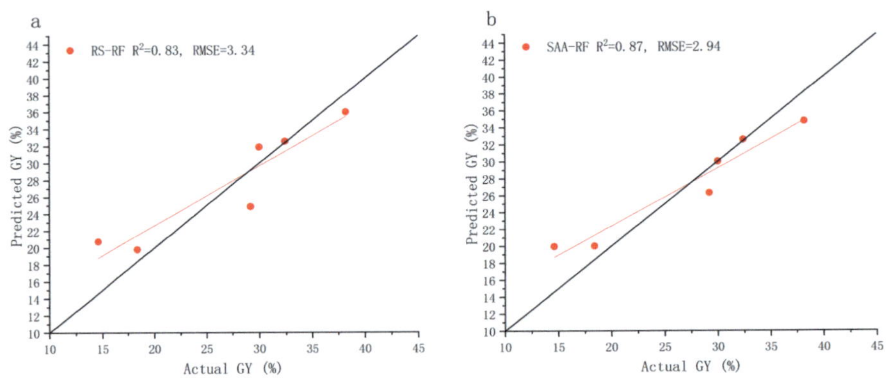

Fig. 2 Test set results. **a** RS-RF, **b** SAA-RF

model clearly outperforms the predictions of its method, and future research can further explore the potential of the simulated annealing optimization algorithm and combine it with other optimization algorithms to improve the model performance.

4 Conclusion

In this study, a Random Forest-based prediction model was developed for predicting the enzymatic glucose yield of pretreated cotton fibers. Thirty-three data sets were identified and obtained by enzymatic digestion of cottonwood samples pretreated with water, acid, and alkali. In order to find appropriate model parameters, the random forest model was tuned using random search, genetic algorithm, particle swarm algorithm, and simulated annealing algorithm. The results show that the random forest model tuned by the simulated annealing algorithm has the best overall effect, with a performance R^2 of 0.90 and RMSE of 2.27 on the training set and a performance R^2 of 0.92 and RMSE of 2.94 on the test set, which is suitable for use. The random forest model tuned by the random search algorithm had the following best overall results, with a performance R^2 of 0.92 and RMSE of 2.03 on the training set and a performance R^2 of 0.83 and RMSE of 3.34 on the test set. In contrast, the model tuned by the genetic and particle swarm algorithms exhibited underfitting. Therefore, the random forest model optimized based on simulated annealing is more suitable for predicting the hydrolyzed glucose yield of kapok fiber.

References

1. Ritchie, H., Roser, M., Rosado, P.: Renewable energy. Our World Data (2024)
2. Abdullah, S.S.S., Shirai, Y., Ali, A.A.M., Mustapha, M., Hassan, M.A.: Case study: preliminary assessment of integrated palm biomass biorefinery for bioethanol production utilizing non-food sugars from oil palm frond petiole. Energy Convers. Manage. **108**, 233–242 (2016). https://doi.org/10.1016/j.enconman.2015.11.016
3. Ali, F., Dawood, A., Hussain, A., Alnasir, M.H., Khan, M.A., Butt, T.M., Janjua, N.K., Hamid, A.: Fueling the future: biomass applications for green and sustainable energy. Discov. Sustain. **5**, 156 (2024). https://doi.org/10.1007/s43621-024-00309-z
4. Conde-Mejía, C., Jiménez-Gutiérrez, A., El-Halwagi, M.: A comparison of pretreatment methods for bioethanol production from lignocellulosic materials. Process Saf. Environ. Prot. **90**, 189–202 (2012). https://doi.org/10.1016/j.psep.2011.08.004
5. Abdulsattar, M.O., Abdulsattar, J.O., Greenway, G.M., Welham, K.J., Zein, S.H.: Optimization of pH as a strategy to improve enzymatic saccharification of wheat straw for enhancing bioethanol production. J. Anal. Sci. Technol. **11**, 17 (2020). https://doi.org/10.1186/s40543-020-00217-7
6. Ajijolakewu, A.K., Leh, C.P., Wan Abdullah, W.N., Lee, C.K.: Optimization of production conditions for xylanase production by newly isolated strain *Aspergillus niger* through solid state fermentation of oil palm empty fruit bunches. Biocatal. Agric. Biotechnol. **11**, 239–247 (2017). https://doi.org/10.1016/j.bcab.2017.07.009
7. Tye, Y.Y., Leh, C.P., Wan Abdullah, W.N.: Total glucose yield as the single response in optimizing pretreatments for *Elaeis guineensis* fibre enzymatic hydrolysis and its relationship with

chemical composition of fibre. Renew. Energy **114**, 383–393 (2017). https://doi.org/10.1016/j.renene.2017.07.040
8. Tye, Y.Y., Lee, K.T., Wan Abdullah, W.N., Leh, C.P.: Second-generation bioethanol as a sustainable energy source in Malaysia transportation sector: status, potential and future prospects. Renew. Sustain. Energy Rev. **15**, 4521–4536 (2011). https://doi.org/10.1016/j.rser.2011.07.099
9. Tye, Y.Y., Lee, K.T., Wan Abdullah, W.N., Leh, C.P.: The world availability of non-wood lignocellulosic biomass for the production of cellulosic ethanol and potential pretreatments for the enhancement of enzymatic saccharification. Renew. Sustain. Energy Rev. **60**, 155–172 (2016). https://doi.org/10.1016/j.rser.2016.01.072
10. Hemashini, T., Lee, C.K., Goh, C.F., Tye, Y.Y., Ismayati, M., H'ng, Y.Y., Leh, C.P.: Elucidating the correlation of lignocellulosic compositions and physicochemical alterations in oil palm (*Elaeis guineensis*) biomass on enzymatic saccharification yield. Bioenerg. Res. (2024). https://doi.org/10.1007/s12155-024-10762-3
11. Djioleu, A., Carrier, D.J.: Effects of dilute acid pretreatment parameters on sugar production during biochemical conversion of switchgrass using a full factorial design. ACS Sustain. Chem. Eng. **4**, 4124–4130 (2016). https://doi.org/10.1021/acssuschemeng.6b00441
12. Ying, T.Y., Teong, L.K., Abdullah, W.N.W., Peng, L.C.: The effect of various pretreatment methods on oil palm empty fruit bunch (EFB) and kenaf core fibers for sugar production. Procedia Environ. Sci. **20**, 328–335 (2014). https://doi.org/10.1016/j.proenv.2014.03.041
13. Tye, Y.Y., Leh, C.P., Lee, K.T., Abdullah, W.N.W.: Non-wood lignocellulosic biomass for cellulosic ethanol production: effects of pretreatment on chemical composition in relation to total glucose yield. J. Jpn. Inst. Energy **96**, 503–508 (2017)
14. Aghbashlo, M., Peng, W., Tabatabaei, M., Kalogirou, S.A., Soltanian, S., Hosseinzadeh-Bandbafha, H., Mahian, O., Lam, S.S.: Machine learning technology in biodiesel research: a review. Prog. Energy Combust. Sci. **85**, 100904 (2021). https://doi.org/10.1016/j.pecs.2021.100904
15. Breiman, L.: Random forests. Mach. Learn. **45**, 5–32 (2001). https://doi.org/10.1023/A:1010933404324
16. Scornet, E., Biau, G., Vert, J.-P.: Consistency of random forests. Ann. Stat. **43**, 1716–1741 (2015). https://doi.org/10.1214/15-AOS1321
17. Biau, G., Scornet, E.: A random forest guided tour. TEST **25**, 197–227 (2016). https://doi.org/10.1007/s11749-016-0481-7
18. Cheng, L., Chen, X., De Vos, J., Lai, X., Witlox, F.: Applying a random forest method approach to model travel mode choice behavior. Travel Behav. Soc. **14**, 1–10 (2019). https://doi.org/10.1016/j.tbs.2018.09.002
19. Bergstra, J., Bengio, Y.: Random search for hyper-parameter optimization. J. Mach. Learn. Res. **13**, 281–305 (2012)
20. Kennedy, J., Eberhart, R.: Particle swarm optimization. In: Proceedings of ICNN'95-International Conference on Neural Networks. pp. 1942–1948. IEEE (1995)
21. Guo, Y., Wen, J.: A new power system reactive power optimization algorithm. Electr. Power Survey and Des. **12**, 66–70 (2005)
22. Li, M., Kang, H., Zhou, P.: Hybrid optimization algorithm based on chaos, cloud and particle swarm optimization algorithm. J. Syst. Eng. Electron. **24**, 324–334 (2013). https://doi.org/10.1109/JSEE.2013.00041

Environmental Impact of Motorized Transportation for School Commutes

Firas Alrawi and Faisal Mohammed

Abstract In Baghdad, recent years have witnessed significant and rapid changes in school trip patterns from non-motorized trips (active transport) to motorized trips due to economic, social, and safety factors. This change led to an increase in gaseous pollutants emitted into the local environment due to vehicle fuel combustion and exceeding the permissible limits, thus its harmful impact on the health of residents. This paper tried to study essential gaseous pollutants affecting the local environment caused by changing the school trip pattern. Two schools have been chosen in Baghdad to conduct a field study. At each school location, two tests were conducted one on a school day and one on the weekend. Measurement has been taken via portable devices (carbon monoxide meter, CO_2 meter, SO_2, NO_2, and O_3 meters). The results found that a high concentration of NO_2, SO_2, and CO_2 gases exceeded the national criteria's limits during the school day compared to the weekend, as it decreased to below those criteria. On the other hand, the concentrations of ozone gas O_3 and carbon monoxide CO were within the permissible limits during the two tests. The air quality index (AQI) was also calculated at the two sites, which gave the same results as the previous ones. However, this indicator rose on the school day, indicating the impact of the change in the school trip pattern on the local environment compared to the weekend, which was within acceptable limits.

Keywords Motorized trip · Active transport · Air pollutants · Air quality index

F. Alrawi (✉) · F. Mohammed
Urban and Regional Planning Centre, University of Baghdad, Baghdad, Iraq
e-mail: dr.firas@uobaghdad.edu.iq; fhamodi@asu.edu

F. Alrawi
School of Geographical Sciences and Urban Planning, Arizona State University, Tempe, AZ, USA

Y. Zhu (ed.), *Proceedings of the 4th International Symposium in Environmental Science and Industrial Ecology*, https://doi.org/10.1007/978-981-96-1578-0_16

1 Introduction

The increasing use of automobiles has led to a range of environmental problems, particularly related to the pollutants emitted into the air from vehicle fuel combustion [1]. These pollutants include nitrogen dioxide (NO_2), sulfur dioxide (SO_2), ozone (O_3), carbon monoxide (CO), carbon dioxide (CO_2), and suspended particles (PM2.5, 10) [2]. Traffic congestion leads to higher concentrations of these pollutants, creating what is known as traffic pollution [3]. School students are among the groups most at risk from exposure to these pollutants, which can lead to long-term health problems, especially respiratory diseases and heart disease, impacting their educational capabilities [4, 5]. Pollutant concentrations vary by season, reaching their highest levels in summer when temperatures rise [6]. The earth's surface temperature and air masses play a role in affecting these pollutants, with cold air masses staying close to the surface and warm air masses rising above them. This creates a barrier that traps pollutants near the surface, increasing their harmful effects on public health [7].

1.1 Factors Influencing Air Pollution

The change in the natural environment, represented by factors such as temperature, relative humidity (RH), clouds, rainfall, wind direction, and speed, influences the concentration of air pollutants either positively or negatively. As poor air quality leads to various health issues, especially for young age groups like school students, affecting their digestive, nervous, circulatory, and respiratory systems, understanding these effects is crucial. Temperature is one of the main factors, as it is directly proportional to the increase of pollutants. The concentration of contaminants rises during the day due to the warmth of the earth's surface from sun-absorbed heat and decreases at night, leading to seasonal variation in pollution levels, especially between winter and summer [8]. Rain plays a significant role in dissolving gaseous pollutants and suspended particles, reducing their concentration in the air, and making rainfall inversely proportional to pollutant intensity [9]. Relative humidity also exhibits an inverse relationship with pollutant concentration, where higher humidity levels reduce airborne pollutants. However, certain gases like SO_2 and NO_2 react with water vapor to form acid rain, causing environmental harm [10]. Wind influences the accumulation of air pollutants, especially suspended particles near the earth's surface. Higher wind speeds decrease the concentration of these pollutants, while slower winds increase it. Wind direction is also crucial, as it transports pollutants from one place to another, making it an important factor when considering the location of facilities or buildings [11].

1.2 Types of Gaseous Pollutants

There are many gaseous pollutants, including nitrogen dioxide (NO_2), which originates from fuel combustion in vehicles and factories. NO_2 primarily impacts public health by contributing to respiratory diseases, with 70–90% of it being absorbed by the body. It also affects the eyes and circulatory system [12]. Sulfur dioxide (SO_2), emitted by vehicles and power plants, affects both the respiratory and nervous systems and also damages plants and buildings due to its reactivity with building materials [13]. Ozone (O_3) from vehicles, power plants, refineries, and petrochemical plants causes respiratory and eye issues and affects crops by causing premature fruit dropping in trees like citrus. Carbon oxides (CO_x), from vehicle emissions, refineries, and power plants, pose significant risks to the nervous and circulatory systems [14].

1.3 Planning, Development, and Environmental Pollution

Poor urban design, sprawl, and inadequate housing contribute to increased exposure to noise and air pollution, primarily from vehicle traffic. Unplanned expansion often leads to greater reliance on private vehicles at the expense of public transportation, exacerbating traffic-related air pollution and noise emissions. Poor management and deficiencies in spatial planning also play a role in this issue. Thus, improvements in urban design, planning, and development are essential in reducing both noise and air pollution, and they can also help address related social and health inequalities [15].

Spatial planning is crucial in enhancing air quality and reducing exposure to pollution. Developing effective local planning policies is vital, as land-use planning can significantly improve local air quality. At a strategic level, spatial planning can create more sustainable transportation links between residential areas and schools. Using spatial planning tools and data wisely can help separate living, working, and commuting areas from polluted zones. Additionally, stricter regulations on air and noise emissions would reduce the adverse health effects on school students and residents [16].

2 Materials and Methods

2.1 Study Area

The study area is represented by the 14th of July community, neighborhood 510, located east of Baghdad within the Rusafa district. It borders the Aadhamiya sector to the north, the 9th of April sector to the south, the Sadr sector to the east, and the rest of the communities of the Rusafa Center sector to the west (Fig. 1) [17]. The population of the study area is 14,052 citizens, and its area is 1,062,913 m^2 [18].

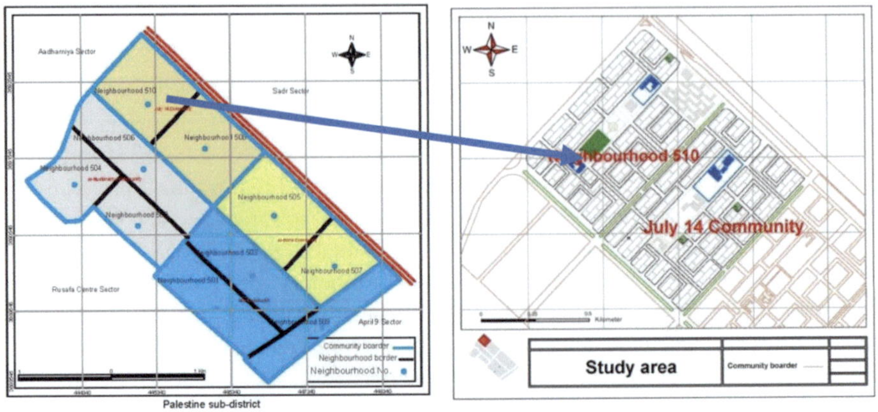

Fig. 1 Study area

2.2 Methodology of Data Collection and Analysis

Portable devices were used to collect data on pollutant concentrations near two selected schools (Fig. 2). Each location had two measurement periods: one on a school day and another on the weekend. The tests measured concentrations of SO_2, NO_2, CO_2, CO, and O_3 over one hour, along with atmospheric elements like temperature and relative humidity, influencing pollutant concentrations (Tables 1, 2, 3 and 4).

M1 and M3 are measuring stations near Al-Tasami Primary School, while M2 and M4 are near Palestine Middle School, with measurements taken on school days (M1, M2) and weekends (M3, M4).

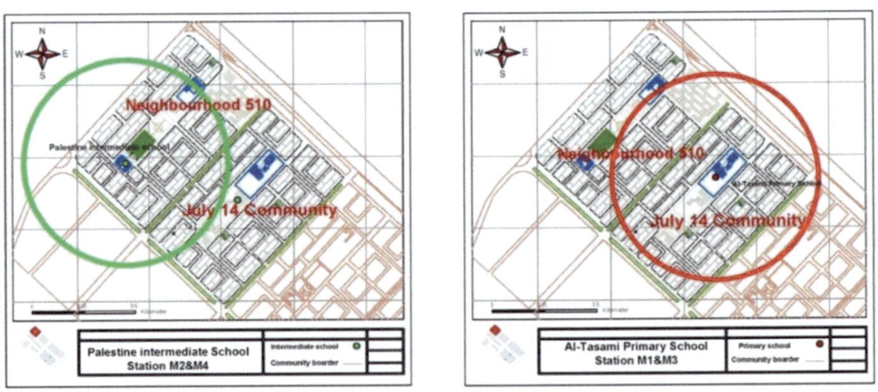

Fig. 2 Testing stations

Table 1 Pollutant concentration in station M1

Station M1							
Pollutant	SO_2	NO_2	CO	O_3	CO_2	Temp. (°C)	R.H.
Concentration (ppm)	0.14	0.27	1.09	0.08	446.5 μg/m^3	22	14%

Table 2 Pollutant concentration in station M3

Station M3							
Pollutant	SO_2	NO_2	CO	O_3	CO_2	Temp. (°C)	R.H.
Concentration (ppm)	0	0	0	0.03	205.2 μg/m^3	29.2	15%

Table 3 Pollutant concentration in station M2

Station M2							
Pollutant	SO_2	NO_2	CO	O_3	CO_2	Temp. (°C)	R.H.
Concentration (ppm)	0.11	0.31	2.1	0.05	417.2 μg/m^3	26.6	19%

Table 4 Pollutant concentration in station M4

Station M4							
Pollutant	SO_2	NO_2	CO	O_3	CO_2	Temp. (°C)	R.H.
Concentration (ppm)	0	0	0	0.02	198.1 μg/m^3	29	7%

3 Results and Discussion

Table 5 and Figs. 3, 4 show the concentration of gaseous pollutants compared to national criteria through field tests carried out over four stations near schools. The first test was on an official working day, and the second was on the weekend. The results indicate that students' use of mechanical transportation to reach school has a significantly negative impact on the local environment.

Table 5 Pollutant concentration in the concerned stations

Pollutant	National criteria	WHO criteria	M1	M3	M2	M4
SO_2 (ppm)	0.04 h	0.01 h	0.14	0	0.11	0
NO_2 (ppm)	0.25 h	0.11 h	0.27	0	0.31	0
CO (ppm)	35 h	9 h	1.0909	0	2.1	0
O_3 (ppm)	0.12 h	0.12 h	0.08	0.03	0.05	0.02
CO_2 (μg/m^3)	250 μg/m^3	250 μg/m^3	446.5	205.2	417.2	198.1

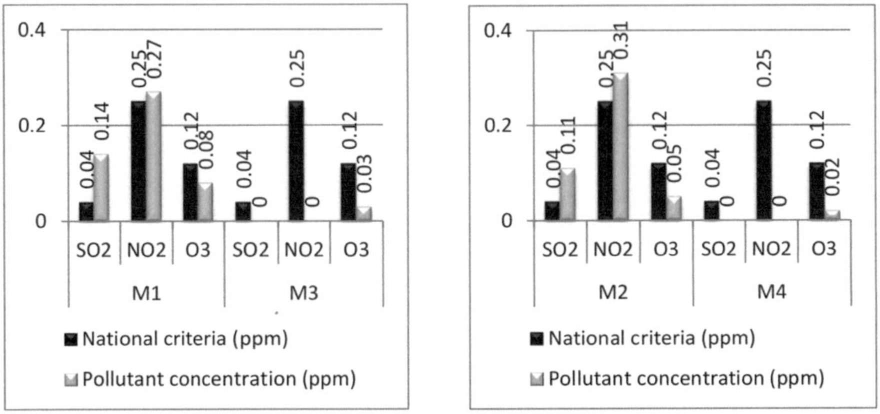

Fig. 3 Pollutant gases (SO_2, NO_2 and O_3) concentration in the concerned stations

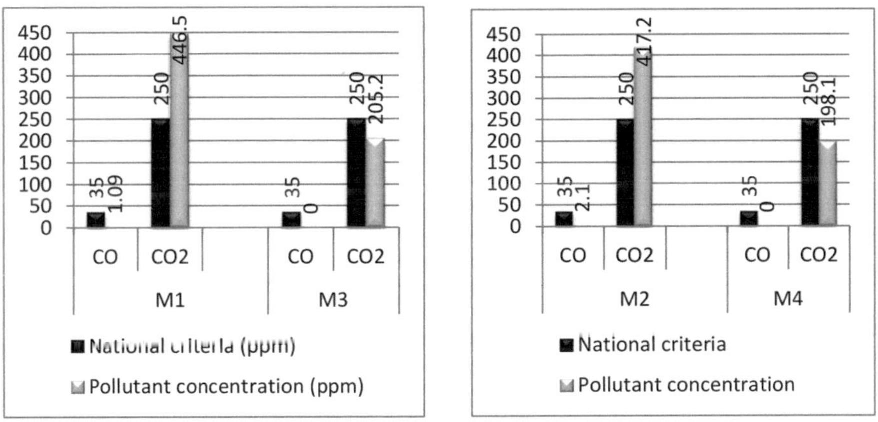

Fig. 4 Pollutant gases (CO, CO_2) concentration in the concerned stations

3.1 Comparison of Air Pollutant Levels with National Standards

Table 5 compares the air pollutant concentrations in the study area with national standards. At M1 and M3 the levels of certain gases exceeded the acceptable standards. The concentration of SO_2 gas was recorded at 0.14 ppm, surpassing the national limit of 0.04 ppm. However, on the weekend, the concentration dropped to zero, indicating it was below the limit of detection. The concentration of NO_2 gas reached 0.27 ppm during the school day, exceeding the national limit of 0.25 ppm, but decreased to zero on the holiday. CO gas was measured at 1.09 ppm on the school day, which meets the national limit of 35 ppm, and it dropped to zero on the weekend. The concentration of CO_2 gas was 446.5 $\mu g/m^3$ on the school day, surpassing the national criteria of 250 $\mu g/m^3$, but decreased to 205.2 $\mu g/m^3$ on the weekend, meeting the national criteria. The concentration of O_3 gas did not exceed the national standard of 0.12 ppm, measuring at 0.08 ppm on the school day and 0.03 ppm on the weekend.

In the second site, M2 and M4, it was found that most air pollutants exceeded the acceptable standards during the school day. For SO_2 gas, the concentration recorded was 0.11 ppm, exceeding the national standard of 0.04 ppm. However, this concentration dropped to zero on the weekend. Similarly, for NO_2 gas, the concentration was 0.31 ppm on a school day, surpassing the national criteria of 0.25 ppm, but it decreased to zero on the weekend. CO gas recorded a concentration of 2.1 ppm during the school day, which is lower than the national criteria of 35 ppm and dropped to zero on the weekend. CO_2 gas concentration on a school day was 417.2 $\mu g/m^3$, exceeding the national criteria of 250 $\mu g/m^3$, but it decreased to 198.1 $\mu g/m^3$ on the weekend, below the national criteria. O_3 gas concentration did not exceed the national criteria of 0.12 ppm on both the school day and the weekend, with concentrations of 0.05 ppm and 0.02 ppm respectively.

3.2 Air Quality Index (AQI) at Stations M1 and M3

The air quality index consists of numbers that correspond to specific categories used by environmental departments to indicate the air quality in a given location. Higher category values indicate greater levels of air pollution, which can harm public health [12]. The air quality index (AQI) was calculated using the following equation [19] and based on the tabular values of pollutants for the city of Baghdad (Table 6) [20].

$$IP\ (AQI) = \frac{I_{HI} - I_{Lo}}{BP_{HI} - BP_{Lo}}(CP - BP_{Lo}) + I_{Lo} \quad (1)$$

where

IP Air quality index
IHI AQI value opposite to BPHI

ILO AQI value opposite to BPLo
BPHI Breakpoint, which is greater than or equal to CP
BPLo Breakpoint which is less than or equal to CP
CP Concentration measured of pollutant.

These colors indicate as shown in Table 7 [21].

On a school day. Table 8 shows that the AQI for sulfur dioxide (SO_2) at station M1 on a school day reached 138.69, falling within the 101–150 range, indicating moderate pollution with a potential impact on public health (orange). Similarly, the AQI for nitrogen dioxide (NO_2) at the same station on the same day was 132.27, also within the 101–150 range, signaling a comparable health risk (orange). In contrast, the AQI for carbon monoxide (CO) on the same day was 12.52, placing it in the 0–50 range, signifying no health risks (green). The AQI for ozone (O_3) was 42.1, also within the 0–50 range, indicating no pollution concerns for this gas (green).

Table 6 Tabular values of air pollutants in Baghdad City

AQI	Health effects	Indicator color	NO_2 (ppm)	SO_2 (ppm)	CO (ppm)	O_3 (ppm)
0–50	Good	Green	0–0.052	0–0.024	0–4.35	0–0.095
51–100	Medium	Yellow	0.053–0.099	0.025–0.035	4.45–8.85	0.10–0.115
101–150	Unhealthy pollution	Orange	0.113–0.359	0.04–0.17	9.10–12.35	0.124–0.163
151–200	Unhealthy	Red	0.360–0.619	0.18–0.307	12.45–15.35	0.164–0.203
201–300	Very unhealthy	Purple	0.620–1.214	0.308–0.607	15.45–30.35	0.204–0.403
301–500	Dangerous	Maroni	1.215–2	0.608–1	30.45–50	0.404–0.603

Table 7 The range and meaning of the air quality index (AQI)

AQI	Scale's degree	Indicator	Effects
0–50	Good	Green	Pollutants in the air do not pose a threat to public health, meaning the air quality is considered satisfactory
51–100	Moderately	Yellow	There is a moderate health concern about some pollutants in the air, which pose a threat to some groups of the population that have particular allergies to air pollution
101–150	Unhealthy pollution	Orange	There is a health concern for population groups that are highly sensitive to air pollution and thus suffer from serious health effects
151–200	Unhealthy	Red	There is a health concern for the population groups that are sensitive to air pollution and thus suffer severe health effects
201–300	Very unhealthy	Purple	Health warning: all groups of society may suffer from serious health effects
301–500	Dangerous	Maroni	Health emergencies: all segments of society may suffer from severe health effects

Table 8 AQI for stations M1 and M3

Station	Pollutants			
	SO_2	NO_2	CO	O_3
M1	138.69	132.27	12.52	42.1
M3	0	0	0	15.78

On a holiday. Table 8 reveals that the holiday AQI at station M3 for sulfur dioxide (SO_2) was zero, indicating no pollution, as it falls within the 0–50 range, marked green for acceptable public health levels. Similarly, nitrogen dioxide (NO_2) and carbon monoxide (CO) also showed no pollution at this station. The AQI for ozone (O_3) on the same day was 15.78, also within the 0–50 range, confirming the absence of pollution and safe levels for public health, indicated by green.

3.3 Air Quality Index (AQI) at Stations M2 and M4

By applying the same equation in stations M1 and M3 and using values from the tabular value in Table 9, the results were obtained.

On a school day. Table 9 shows that the AQI at station M2 on the school day for sulfur dioxide (SO_2) was 127.38, placing it within the 101–150 range, indicating significant pollution that affects public health, marked in orange. The AQI for nitrogen dioxide (NO_2) at the same station on the same day was 140.24, also within the 101–150 range, reflecting a similar public health concern, indicated in orange. For carbon monoxide (CO), the AQI was 24.13, within the 0–50 range, suggesting no pollution and acceptable levels for public health, marked in green. Ozone (O_3) levels at the same station reached 26.31, also in the 0–50 range, indicating no pollution and acceptable concentrations for public health, marked in green.

On a holiday. From Table 9, the holiday AQI at station M4 for sulfur dioxide (SO_2) was zero, placing it in the 0–50 range, indicating no pollution, with concentrations acceptable for public health, marked in green. The same applies to nitrogen dioxide (NO_2) and carbon monoxide (CO) gases. For ozone (O_3) at the same station, the AQI was 10.52, also in the 0–50 range, showing no pollution and acceptable levels for public health, marked in green. These results highlight the negative environmental

Table 9 AQI for stations M2 and M4

Station	Pollutants			
	SO_2	NO_2	CO	O_3
M2	127.38	140.24	24.13	26.31
M4	0	0	0	10.52

impact of mechanical school transportation, as shown by the differences between school day and holiday measurements at the four stations.

4 Conclusion

The comparison of air pollutant concentrations near schools with national criteria, alongside AQI measurements, provides a clear understanding of air quality in the school environment and identifies the most impactful pollutants. This study highlights how the shift from non-motorized to motorized school trips has significantly increased pollutants, negatively affecting the local environment. The results showed that ozone (O_3) and carbon monoxide (CO) levels were within acceptable limits on both school days and weekends. However, nitrogen dioxide (NO_2), sulfur dioxide (SO_2), and carbon dioxide (CO_2) levels were higher on school days due to vehicle emissions, surpassing local permissible limits. These concentrations dropped below limits during the weekend, indicating that motorized school transportation is a key contributor to pollution. This situation calls for city planners and other stakeholders to implement protective measures, such as separating pedestrian and vehicle traffic near schools. Reducing motorized transport and enhancing pedestrian-friendly infrastructure could mitigate the environmental impact and improve air quality for students and surrounding communities.

References

1. Yang, F.: The research of long-term haze pollution in Shanghai, China. Analysis, plans and scenarios. M.Sc. thesis, Royal Institute of Technology, Sweden, p. 4, 2014
2. The United States Environmental Protection Agency (EPA): Near Roadway Air Pollution and Health: Frequently Asked Questions. Office of Transportation and Air Quality, 420-F-14-044, p. 1 (2014)
3. McAuley, T.R., Pedroso, M.: Safe Routes to School and Traffic Pollution. Technical Report, pp. 4–5. American Public Health Association (2012)
4. Figueiredo, A., et al.: The Mayor's School Air Quality Audit Programme. Programme Report, p. 1. Greater London Authority (2018)
5. Broekstra, N., Luck, A., Gordeljevic, V.: Healthy Air, Healthier Children, Health and Environment Alliance (HEAL), p. 2 (2019)
6. Al-Waeely, A.A., Salman, S.D., Abdol-Reza, W.K., Chaichan, M.T., Kazem, H.A., Al-Jibori, S.H.: Evaluation of the spatial distribution of shared electrical generators and their environmental effects. Int. J. Eng. Technol. **14**(02), 16–23 (2014)
7. Aenab, A.M., Singh, S.K., Lafta, A.J.: Air quality assessment: a statistical approach to stationary air monitoring stations. Int. J. Adv. Res. **3**(3), 68–80 (2015)
8. Lu, L.S., Lin, B.H., Wong, T.C.: Air Pollution in Asia. Interactive Qualifying Project, p. 24 (2006)
9. Santana, J.C., Miranda, A.C., Yamamura, C.L., Silva Filho, S.C., Tambourgi, E.B., Lee Ho, L., Berssaneti, F.T.: Effects of air pollution on human health and costs. Sustainability **12**(4875), 1–21 (2020)

10. Krzyzanowski, M.: Health Effects of Transport-Related Air Pollution, p. 59. The Regional Office for Europe of the World Health Organization (2005)
11. Chaichan, M.T., Kazem, A.H., Abed, T.A.: Traffic and outdoor air pollution levels near highways in Baghdad, Iraq. Environ. Dev. Sustain. **18**(6) (2016). In: A Multidisciplinary Approach to the Theory and Practice of Sustainable Development. Springer Science Business Media, Dordrecht
12. Javadekar, P., Lavasa, A., Kumar, S.: National Air Quality Index, pp. 17–24. Central Pollution Control Board, India (2014)
13. Lafteh, S., Jaber: Utilizing "environmental impact assessment" in identifying effects of industrial projects in city's master plans. M.Sc. thesis in Urban & Regional Planning, Centre of Urban & Regional Planning for Post Graduate Studies, Iraq, p. 24, 2006
14. Ministry of Health: Technical Affairs Directorate, Pollution Control Division (2021)
15. Addison, L., Lyons, G., Harris, J., Parker, J., Mitchell, K., Taylor, N., Kane, N., Jones, P., Ciaburro, T., Cookson, J., Matthews, S., Higgins, L.: Better Planning, Better Transport, Better Places, p. 8. Chartered Institution of Highways & Transportation (CIHT), UK (2019)
16. Moorcroft, S., Barrowcliffe, R.: Land-Use Planning & Development Control: Planning for Air Quality, p. 8. Institute of Air Quality Management, UK (2017)
17. Baghdad Municipality: Design Directorate, GIS Department (2021)
18. Ministry of Planning, Central Statistics Organization, Directorate of Publication and Relations: Iraq Population Estimates for 2019. Iraq (2021)
19. Rabee, A.M.: Estimating the Health Risks Associated with Air Pollution in Baghdad City, pp. 1–12. Springer International Publishing Switzerland (2015)
20. The Ministry of Science and Technology: Environmental and Water Technology Research Department. Iraq (2021)
21. The United States Environmental Protection Agency (EPA): Air Quality Index: A Guide to Air Quality and Your Health, pp. 2–3 (2003)

Ecological Assessment of Artificial Landscapes as a Basis for Their Sustainability Management

V. Tatarintsev, L. Tatarintsev, L. Lebedeva, and S. Makenova

Abstract The article presents materials on the study of the environmental sustainability of artificial landscapes, which depends on many natural conditions that limit the productivity of agrocenoses at the level of an agricultural enterprise. The conditions were assessed using the example of agricultural land use located in the forest-steppe zone of the Altai Krai (RF) with an area of more than 17.2 thousand hectares. The sustainability of agrolandscapes of the agricultural territory was assessed based on more than 20 indicators of the qualitative state of agricultural land. It was found that all conditions for manageability are divided into manageable, regulated, partially regulated and unregulated. The efficiency of using agrolandscapes is affected by physical and geographical, natural and climatic, geomorphological, soil conditions, natural and anthropogenic factors. For effective design of agro-, phyto-, and melioration measures in various agrolandscapes, they were differentiated from the point of view of the influence of limiting effects on biological objects. We compiled a complex of cartographic materials showing the differences in economic conditions at the local management level. The authors have carried out an ecological classification of agricultural lands, on the basis of which zoning and modeling of artificial landscapes were carried out. With the help of unique cartographic material, the spatial differentiation of agricultural landscapes by limiting effects is shown. The presented results are recommended for use in organizing new agricultural enterprises and optimizing existing ones.

Keywords Classification and typification of landscapes · Assessment of environmental conditions · Assessment and design of sustainability of agricultural landscapes · Management of sustainability of agricultural landscapes · Effective use of agricultural landscapes in agricultural production

V. Tatarintsev (✉)
National Research Tomsk State University, Tomsk, Russia
e-mail: kafzem@bk.ru

L. Tatarintsev · L. Lebedeva
Altai State Agrarian University, Barnaul, Russia

S. Makenova
S. Seifullin Kazakh Agrotechnical Research University, Astana, Kazakhstan

Y. Zhu (ed.), *Proceedings of the 4th International Symposium in Environmental Science and Industrial Ecology*, https://doi.org/10.1007/978-981-96-1578-0_17

1 Introduction

Modern agricultural landscapes in the world are subject to extreme anthropogenic load, which has a catastrophic effect on their sustainability and leads to land degradation [1, 2]. At the interstate level, strategic program documents have been adopted that promote the optimization of the use and protection of natural resources, including land [3, 4]. Modern trends related to food security and quality of life are based on sustainable agricultural production, the basis of which is land resources. An innovative direction for increasing agricultural productivity is the environmental assessment of artificial landscapes (agrolandscapes). However, at present, innovative proposals often do not find support from agricultural producers for many reasons. For example, differences in environmental conditions and factors affecting the productivity and sustainability of agricultural land use are difficult to model and implement in the production process [5–8]; agro-, phyto, and melioration measures developed for use in a particular natural zone (subzone) are ineffective due to local differences inherent in the agricultural territory of the economic entity [9–12], etc.

Therefore, the purpose of this study was to assess the environmental conditions of agrolandscapes of an agricultural enterprise to manage their sustainability at the local level (enterprise level). Research objectives: assessment of environmental conditions affecting the sustainability of artificial landscapes, using the example of the forest-steppe zone of the Altai Territory (Russian Federation); analysis of the use of agricultural land and identification of problems; to carry out an agroecological typification of lands and to model scenarios of agricultural land use with different sustainability.

2 Methods and Materials

The research is based on general scientific methods [13], such as observation, description, comparative, historical, modeling, system analysis, etc. Land typing was carried out using methods of analysis and synthesis [14, 15]. Special cartographic materials were compiled based on a field survey of the territory and a cartographic method at a scale of 1:100,000 by the authors of the scientific study. The scientific work used materials from design and scientific organizations that participated in the study of landscapes in different time lags, which are freely available [16, 17]. The work also involved archival data from professors Tatarintsev and Tatarintsev, who collectively devoted more than eighty years to the study, assessment, monitoring and analysis of landscape elements in the south of Western Siberia [18–20].

3 Research Results

The local level of land management (the level of an agricultural enterprise) is one of the most problematic. This is due to many reasons: the labor intensity and cost of scientific research in the territories used in agriculture; heterogeneity and lag dynamics of natural, climatic, soil, morphological and other conditions of agricultural territories; the lack of an appropriate regulatory framework for implementation at the local level; insufficient motivation of the agricultural producer to implement such project solutions.

We have developed a technology for the environmental assessment of artificial landscapes (agrolandscapes) at the level of agricultural enterprise management (local level), which includes the stages presented in Fig. 1.

We have studied artificial landscapes of an agricultural enterprise located in the middle forest-steppe of the Altai Krai, on an area of more than 17.2 thousand hectares. A landscape map was compiled, on which six landscape areas were identified. During the survey, it was found that almost 90% of the territory is actively used for growing agricultural crops. Agrolandscapes are significantly exposed to erosion processes, as evidenced by the compiled geomorphological profile of the agricultural enterprise (Fig. 2) and the slope steepness map (Fig. 3).

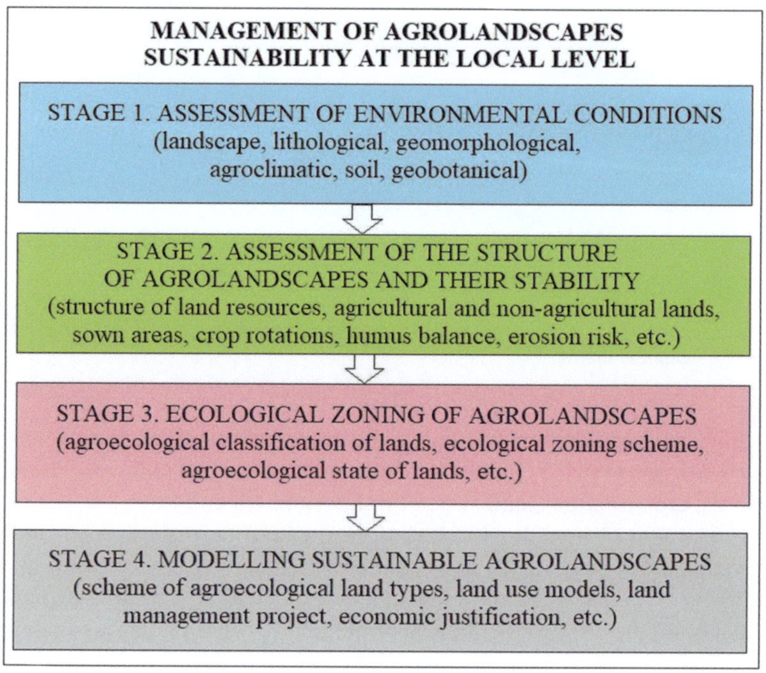

Fig. 1 Stages of ecological assessment of agricultural landscapes for managing their sustainability at the local level

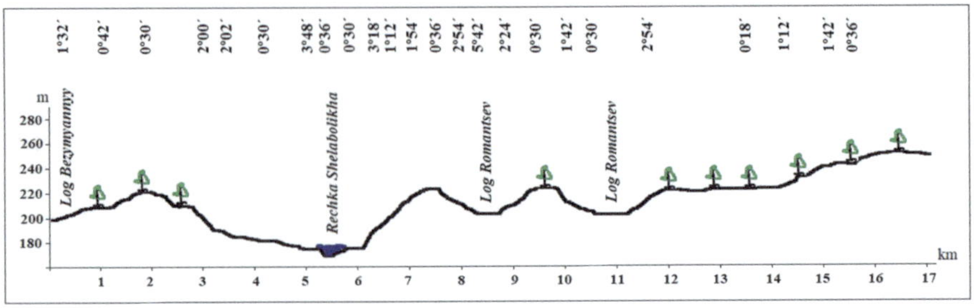

Fig. 2 Geomorphological profile

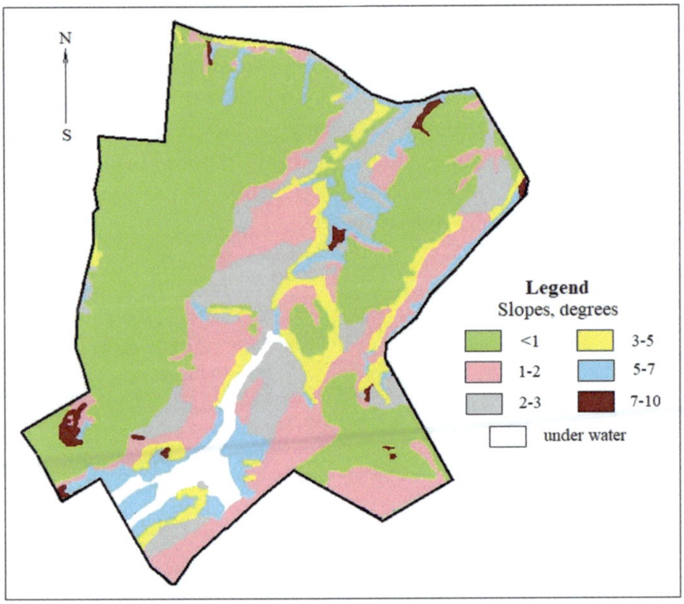

Fig. 3 Slope steepness map

The relief and its morphometric characteristics are taken into account when differentiating land use systems, which will affect the ratio of lands, the structure of agrocenoses, crop rotations, and agricultural technologies.

The assessment of agroclimatic resources was carried out using ten indicators characterizing the quantity and quality of precipitation, temperature conditions. Soil conditions were assessed for the manifestation of erosion processes, which were differentiated by the strength of their distribution, shown on a unique map and the limiting effect of erosion and deflation on the productivity of agricultural crops. It was determined that erosion and deflation processes develop in the non-floodplain left-bank part of the study area, the share of which is three quarters of the total

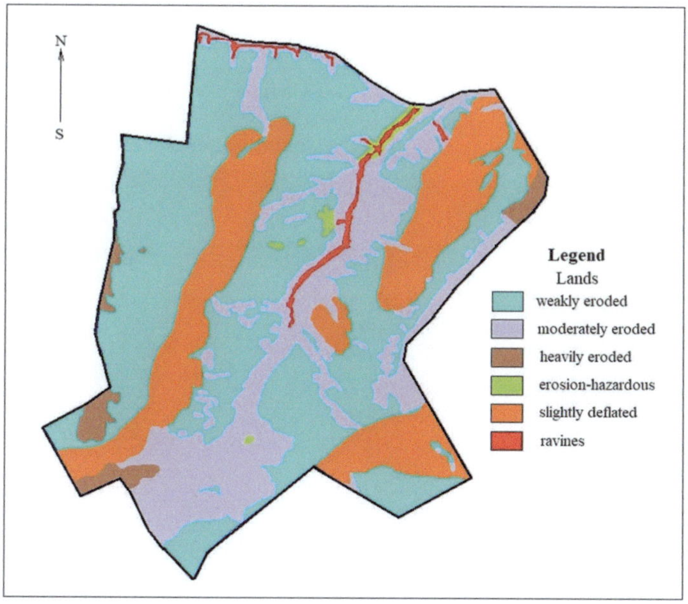

Fig. 4 Soil erosion map

area. Eroded soils are common on coastal slopes, which are characterized by higher dissection and steepness. Deflated soils are identified at the highest watershed spaces. The compiled unique soil map and soil erosion map (Fig. 4) made it possible to establish the deficiencies of agricultural landscapes on the ground.

Differences in soil cover significantly affect effective fertility, as well as variations in the cost of agricultural products, which increases due to overcoming the stress impact on biological objects.

The use of agricultural land from the point of view of its commercial organization at the present stage is presented in Fig. 5. Having assessed the ecological state of the agricultural territory using 11 indicators, it was established that it has a very high degree of agricultural development and plowing, which leads to the development of erosion processes. Environmental damage from irrational use of land exceeds 15%, leveling the intensification of the technological process.

Having assessed the above-mentioned land areas using more than 20 indicators and having calculated the share of the land area (%) that changed its quality condition under the influence of the factor from the total area of the agricultural landscape, agricultural land, arable land, and forage land, it was found that the area under study consists of 20% very unstable agricultural landscapes, 62% unstable, 12% moderately stable, and 6% very stable.

Subsequently, the analysis of the field crop rotation system cultivated on the territory showed their very low anti-erosion organization. The calculation of the humus balance made by the authors revealed that a positive humus balance is observed

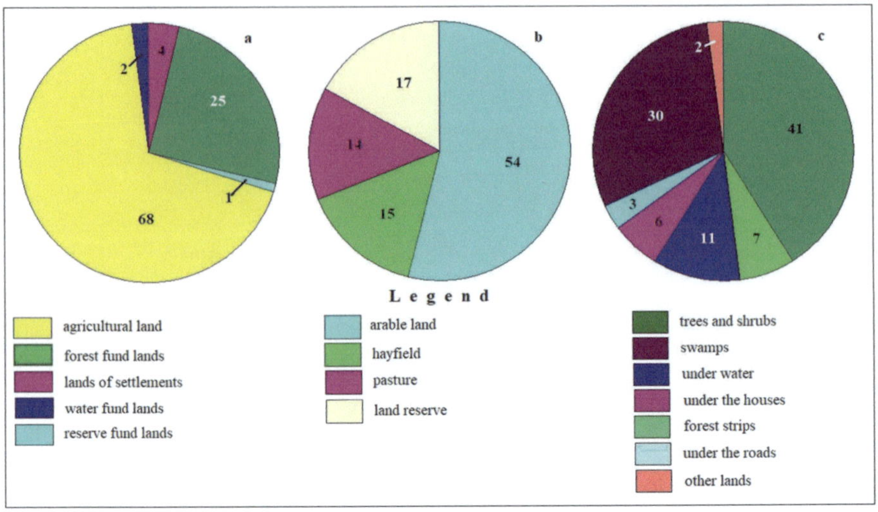

Fig. 5 Structure of land fund (**a**), agricultural (**b**), non-agricultural (**c**) lands

only in the grain-fallow four-field and forage six-field crop rotations. In all other crop rotations, a loss of organic matter is observed.

Next, we performed an agroecological grouping of lands based on their classification by agroecological indicators: groups, subgroups, classes, categories, genera, subgenera [21–23]. Species and subspecies have not been identified due to a lack of factual material, but work in this direction is underway. Four agroecological groups were established on the territory, within which 17 agroecological subgroups were objectively identified. Based on the agroecological grouping of lands, we compiled a map of ecological zoning of the study area and a map of agroecological land types (Fig. 6).

In the study area, among the limiting factors, controllable ones predominate—lack of mineral nutrition elements; regulated ones—humus content, lack of moisture, which are caused by the influence of deflation in watershed areas and erosion on slope lands with slopes of up to 3°. The second category of lands [24, 25] can be used in field (grain-fallow, grain-row, grain-fallow-grass, or field with a grass output field) crop rotations quite intensively.

The remaining agricultural landscapes require additional measures aimed at reducing the negative impact of degradation processes. Those territories that occupy slope lands and bottom ravines are not suitable for effective use in agriculture. Their resistance to anthropogenic activity is extremely low or completely absent.

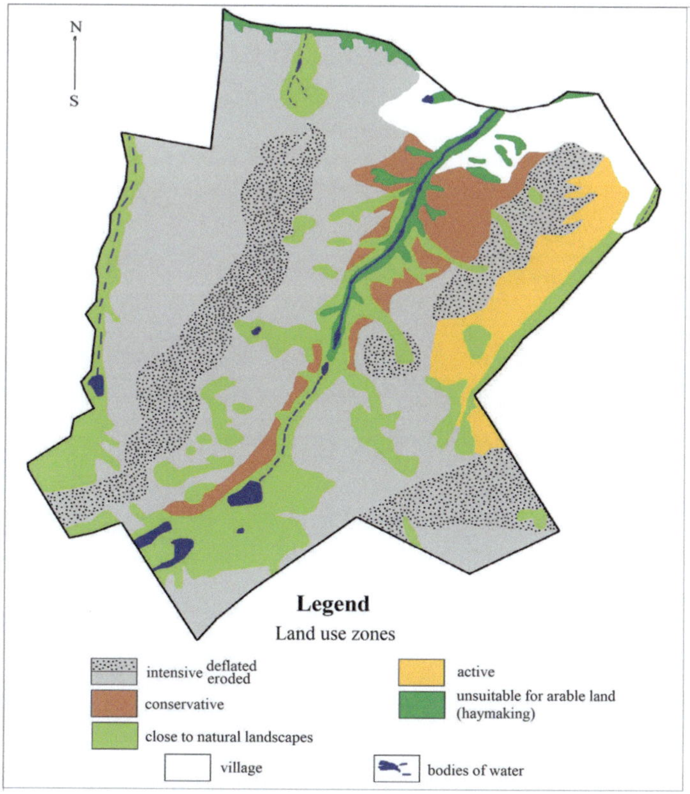

Fig. 6 Agroecological land type map

4 Discussion of Findings

The study could have been stopped at this point, but agricultural producers set different goals when farming the land, taking into account capital investments and production costs attracted or planned for the medium or long term. The area of land use, the availability of capital investments, the time lag and other elements of the organization of agricultural land use have a significant impact on the farming process. Therefore, it is important to show farmers what models of using agricultural landscapes are possible from the point of view of managing their sustainability. In Fig. 7, we show agricultural land use models with different levels of agricultural landscape sustainability.

The authors understand the sustainability of agrolandscapes as a state of the territory close to natural. There are not many opportunities in the practice of modern agriculture to achieve the cherished goal—high yield, low cost and environmentally friendly products. However, by varying the main organizational elements—arable land, hayfields, pastures, the desired balance can be achieved. This effect will not

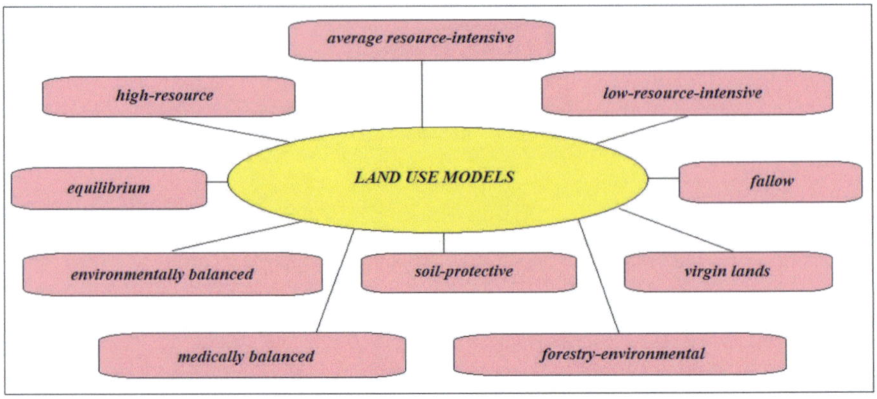

Fig. 7 Models for managing the sustainability of agrolandscapes

occur immediately, but in the medium and long term, we will all feel it in the entire ecosystem in the form of healthy and high-quality products, raw materials and feed.

5 Conclusion

In conclusion of the scientific study, it should be noted that the assessment of the ecological potential of anthropogenic landscapes is extremely necessary when organizing new and optimizing existing agricultural production. As a result of such work, a set of limiting ecological conditions will be known that affect the effective fertility and, as a consequence, the sustainability of anthropogenic landscapes. By means of land use modeling, it will be possible to accurately predict the long-term productivity of 1 ha of agricultural land, the amount of chemicals, seed material and other elements of the complex technological process of organizing agricultural landscapes.

References

1. Dale, P. (Group Leader): Guidelines for Land Administration, 150 pp. UNECE, New York, Geneva (1996)
2. Sustainable Development Summit. Transforming Our World for People and Planet, 25–27 Sept 2015 [Electronic resource]. United Nations [Official website]. URL: http://www.un.org/sustainabledevelopment/ru/wp-content/uploads/sites/5/2015/08/0verview_Sustainable_Development_Summit.pdf
3. The 2030 Agenda for Sustainable Development. The Program was Adopted on September 25, 2015 at the UN Summit [Electronic resource] (2018). URL: https://ec.europa.eu/europeaid/policies/european-development-policy/2030-agenda-sustainable-development
4. Voluntary Guidelines on the Responsible Governance of Tenure of Land, Fisheries and Forests in the Context of National Food Security, 416 pp. FAO, Rome (2013)

5. Tatarintsev, V.L., Merzlyakov, O.E., Ozeranskaya, N.L., Shakenova, Z.K.: Analysis of the qualitative state of agricultural land as the basis for the sustainability of agricultural land use. Sustain. Dev. Mt. Territ. **14**(4), 644–656 (2022). https://doi.org/10.21177/1998-4502-2022-14-4-644-656
6. Tursynbaev, N.A., Kuzhevskaya, I.V., Tatarintsev, L.M., Ospanova, A.A.: Assessment of the environmental state and sustainability of agrolandscape at the local level. Sustain. Dev. Mt. Territ. **15**(3), 662–671 (2023). https://doi.org/10.21177/1998-4502-2023-15-3-662-671
7. Duisenova, D.D., Tireuov, K.M., Tatarintsev, L.M., Lebedeva, L.V.: Ecological typization of agricultural land and modeling of sustainable agrolandscape at the local level. Sustain. Dev. Mt. Territ. **16**(1), 94–110 (2024). https://doi.org/10.21177/1998-4502-2024-16-1-94-110
8. Tatarintsev, V.L., Lebedeva, L.V., Syzdykova, G.D., Inkarov, D.S.: Management of agricultural land use sustainability at the local level using GIS technologies and special map material. Sustain. Dev. Mt. Territ. **15**(4), 864–876 (2023). https://doi.org/10.21177/1998-4502-2023-15-4-864-876
9. Asanova, G.A., Merzlyakov, O.E., Tatarintsev, V.L., Unysheva, N.K.: Analysis of the sustainability of agricultural landscapes in the forest-steppe zone of the Krasnoyarsk Territory and measures for their protection. Sustain. Dev. Mt. Territ. **15**(2), 264–274 (2023). https://doi.org/10.21177/1998-4502-2023-15-2-264-274
10. Nurpeisov, M.N., Ruchkina, K.V., Androkhanov, V.A., Makenova, S.K.: Assessment of geoecological factors and conditions affecting sustainability of agricultural land use in the forest-steppe zone of Altai Krai. Sustain. Dev. Mt. Territ. **15**(2), 419–430 (2023). https://doi.org/10.21177/1998-4502-2023-15-2-419-430
11. Akhmetov, B.Z., Kasenova, A.Z., Merzlyakov, O.E., Tatarintsev, L.M.: Modeling of sustainable agricultural land use based on agro-ecological assessment of the territory. Sustain. Dev. Mt. Territ. **15**(1), 17–26 (2023). https://doi.org/10.21177/1998-4502-2023-15-1-17-26
12. Tatarintsev, V.L., Tatarintsev, L.M., Matsyura, A.V., Bondarovich, A.A.: Organization of sustainable agricultural land use in the Altai Territory using landscape analysis. Sustain. Dev. Mt. Territ. **12**(3), 339–349 (2020). https://doi.org/10.21177/1998-4502-2020-12-3-339-348
13. Afanasyev, V.V., Gribkova, O.V., Ukolova, L.I.: Methodology and Methods of Scientific Research, 164 pp. Yurait, Moscow (2023)
14. Lebedev, S.A.: Methodology of Scientific Knowledge, 154 pp. Yurait, Moscow (2023)
15. Mokiy, M.S., Nikiforov, A.L., Mokiy, V.S.: Methodology of Scientific Research, 260 pp. Yurait, Moscow (2024)
16. Public cadastral map of Russia [Electronic resource]. URL: https://pkk.rosreestr.ru/
17. Forecast of Scientific and Technological Development of the Agro-Industrial Complex of the Russian Federation for the Period Up to 2030, 140 pp. Ministry of Agriculture of Russia, National Research University Higher School of Economics, HSE, Moscow (2017)
18. Tatarintsev, V.L., Tatarintsev, L.M., Makenova, S.K., Shostak, M.M.: Geoecological assessment of landscapes as a basis for organizing sustainable agricultural land use. Sustain. Dev. Mt. Territ. **13**(4), 485–497 (2021). https://doi.org/10.21177/1998-4502-2021-13-4-485-496
19. Tatarintsev, V.L., Tatarintsev, L.M., Ermekov, F.K., Lisovskaya, Y.S.: Assessment of the agroecological state of agrolandscapes to improve their sustainability. Sustain. Dev. Mt. Territ. **14**(1), 76–87 (2022). https://doi.org/10.21177/1998-4502-2022-14-1-76-86
20. Tatarintsev, V., Lisovskaya, Yu., Tatarintsev, L.: Agricultural landscape quality as a key factor fostering environmentally safe agricultural land use in the arid steppe of the Altai Region. IOP Conf. Ser. Earth Environ. Sci. **670**, e012036 (2021). https://iopscience.iop.org/issue/1755-1315/670/1
21. Kiryushin, V.I., Ivanov, A.L.: Agroecological Assessment of Land, Design of Adaptive Landscape Systems of Agriculture and Agricultural Technologies, 784 pp. Rosinforagrotekh, Moscow (2005)
22. Ramensky, L.G.: On the principles of installation, basic concepts and terms of production typology of lands, geobotany and ecology. Sov. Botany (4) (1938)
23. Vaseneva, I.I. (ed.): Methodology of Agroecological Typification of Lands in the Agricultural Landscape, 102 pp. Rosselkhozakademiya, Moscow (2004)

24. Tatarintsev, V.L., Lisovskaya, Y.S., Tatarintsev, L.M.: Agricultural land protection as a basis of sustainable land management in the dry steppes of Altai Krai. Sib. J. Life Sci. Agric. **14**(2), 338–355 (2022). https://doi.org/10.12731/2658-6649-2022-14-2-338-355
25. Tatarintsev, V.L., Shostak, M.M., Tatarintsev, L.M.: Organizing sustainable agricultural land management in Altai Krai: a geo-ecological prospect. Sib. J. Life Sci. Agric. **14**(2), 356–372 (2022). https://doi.org/10.12731/2658-6649-2022-14-2-356-372

Comparative Analysis of Activity-Based and Spend-Based Methods for Scope 3 Greenhouse Gas Emissions Calculation

Chufan Liu, Lili Yin, Yuanzhe Li, and Yu Hao

Abstract This paper presents a comparative analysis of two primary methodologies for calculating Scope 3 greenhouse gas (GHG) emissions: the activity-based approach and the spend-based approach. Scope 3 emissions, encompassing all indirect emissions within a company's value chain, pose significant challenges in data acquisition and accuracy. The activity-based method provides a detailed calculation by directly measuring emissions from all activities in the value chain, thus offering granular insights into environmental impacts but requiring extensive data collection. In contrast, the spend-based method estimates emissions based on financial expenditures on goods and services, simplifying data collection but potentially sacrificing specificity for convenience. This study assesses the efficacy, challenges, and suitability of each approach depending on various industry needs, the complexity of supply chains, and regulatory compliance requirements, such as those from the Corporate Sustainability Reporting Directive (CSRD) and the Science Based Targets initiative (SBTi). The findings suggest a potential hybrid approach, leveraging the detailed accuracy of the activity-based method where feasible, supplemented by the broader overview provided by the spend-based method for initial assessments and less critical areas.

Keywords GHG Scope 3 emissions · Activity-based method · Spend-based method · Corporate Sustainability Reporting Directive · Emission reduction strategy

---

C. Liu · Y. Li (✉)
School of Civil and Environmental Engineering, University of Auckland, Auckland, New Zealand
e-mail: yuanzhe001@e.ntu.edu.sg

L. Yin
Murdoch Business School, Murdoch, WA, Australia

Y. Hao
NUS College of Design and Engineering, National University of Singapore, Singapore, Singapore

Y. Zhu (ed.), *Proceedings of the 4th International Symposium in Environmental Science and Industrial Ecology*, https://doi.org/10.1007/978-981-96-1578-0_18

1 Introduction to Scope 3 GHG Emissions

The escalating global climate crisis has intensified scrutiny on corporate environmental impacts, pushing companies to extend their responsibility beyond direct (Scope 1) and energy-related (Scope 2) GHG emissions. To fully capture a company's carbon footprint, Scope 3 emissions—which encompass all other indirect emissions throughout a company's value chain—must be considered [1]. These emissions arise from upstream activities like the extraction of raw materials and downstream activities, including the use and disposal of sold products [2]. Often the largest component of a company's total GHG emissions, Scope 3 can account for more than 70% of the overall impact, depending on the industry [3]. Scope 3 covers a broad range of activities, including transportation, distribution, business travel, waste, and leased assets. It is a critical area for companies aiming to reduce their carbon footprint and align with global sustainability commitments and regulatory requirements [4]. Standards such as CSRD, the Global Reporting Initiative (GRI), and SBTi mandate transparent reporting and active management of these emissions. Despite the significant challenges of data collection and measurement uncertainties, managing Scope 3 emissions offers opportunities to identify inefficiencies, drive innovation, and enhance resilience across value chains through collaborative efforts with suppliers, customers, and other stakeholders [5, 6]. Given the critical role of Scope 3 in corporate sustainability, this review explores its significance within environmental accountability frameworks. It emphasizes the importance of integrating comprehensive GHG accounting into corporate decision-making, enabling companies to achieve compliance and contribute to global efforts to mitigate climate change [6–8].

2 Foundations of GHG Emission Calculations

2.1 Simplification in Scope 1 and 2 Calculations and Challenges in Scope 3 Emissions

Calculations of Scope 1 and Scope 2 emissions are relatively straightforward due to Calculations of Scope 1 and Scope 2 emissions are relatively straightforward due to the direct access to operational data. Scope 1 includes direct emissions from sources owned or controlled by the organization, such as vehicles and on-site fuel combustion. For instance, emissions from company-owned boilers can be directly measured through fuel consumption records. Scope 2 emissions involve indirect emissions from purchased energy, which are often derived from utility bills and metering systems, allowing for accurate reporting [9–11]. However, Scope 3 emissions, which include all other indirect emissions within a company's value chain, present a far more complex challenge. These emissions cover a wide range of activities, such as raw material production, business travel, and the use of sold products.

The GHG Protocol outlines 15 categories for Scope 3 emissions, each requiring distinct data and emission factors [12]. For example, a consumer goods company may face difficulties estimating emissions from purchased goods, especially when products are distributed widely and used under varying conditions. The complexity of data quality across the supply chain often leads to significant reporting challenges [13, 14].

2.2 Methodological Approaches to Scope 3 Emissions Calculation

Activity-Based Approach

The activity-based approach involves collecting detailed data from a company's value chain, such as transportation, production processes, energy use, and waste generation. This method provides precise emissions calculations based on actual operational data. For instance, a logistics company may track its fuel consumption to calculate emissions from its fleet. A steel plant producing 100,000 metric tons of steel may estimate its emissions at 1.9 metric tons of CO_2e per ton, resulting in total emissions of 190,000 metric tons of CO_2e [15]. While this approach allows companies to target high-emission activities, it requires substantial data collection efforts, including monitoring equipment and tracking waste outputs. This can be resource-intensive, demanding investments in data systems and collaboration across departments or external partners [16, 17].

Spend-Based Approach

The spend-based approach estimates emissions based on the monetary value of goods and services purchased, offering a broader but less detailed view. For example, a company spending $10 million on raw materials might estimate emissions at 5000 metric tons of CO_2e using an average factor of 0.5 kg CO_2e per dollar [18, 19]. This method simplifies the process, making it suitable for companies lacking detailed activity data. However, it relies on generalized emission factors, which may not accurately reflect specific products or suppliers. It is particularly useful for identifying emission hotspots and guiding further data collection efforts [20].

3 Strategic Importance of Comprehensive Scope 3 Emission Calculation

3.1 Activity-Based Approach

The activity-based approach to Scope 3 emissions calculation focuses on precision and detailed assessment of emissions across various stages of a company's value chain, such as raw material acquisition, transportation logistics, production processes, waste management, and disposal [21, 22]. This method provides granular insights, allowing companies to devise targeted emission reduction strategies based on actual activity data.

Comprehensive Data Requirements

Implementing this approach requires extensive data collection throughout the value chain, including tracking raw material volumes, transportation routes, and waste management. For instance, a chemical company might track emissions from raw material extraction, such as oil and gas, amounting to thousands of metric tons of CO_2e annually [23]. The accuracy of data is crucial; for example, air freight emissions can be significantly higher than ocean shipping, highlighting the need for precise data to select the right reduction pathways. In sectors like automotive manufacturing, companies use this approach to calculate emissions at a component level, enabling them to effectively target high-emission sources [24, 25].

Scope of Emissions Analysis

This approach allows detailed examination of emissions within the 15 Scope 3 categories defined by the GHG Protocol, such as purchased goods, services, and end-of-life product treatment. By categorizing emissions, companies can prioritize interventions where they will have the most impact. For instance, semiconductor production in the electronics industry can account for up to 60% of a product's carbon footprint due to its energy-intensive nature [4, 26]. While effective, the approach demands substantial data management, supplier engagement, and verification efforts, especially in industries with fragmented supply chains [7, 27]. Advanced digital tools, such as blockchain for transparency or IoT devices for real-time tracking, can help mitigate these challenges but add cost and complexity [21, 28].

Critical Assessment

As summarized in Table 1, the activity-based approach requires highly granular data from various stages of the value chain, ensuring precise emissions measurements but demanding extensive data collection efforts. This detailed level of analysis allows companies to capture specific emissions data from processes and components, providing a more accurate environmental impact assessment [11, 14, 29]. However, implementing this approach necessitates significant investment in resources, time, and expertise, including advanced data management systems, dedicated personnel, and complex tracking mechanisms. Data integration from multiple sources, such as suppliers, logistics partners, and internal operations, is complex and requires robust coordination and advanced analytical tools to ensure consistent emissions reporting. Active collaboration with suppliers is crucial, as companies rely on detailed information about production practices and material use, which can be challenging to obtain [30]. Standardizing measurement techniques across the supply chain is essential to maintain data consistency, as variations can lead to discrepancies. Regulatory compliance and verification further complicate the process, as companies must ensure their emissions data meets industry standards, often requiring third-party validation, adding to costs and effort [13]. Advanced technologies like sensors for real-time monitoring or blockchain for transparency can enhance data accuracy, but these solutions demand substantial investment and careful integration, adding to the challenges of the activity-based approach [30].

Table 1 Critical factors must be considered to ensure the accuracy and relevance of emissions calculated using the activity-based approach

Consideration	Description	Example
Data granularity and precision	Requires detailed data collection from various value chain stages for accurate emissions measurement	Tracking emissions per ton in steel production
High resource and time investment	Demands significant time, financial resources, and expertise, including data systems and dedicated teams	Investing in advanced tracking systems
Complexity of data integration	Integrating data from suppliers, logistics, and internal operations is complex and requires coordination	Merging data to report logistics emissions
Supplier engagement and collaboration	Requires active collaboration with suppliers to collect production and material data	Working with suppliers to track energy use
Variability in measurement techniques	Inconsistent measurement methods across the supply chain affect data accuracy; standardization is crucial	Variations in waste management emissions reporting
Regulatory compliance and verification	Compliance with regulatory standards adds complexity and cost, often needing third-party verification	Meeting CSRD or SBTi standards
Technological integration and monitoring	Advanced technologies enhance data precision but require significant investment and integration	Using sensors for real-time emissions tracking

3.2 Spend-Based Approach

Advantages of Simplified Data Collection

The primary advantage of the spend-based approach is its streamlined data collection process, leveraging existing financial and procurement data to bypass the complex data gathering required by activity-based methods. For example, emissions from $10 million spent on steel can be estimated using a factor of 2.1 kg CO_2e per dollar, resulting in 21,000 metric tons of CO_2e [17]. This approach is suitable for industries with complex supply chains, such as retail or electronics, where tracking emissions across numerous suppliers is impractical. It allows companies to quickly identify major emissions categories, offering rapid, high-level insights, particularly useful during initial assessments or when resources are limited [14, 31].

High-Level Emission Insights

By translating expenditures into emission estimates, the spend-based approach provides a broad overview of potential hotspots, guiding strategic decisions on where to focus reduction efforts. For instance, a food and beverage company might find that 60% of its emissions stem from agricultural products, prompting a shift to more sustainable sourcing. However, this method offers less detailed insights compared to activity-based calculations and may overlook specific emission drivers, such as variations in supplier efficiency or waste management practices, leading to discrepancies between estimated and actual emissions [15, 32].

Key Considerations for Emission Factors

As summarized in Table 2, market average emission factors used in the spend-based approach often rely on generalized data that may not capture supplier-specific efficiencies, potentially leading to inaccuracies. External factors like inflation and exchange rates complicate the process, as changes in financial expenditures don't directly reflect actual emissions variations, making accurate correlations difficult [20]. Price fluctuations can further distort emissions estimates, as rising costs might falsely indicate increased emissions. Supply chain transparency and data quality are crucial, as insufficient visibility into supplier operations can lead to significant estimation errors. Regional differences in emissions factors and dynamic changes in supply chains further reduce the spend-based approach's effectiveness for real-time emissions management [32, 33].

Table 2 Critical factors must be considered to ensure the accuracy and relevance of emissions calculated using the spend-based approach

Consideration	Description	Example
Market average emission factors	Industry averages may not reflect specific supplier practices, leading to estimation errors	Averages might not account for renewable energy use by suppliers
Impact of inflation and exchange rates	Inflation and currency shifts can misalign spending changes with actual emissions	Inflation may increase costs without corresponding emissions rise
Price variability effects	Price changes not linked to operations can mislead emissions calculations	Higher prices might falsely indicate increased emissions
Supply chain transparency and data quality	Lack of transparency in complex supply chains can lead to inaccurate estimates	Poor data quality skews emissions calculations
Regional differences in emission factors	Emission factors vary by region due to different energy sources and practices	Products made in cleaner energy regions may have lower emissions
Technological advancements and efficiency	Spend-based data often lags in reflecting recent process improvements	Efficiency upgrades may reduce emissions not captured in estimates
Dynamic supply chain changes	Frequent changes in suppliers or processes can alter emissions profiles undetected	Switching to lower-emission suppliers may not be reflected promptly

4 Considerations for Method Selection in Scope 3 Emissions Calculation

4.1 Resource Allocation and Precision

The choice between activity-based and spend-based methods depends on available resources and the desired precision in emissions data. The activity-based approach is resource-intensive, requiring significant investments in tools, software, and personnel training. This approach provides detailed emissions insights, such as identifying that one ton of steel production contributes about 1.9 metric tons of CO_2e [18, 32]. Conversely, the spend-based approach relies on financial data, making it less costly and ideal for initial assessments or smaller companies with limited budgets. While it lacks the precision of the activity-based method, it allows for a quick overview of major emissions sources without heavy resource commitment [33].

4.2 Data Collection and Engagement with Stakeholders

The activity-based approach requires active engagement with stakeholders, including suppliers and logistics providers, to collect specific data on material sourcing and energy use. This collaboration is essential for high-quality emissions data but can be complex and time-consuming [29]. In contrast, the spend-based approach simplifies data collection by using financial transactions, making it easier to implement but often relying on generalized emission factors that may not reflect specific supplier practices [8, 31].

4.3 Integration of Methods for Comprehensive Insights

Companies can optimize Scope 3 emissions calculations by integrating both activity-based and spend-based approaches. The GHG Protocol allows using different methods across impact categories, enabling companies to balance accuracy with resource constraints. For example, a company might use the activity-based approach for high-impact categories like agricultural inputs and the spend-based method for less critical areas like office supplies, ensuring precise quantification of key emissions while conserving resources [14].

4.4 Strategic and Regulatory Considerations

Method selection is also influenced by strategic sustainability goals and regulatory standards. Increasingly, detailed emissions reporting is mandated, pushing companies toward more rigorous methods like activity-based calculations, particularly for high-impact categories [18]. Companies must balance precision with resource efficiency, considering their sustainability journey and regulatory obligations. For instance, a company aiming for carbon neutrality may prioritize detailed data, while one just beginning emissions tracking might start with the spend-based approach to establish a baseline [20].

5 Conclusion

This paper has examined the activity-based and spend-based methods for estimating Scope 3 GHG emissions, highlighting their benefits, drawbacks, and suitability across various industries and regulatory contexts. The activity-based approach offers precision and depth, ideal for sectors with complex supply chains where detailed emissions tracking is essential for compliance and management. However, its high resource demands and need for stakeholder engagement make it suitable primarily for organizations with substantial resources. Conversely, the spend-based method provides a simpler, less detailed overview of emissions, making it suitable for initial assessments and companies at the beginning of their sustainability journey. While it relies on financial data and may not fully capture the complexity of an organization's environmental impact, it is easier to implement. By integrating both methods, companies can mitigate the limitations of each approach, tailoring their strategy to the significance of different impact categories. This hybrid approach enables businesses to meet compliance standards and strategic goals, balancing accuracy with resource allocation. Ultimately, the choice of method depends on the company's unique needs, supply chain complexity, and regulatory environment. To support broader sustainability goals like achieving Net-Zero emissions and aligning with standards such as CSRD and SBTi, companies must carefully select or combine these approaches to enhance Scope 3 emissions management. This integrated approach not only supports regulatory compliance but also strengthens the company's sustainability profile and provides a clearer visualization of its environmental impact.

References

1. Voss, G.B., Sirdeshmukh, D., Voss, Z.G.: The effects of slack resources and environmental threat on product exploration and exploitation. Acad. Manag. J. **51**(1), 147–164 (2008)
2. Wang, Y., Hao, Y., Hou, Y., Quan, Q., Li, Y.: Optimizing scope 3 emissions in the automotive manufacturing industry: a multidisciplinary approach. Carbon Res. **3**(1), 49 (2024)
3. Babiack, K., Trendafilova, S.: CSR and environmental responsibility: motives and pressures to adopt green management practices. Corp. Soc. Responsib. Environ. Manag. **18**, 11–24 (2011)
4. Chen, J., Zhang, F.: Does environmental responsibility matter in cross-sector partnership formation? A legitimacy perspective. J. Environ. Manage. **231**, 612–621 (2019)
5. Clarkson, P.M., Li, Y., Richardson, G.D., Vasvari, F.P.: Does it really pay to be green? Determinants and consequences of proactive environmental strategies. J. Account. Public Policy **30**(2), 122–144 (2011)
6. Meng, X.H., Zeng, S.X., Shi, J.J., Qi, G.Y., Zhang, Z.B.: The relationship between corporate environmental performance and environmental disclosure: an empirical study in China. J. Environ. Manage. **145**, 357–367 (2014)
7. Zhang, L.: Understanding Chinese Corporate Governance: Practical Guidance for Working with Chinese Partners. Taylor & Francis (2023)
8. Zhong, J., Wang, Y., Quan, Q., Li, Y.: Biofilm dynamics in fluoride-based wastewater systems: a mini-review on *Pseudomonas* spp. and *Bacillus* spp. biofilms in semiconductor manufacturing WWTP. Korean J. Chem. Eng. 1–15 (2024)
9. Corte, A.E.R.: Promoting pro-environmental attitudes and behaviors to reduce the consumption of illegal wildlife products in China. Doctoral dissertation, 2015
10. McGuire, J.B., Sundgren, A., Schneeweis, T.: Corporate social responsibility and firm financial performance. Acad. Manag. J. **31**(4), 854–872 (1988)
11. McWilliams, A., Siegel, D.: Corporate social responsibility: a theory of the firm perspective. Acad. Manag. Rev. **26**(1), 117–128 (2001)
12. Crisostomo, V.L., Freire, F.D.S., Vasconcellos, F.C.: Corporate social responsibility, firm value and financial performance in Brazil. Soc. Responsib. J. **7**(2), 295–309 (2011)
13. Keupp, M.M., Gassmann, O.: The past and future of international entrepreneurship: a review and suggestions for developing the field. J. Manag. **35**(3), 600–633 (2009)

14. Leung, T.C.H., Guan, J., Lau, Y.Y.: Exploring environmental sustainability and green management practices: evidence from logistics service providers. Sustain. Account. Manag. Policy J. (2023)
15. Darnall, N., Kim, Y.: Which types of environmental management systems are related to greater environmental improvements? Public Adm. Rev. **72**(3), 351–365 (2012)
16. Orlitzky, M., Louche, C., Gond, J.-P., Chapple, W.: Unpacking the drivers of corporate social performance: a multilevel, multistakeholder, and multimethod analysis. J. Bus. Ethics **144**(1), 21–40 (2017)
17. Pava, M.L., Krausz, J.: The association between corporate social-responsibility and financial performance: the paradox of social cost. J. Bus. Ethics **15**(3), 321–357 (1996)
18. Qian, Y., Li, Y., Hao, Y., Yu, T., Hu, H.: Greenhouse gas control in steel manufacturing: inventory, assurance, and strategic reduction review. Carbon Res. **3**(1), 27 (2024)
19. Flammer, C.: Does corporate social responsibility lead to superior financial performance? A regression discontinuity approach. Manage. Sci. **61**(11), 2549–2568 (2015)
20. Garcia-Pont, C., Nohria, N.: Local versus global mimetism: the dynamics of alliance formation in the automobile industry. Strateg. Manag. J. **23**(4), 307–321 (2002)
21. Hayes, A.F.: Introduction to Mediation, Moderation, and Conditional Process Analysis: A Regression-Based Approach. The Guilford Press (2018)
22. Gola, M., La Milia, D., Cadeddu, C., Bardini, F., Bianconi, B., Bisceglia, R., et al.: Buffer spaces in healthcare facilities: strategies for managing and designing strategic areas. Popul. Med. **5**(Supplement), 171–172 (2023)
23. Griffin, A., Price, R.L., Maloney, M.M., Vojak, B.A., Sim, E.W.: Voices from the field: how exceptional electronic industrial innovators innovate. J. Prod. Innov. Manag. **26**(2), 222–240 (2009)
24. Smith, W.A., Passmore, J., Turner, E., Lai, Y.L., Clutterbuck, D. (eds.): The Ethical Coaches' Handbook: A Guide to Developing Ethical Maturity in Practice. Taylor & Francis (2023)
25. Harjani, T., He, H., Chao, M.M.: The moral foundations of vaccine passports. J. Bus. Ethics 1–29 (2023)
26. Hobday, M.: East Asian latecomer firms: learning the technology of electronics. World Dev. **23**, 1171–1193 (1995)
27. Hobday, M.: Innovation in East Asia: The Challenge to Japan. Edward Elgar (1995)
28. Yeung, S.S.: The cultural logics of mothering practices: agency capacity of the university-educated Chinese mothers in Singapore and Hong Kong. Soc. Transform. Chin. Soc. (2023)
29. Jalilvand, A., Kim, S.M.: Matching slack resources and investment strategies to achieve long-term performance: new perspectives on corporate adaptability. J. Econ. Asymmetries **10**, 38–52 (2013)
30. Jung, J., Lai, M., Li, X.: Part-time master's students' attitudes towards study and work. Stud. Contin. Educ. 1–18 (2023)
31. Liu, H., Guo, K., Li, Y.: Impact of technical barriers to trade on sulfur dioxide and nitrogen oxide emissions in China's manufacturing sector. Environ. Pollut. Bioavailab. **36**(1), 2387682 (2024). https://doi.org/10.1080/26395940.2024.2387682
32. Mangili, S., Ferraguzzi, G., Capolongo, S.: Assessing the quality of the built environment in dementia: a framework to evaluate long-term care facilities. Popul. Med. **5**(Supplement), 55–55 (2023)
33. Rettig, L.A.: The Analysis and Characterization of the Cultural Dimensions of Food Waste Behavior in Hong Kong's Luxury Hotels. Hong Kong University of Science and Technology, Hong Kong (2021)

Risk Assessment and Its Prediction of Flood Risk Zone Using Multi-Criteria Decision-Analysis (MCDA) and Geographical Information System (GIS) Techniques: A Case Study of the Brahmani Watershed, Odisha, India

Abhijeet Das

Abstract Brahmani River Basin of eastern Odisha, India, has frequently seen flash floods over the past three decades despite being located near semi-arid/arid regions. However, in situations where there is a dearth of flood-related data, identifying a basin's flood risk zone is crucial. Therefore, the research sought to assess the watershed's flood risks and high-risk zones using Multi-Criteria Decision-Analysis (MCDA) and Geographical Information System (GIS). For this approach, the present study proposed the MCDA theory to list the five primary criteria that were taken into account while utilizing a spatial database to locate areas at risk of flooding, consisting of agents such as, number of households (NH), population (P), literacy (L), land use (LU), and infrastructure vulnerabilities (IV). Meanwhile, population, flood hazard zone and land use were given careful consideration when mapping this watershed's flood risk area. A weighted overlay analysis tool has been applied to the integration of the thematic layers in order to determine the zones of flood risk and hazard. The outcomes of the geospatial method were contrasted with the results derived from MCDA theory. Based on the observation, the results showed that this theory can also accurately and precisely define a basin's flood danger zone. In this chapter, around 1850 villages corresponds to high infrastructural vulnerability zones, whereas 75 villages is observed to have a moderate-high social vulnerability zone. About 48 villages of the study area, consisting of villages, where agriculture is observed to have a primary source of subsistence, that exhibits high vulnerability zones. These results underline the necessity of regularly and sustainably implementing flood prediction, early warning systems, and efficient management techniques.

A. Das (✉)
Department of Civil Engineering, C.V. Raman Global University (CGU), Bhubaneswar, Odisha, India
e-mail: das.abhijeetlaltu1999@gmail.com; abhijeetlaltu1994@gmail.com

Y. Zhu (ed.), *Proceedings of the 4th International Symposium in Environmental Science and Industrial Ecology*, https://doi.org/10.1007/978-981-96-1578-0_19

Keywords Brahmani · Flood hazard · MCDA · GIS · Overlay · Vulnerability · Agriculture · Management · Sustainable

1 Introduction

Flood is among the most frequent natural disasters that causes property damage, infrastructure damage, and fatalities [1]. Floods cannot be avoided, as seen by the numerous extreme catastrophes that occur, but their harmful effects might be minimized with cautious preparation and administration. However, in India, in particular, has been grappling with the recurring issue of floods, which profoundly impact both lives and livelihoods [2]. Thus, it experiences a multitude of natural hazards, with floods being the most prominent among them, resulting in extensive damage to infrastructure and agriculture, as well as tragic loss of life. The frequency of flood events has significantly increased globally over the past three decades. In certain instances, the frequency of floods with greater size has increased globally over the past 20 years as a result of climate change [3]. The proliferation of artificial surfaces can significantly impact runoff characteristics, leading to an increase in submerged areas during floods. The alternation in natural processes poses a particular challenge for developing countries, which often have limited resources for recovery and repairs in the aftermath of natural disasters [4].

Meanwhile, land use, land cover, rainfall, and population are examples of dynamic elements that vary over time. It is necessary to regularly analyze the changes in these dynamic elements using a variety of methods. Therefore, floods have the potential to result in both loss of life and substantial disruption to economic operations [5]. The rise in global flooding is attributed to various factors, such as changes in rainfall patterns, an increase in extreme weather events, alterations in land use, and population growth. Growing agricultural operations have put strain on surface water resources for irrigation over the past few decades, mostly in areas of the world that are prone to flooding. In the meantime, human activities like deforestation and urbanization in flood plains are likely to result in climate change, catastrophic weather events like floods, and regular variations in the flow of rivers [6]. Additionally, rapid industrialization and urbanization without proper planning has resulted in the occupation of unsuitable areas, including river banks and flood plains, which are prone to flooding. Subsequently, the increased anthropogenic activities along the majority of the Indian subcontinent's river basins are one factor that greatly increases its susceptibility to flooding and the associated alterations.

Evidently, literature demonstrates that by appropriately mapping the flood-prone areas, flood mapping is essential to reducing the risk of flooding and the susceptibility to flooding [7]. Therefore, flood mitigation is necessary. Also, in this study, early flood event can be discovered by applying optimization and computational intelligence techniques. While, it may not be possible to prevent these hazards entirely, implementing appropriate mitigation measures can lessen the magnitude of the calamity and its effects. The identification of flood risk zones is essential for this reason [8].

The assessment of flood disaster risk is impacted by variables such as population, land utilization, and infrastructure. Hence, it is crucial to address these challenges and find sustainable solutions to impact the impact of floods in the country, protect communities, and minimize the destruction caused by these natural disasters.

Subsequently, making use of the hydrodynamic and hydrological models, it takes a lot of work and rigorous calibration to accurately identify flood risk zones [9]. The most recent methods for decision-learning and statistics are helpful for locating areas that are prone to flooding, but no single technique is adequate for mapping a basin's flood susceptibility. The integration of multi-criteria decision analysis (MCDA) with Geographical Information System (GIS) can be used to determine which places are most at danger from flooding by taking into account a number of separate factors. However, GIS and Remote Sensing (RS) technologies provide prompt access to satellite data and effective handling of substantial volumes of data in order to recognize distinct signs across a range of dimensions [10]. Managers and policy makers frequently integrate MCDA with GIS since it enables the geographical analysis of multi-criteria decision support systems. Whereas the second group uses the judgment of experts, decision-makers, and practitioners from various agencies in a particular area to decide the weights in MCDA, the first category assigns the weights based on statistical analysis of the data of water quality parameters [11]. Several studies used MCDA to establish the weights and determine the same conditions for the creation of flood-prone zones.

In this perspective, MCDA is the most generally used method because it incorporates framework concerns, consistency, and accessibility into the study, which can integrate several assessment variables. Compared to more conventional techniques that frequently required specialized knowledge, such as drilling, resistivity surveys, hydrogeological, geophysical, and ground-penetrating radar surveys, this method is more productive and economical [12]. On the contrary, GIS-MCDA yields the best instruments for analyzing environmental issues. However, this mechanism has the capability to evaluate these variables and their respective effects. For instance, it is described as the processes for gathering, transforming, and integrating spatial data in order to determine which alternative solution, out of several options, would best support decision-making. Because this approach determines the flood risk zone by consulting experts, it could be vulnerable to subjectivity and uncertainty-based cognitive limitations. After that, while uncertainty during the assessment of flood risk cannot be completely eliminated, it can be reduced [13].

Furthermore, the combination of GIS and MCDA, is very successful at assessing, tracking, and managing surface water in space and time within a predetermined window of time. In Odisha, the majority of the population works in agriculture, and the river helps with irrigation. So, it modifies the decision-making process by calculating the various contributing aspects. Hence, MCDA has gained recognition on a global scale as a very useful instrument for handling difficult decision-making. Likewise, these technologies have become commonly used in investigating multi-dimensional issues like mapping areas prone to flood-related problems, considering both spatial and temporal dimensions. However, the river floods during the monsoon season, destroying houses and croplands in the low-lying areas [14]. To address the vulnerability of certain areas to flooding, an integrated approach combining RS and GIS can be employed, that has resulted in notable advancements over conventional map overlay techniques. In India, some scholars have conducted studies on flood hazards, vulnerability, and risk analysis using geospatial technologies like RS and GIS techniques in Indian river basins.

Brahmani River, Odisha, is a perennial river and has experienced recurrent flooding, leading to significant economic losses, and impacting numerous individuals. The majority of Odisha state's villages as well as agricultural land were harmed by the flood. The process of identifying a river basin's flooding zones gets difficult when there is a dearth of available data. There are currently no comparative studies between MCDA and GIS for a feature selection in the flood dataset, despite multiple trials utilizing those approaches. Additionally, It has been noted that policymakers pay less attention to flood preparedness and management in middle- to small-sized river basins. Despite the lack of previous investigations on flood hazards and risk zones in the Brahmani watershed, this study aims to fill the gap by specifically focussing on this area. However, there is little study on determining the accuracy of algorithms. In addition, the relatively new GIS-based multi-criteria framework for flood-risk assessment has not been applied in many studies. More precisely, an appropriate method for classifying the experiment's input data needs to be created. In order to identify the best alternative with the maximum accuracy for choosing the best flood quality characteristic, they were employed to address these problems in the input data sorting process.

Indeed, the objective of this study is to utilize GIS, RS, and MCDA techniques to detect flood affected areas and zones of risk in the Brahmani Watershed. The MCDA method offers a variety of alternative potential solutions so that decision-makers can choose the best one. To attain this objective, thematic layers of criterion that influence flood hazard and risk, this study examined five key factors such as social that contains number of households, population, literacy, land use, and infrastructure vulnerabilities, mainly roads. The main highlights of this study are listed as follows: (1) to evolve flood hazard zonation map from satellite data, (2) to classify the basin into several flood risk zones, depending on the severity of the danger, (3) A GIS-MCDA analysis tool will be employed. Our geographical focus will be on the semi-arid areas of Odisha and (4) We will investigate the flood regions, as a potential resource for the selected studied basin. As per authors knowledge, there is no literature on applying the multi-criteria theory to assess a river basin's flood risk. This theory offers a framework, accountability, transparency, and resilience to judgments. Therefore, this study will be useful for decision-makers and managers in designing early warning systems and flood hazard reduction measures.

2 Description of Study Area

The Brahmani River is the 2nd largest river in the state of Odisha, eastern part of India, after Mahanadi and has a length of around 799 km. Its basin occupies 15% of the total state geographical area. It is produced by the meeting of the Sankh and Koel rivers, situated at the latitudinal range of 21–23° North and the longitudinal range of 84–87° East, and the river often, travels southward and passes through the districts of Odisha namely, Sundargarh, Deogarh, Angul, Dhenkanal, Cuttack, Jajpur, and Kendrapara [15]. The basin area of river Brahmani in Odisha, includes 41 sample monitoring stations, thereby making up 57.63% of the basin's overall area. The basin gets an average of yearly monsoonal precipitation of 1400 mm, which is compatible with a climate that is semi-arid to arid. The maximum mean temperature varies from 38 to 43 °C, that prevails inside the catchment. The river flows through numerous industrial facilities and a substantial section of an agricultural area before joining Bay of Bengal. The lower portion of the basin receives less rainfall on average each year, but the higher and sections of the middle portion of the basin receive significant amounts of rain during the monsoon season [16]. The annual rainfall depending on the amount of rain that fell during the south-west (SW) monsoon season, that fluctuate among June till September. Subsequently, the flood risk is greater during the SW monsoon, although exceptional occurrences can also result in flooding during other seasons. The treatment facilities are extremely limited. The water quality of various locations suffers high level of contamination. Due to huge discharge of industrial wastes into the river, the water quality is getting polluted day by day [17]. The river Brahmani, is being polluted by the effluents from steel plants and chemical industries in the river bed. The major selected places in the study, are represented by Fig. 1.

3 Data Collection and Database Properties

In this scenario, the present study utilizes a collection of satellite data from multiple time periods, consisting of images from Landsat VII and VIII, specifically focusing on this selected sample of the studied river basin. Using Google Maps, the sample source's spatial positions were noted, and further information was transferred in the GIS software [18]. To assess the flood hazard in the Brahmani Watershed, researchers examined five key factors: number of households, population, literacy, land use, and infrastructure vulnerabilities. The Census of India 2011, is utilized for the extraction of socio-economic statistics [19]. Then, for the mapping, an spatial analyst technique was applied in order to rasterize the spatial condition in the GIS program. The creation of thematic maps for each parameter was done using the Inverse Distance Weighted (IDW) technique [20]. A weighted overlay technique combined all the parameters and generated the flood hazard zone. The calculation of the data were analyzed by the use of Statistical Package for Social Sciences (SPSS software Version 20.0). For certain purposes, this technique is a highly flexible method of interpolation that

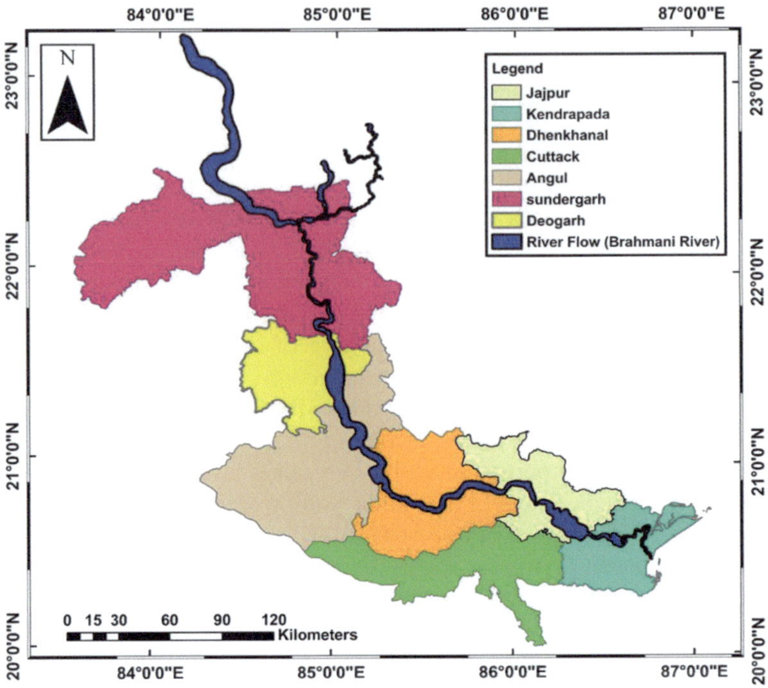

Fig. 1 Study area along the Brahmani River, Odisha, India

measures the data of nearby unknown sites and is objectively accurate under a variety of circumstances. The findings were exported as themed maps following the study [21]. The satellite data for the research region was acquired during a duration of 10 years, encompassing the years 2013–2023.

4 Methods

The current study used two different methodologies for flood risk analysis. The MCDA theory was employed to evaluate the danger of flooding and validate the results obtained from the evaluation of the flood hazard zonation layer. The MCDA theory is a system-integrated method for mapping flood risk that was presented in this study. Therefore, the flood risk analysis is referred as the combination of an examination of flood risk and flood vulnerability that may be divided into two parts [22]. However, in order to determine the river basin's flood risk zone, the current study used decision-making analysis. The following topics were covered in discussion: determining the flood risk area.

4.1 Flood Hazard Zone Analysis

To construct a mathematical model employing this decision-making process, the first step is to recognize the structural components and comprehend the influence of each river basin's flood risk system [3]. In this watershed, flooding watersheds were considered to be among the elements that raise the possibility of floods. To better understand this, the flood hazard zones were categorized according to their vulnerability to flooding risk. Two major sub-systems are found for flood risk systems, but there are also a number of sub-systems found that may have an impact on the sub-systems. The layer with a very high flood hazard was given a ranking of higher score, indicating its susceptibility to risk of flooding. Conversely, the layer with a very low flooding hazard has been ranked as 1, indicating its low susceptibility to flooding risk [8]. The layer of flood hazard zonation was created by utilizing spatial multiple temporal satellite datasets collected during flood-prone seasons, from a duration of 10 years i.e., between 2013 and 2023. However, each flood occurrence is accompanied by a corresponding inundation layer, and these layers were aggregated to create the annual flood inundation layer. A basin's flood risk can be decreased through effectively overseeing flood vulnerability and flood hazard, the two subsystems of a flood risk system [7]. Subsequently, the river basin generates a flood hazard zone outcomes by combining all the annual inundation layers.

4.2 Flood Risk Zone Analysis

The MCDA is a decision-making approach that considers multiple objectives and criteria. It uses a pairwise comparison procedure to determine preferences among different groups. Thus, MCDA quantifies individual preferences or judgements using absolute numbers on a scale [1]. The current study thoroughly assesses the flood risk in the BRB using the MCDA technique, including factors like social (number of households, population, literacy), land use, and infrastructure (roads) vulnerabilities. Therefore, assessment of flood risk is essential for estimating the damages caused owing to flooding by implementing spatial analysis considering space and time as a reference frame. In addition, to analyze the risk of flooding in this area, the different land use classes, such as wetland, built-up, farmland, barren land, agricultural land, scrub forest, and agroforestry, were grouped based on their vulnerability to flooding. A common scale was used to reclassify the types of land use in the sub-basin according to their sensitivity to flood risk [4]. A vulnerability index for the current study was developed based on the weight assigned to each variable according to its relative value to society. Thus, a flood risk assessment is defined as the result of the study's total vulnerabilities plus the flood threat, that includes social, infrastructure and land use parameters. The internal driving forces that create flood risk are known as flood hazard factors, whereas the factors that make a person vulnerable to flooding are known as flood risk resistance. Similar to this, urbanization causes a quicker peak and larger runoff volume, however the presence of vegetation and forests reduces the size of the runoff peak, indicating the impact of land use land cover (LULC) on runoff [5]. However, understanding LULC is crucial for assessing surface runoff and potential flooding in a particular catchment. The presence of water leads to urbanization, which highlights the interdependence of urbanization and its separation from rivers. Similarly, soil erosion can also cause a change in slope during runoff. When rainfall happens near a river, the rapid runoff into the river causes an instantaneous increase in river discharge, which increases the risk of flooding [23]. Different types of land use affect the rate of infiltration, with vegetated areas promoting infiltration, while residential and pastre areas hinder infiltration due to their impervious cover, leading to increased runoff and a higher risk of flooding. Assume that any factor that influences a zone's flood risk is not included in the current study. If enough information about that aspect for a study region is available to include in the final results, then that can be taken into account in subsequent research. Furthermore, the realization of risk assessment involves the multiplication of hazard and the summation of vulnerabilities derived from numerous indicators as represented in Eq. (1) by:

$$\text{Flood Risk} = \text{Hazard} * \text{f(Social, Infrastructure, Land use Vulnerabilities)} \quad (1)$$

5 Results and Discussions

The Floods occur often and have turned into a regular occurrence in this watershed flood plain and other places that get a lot of rain. The adverse effects of flooding typically transpire in an area that is at a lower elevation when it happens [6]. Flood risks are greater in lower elevation, flat areas. Again, the slope is one of the important elements that causes flooding since it regulates the water's velocity and the process of infiltration [11]. Thus, the integration of GIS and Remote Sensing (RS), a cost effective and efficient tool for systematically analyzing several surface features helpful in identifying flood susceptibility zones. The zones of flood was generated based on five factors. The weight of each factor was determined using MCDA approach [15]. For sustainable economic development, in the natural sciences, assessing flood risk is of utmost importance because it is crucial for analysis.

5.1 Flood Hazard Vulnerability

The Brahmani watershed flood hazard map categorized areas based on their flood susceptibility level. Depending on the severity of the flood threat, the flood hazard layer is distinguished into three zones, assessed as low, medium, or high. The findings revealed the following flood risk zones: in the basin, about 1850 villages of the area was characterized as a low flood hazard zone, while, 75 villages was considered as moderate flood hazard zones. Conversely, the regions that indicate minimal flood threats are mainly located in the southern, southeast, central, and eastern regions. They were distinguished by their steep slopes, higher elevation, high drainage density, lower rainfall, agroforest, and soil types of eutric cambisols, which resulted in low infiltration and high runoff. The areas with moderate flood hazard zones frequently characterized the eastern northeastern, southwestern, central parts, and some parts of the western watershed. The analysis again showed that approximately, 48 villages are recognized under high flood hazard zones category. The high zones of vulnerability are distinguished by their gentle slope, lower elevation, lower drainage density, high rainfall, and specific soil types like pellic vertisols and chromic vertisols. However, wetland, built-up, and barren land LULC types further contributed to very high and high flooding in the watershed. Notably, based on the spatial study findings, the areas of the selected river basin that are affected by flooding are namely, the districts of Jajpur, Cuttack, and Kendrapara, which are accounted for the vast bulk of the flooded villages. These zones have mostly been found in the western, northwest, and a few areas in the center.

5.2 *Flood Risk Assessment*

A flood hazard zone is the most important criterion for an effective flood risk assessment. As a result, it serves as a comprehensive performance management system. It is observed to compute the system's performance and considers the interaction between the subsystems at all stages of the assessment procedure. However, if social, infrastructure, and land use vulnerabilities are integrated with flood hazards, the accuracy of the flood risk assessment can be further increased. This innovative study found that 18% of the river basin settlements are grouped in the flood danger zone. Most importantly, the flood risk map shows that very high flood risk is primarily concentrated in the primary communities, which are found in the selected studied area are namely, Kendrapara, Jajpur, and Cuttack respectively. The region with a very high flood risk is distinguished by a very high population density, flooding, and built-up. Due to the plain land, croplands and key infrastructure are located in high flood-risk locations, where people prefer to dwell. The LULC pattern affects the flood risk since places with forest cover have a low risk of flooding whereas urban built-up areas have a high risk. However, because of its low population density, agroforest, scrub forest, and low flood danger, the bulk of the area under study was categorized as possessing a low flood risk. Additionally, a region's flood hazard is inversely correlated with its distance from rivers because areas near rivers are more vulnerable to flooding.

Similarly, from the results, LULC, distance from food shelters, effective drainage network, and literacy rate sub-systems are found to be on the higher end of the spectrum. Most prominently, the decision-makers ought to offer remedies to lessen the consequences of LULC, distance from food shelters, efficient drainage network, and literacy rate sub-systems by constructing food shelters, limiting construction in the flood plain, and ensuring efficient drainage in the basin. These factors will be diminished as their impact on the susceptibility to floods reaches its minimum. Thus, decision-makers ought to offer ways to lessen the worst effects of conditioning variables on the Brahmani River Basin's flood risk. The basin's flood risk will go down as long as the flood hazard and vulnerability subsystems work effectively.

6 Conclusions

Different processes such as Flood risk, flood hazard, and flood vulnerability analysis are frequently employed to improve river basin flood management. The research successfully utilized MCDA and GIS data to assess the risk zones and flood threats in the Brahmani Watershed. Multi-temporal historical satellite data has made village-level flood hazard zone estimation possible in locations with insufficient or unavailable data. This innovative method integrates hazard zones with social, infrastructural, and land-use vulnerabilities at the village level, for flood vulnerability assessment of the river basin. The assessment standards allocate weights to social, infrastructure, and land-use factors under the premise that densely populated regions possess greater economic resources and livelihood opportunities, rendering them more susceptible to the risk of flooding. On a catchment scale, it is found that the flood risk zone can be identified using the MCDA theory. Since the method considers both the individual performance of the sub-systems and their interaction with one another, the results are believed to be more accurate than those derived from other similar methods. The flood hazard investigation of the watershed has provided valuable insights into the distribution of flood hazards across the area. The flood hazard, vulnerability, and risk are divided into three categories: medium, high, and low. Notably, this study shows that 1850 villages was categorized as high infrastructural vulnerability zones, moderate-high social vulnerability zones covers 75 villages, and approximately 48 villages, where agriculture is the primary source of subsistence, are present in high vulnerability zones. Based on flood risk assessment theory, around 18% of villages are situated in areas that are classified as high flood risk zones. The part of the basin consisting of three places namely, Kendrapara, Jajpur, and Cuttack, is highly vulnerable to flooding as a result of its location in a plane area, urbanization, dense population, inadequate drainage system, and lack of flood shelters. Remaining percentage of areas are prone to low flood hazard zones. The lowest population density, forest cover, rapidly declining slope, and higher elevation are all represented by the low flood zones. The results provide important information for understanding and managing flood risks. Therefore, decision-makers can make use of these findings to set up safeguards against flooding in this basin.

The limitation of this methodology involves the availability of datasets with a respectable level of accuracy, the combination of expertly assigned weights, and the need for strong computational capabilities to determine the numerical value for every pixel. For more precise results, the weights assigned by various experts might be further refined. The farmers of these villages should be encouraged to grow flood-resistant crops. Industry startup policies like flood insurance. Additional insurance for low-risk flood areas. Utilizing the MCDA, this algorithm can be improved even more to offer potential alternative management strategies on a catchment size, allowing decision-makers to play a more important part in improving flood risk management. By identifying specific areas with flood hazards and risks, appropriate measures can be taken to mitigate and reduce potential damages. Considering these findings, when planning and implementing interventions is crucial to ensure the safety and well-being of the communities living in the watershed.

Acknowledgements I would like to acknowledge Indian Meteorological Department (IMD), Bhubaneswar, Odisha, for providing relevant data of the study area. The lab facilities and investigation was supported by C.V. Raman Global University (CGU), Bhubaneswar, Odisha. The corresponding author would like to thank Ranu-Prativa Research Group, Ruchika Square, Bhubaneswar, for supporting the completion of this research study. We appreciate the editor and anonymous reviewers for their constructive comments which greatly helped to improve our manuscript. The datasets used and/or analysed during the current study are available in the article/from the corresponding author on request.

References

1. Burayu, D.G., Karuppannan, S., Shuniye, G.: Identifying flood vulnerable and risk areas using the integration of analytical hierarchy process (AHP), GIS, and remote sensing: a case study of southern Oromia region. Urban Clim. **51**, 101640 (2023). https://doi.org/10.1016/j.uclim.2023.101640
2. Mokhtari, E., Mezali, F., Abdelkebir, B., Engel, B.: Flood risk assessment using analytical hierarchy process: a case study from the Cheliff-Ghrib watershed, Algeria. J. Water Clim. Change **14**(3), 694–711 (2023). https://doi.org/10.2166/wcc.2023.316
3. Nair, P.G., Das, S., Medhe, R, Chatterjee, U., Kumar, S., Karuppannan, S.: Flood susceptibility mapping in the Periyar River Basin of Kerala integrating GIS-AHP and Google Earth Engine (2023). https://doi.org/10.2139/ssrn.4549889
4. Senan, C.P.C., Ajin, R.S., Danumah, J.H., Costache, R., Arabameri, A., Rajaneesh, A., Sajinkumar, K.S., Kuriakose, S.L.: Flood vulnerability of a few areas in the foothills of the Western Ghats: a comparison of AHP and F-AHP models. Stoch. Environ. Res. Risk Assess. **37**(2), 527–556 (2023). https://doi.org/10.1007/s00477-022-02267-2
5. Raufu, I.O., Mukaila, I., Olaniyan, K., Awodele, Z.: Application of remote sensing and geographical information system (GIS) in flood vulnerability mapping: a scenario of Akure South, Nigeria. IJEGEO **10**(1), 90–99 (2023). https://doi.org/10.30897/ijegeo.1073697
6. Melkamu, T., Bagyaraj, M., Adimaw, M., Ngusie, A., Karuppannan, S.: Detecting and mapping flood inundation areas in Fogera-Dera Floodplain, Ethiopia during an extreme wet season using Sentinel-1 data. Phys. Chem. Earth Parts A/B/C **127**, 103189 (2022). https://doi.org/10.1016/j.pce.2022.103189
7. Kamaraj, P., David Thangapandian, I., Karuppannan, S., Garo, T.: A statistical-based geospatial approach to prioritize the watersheds for soil erosion conservation in the Upper Awash Basin (Upstream Koka), Ethiopia. Kuwait J. Sci. **51**(2), 100198 (2024). https://doi.org/10.1016/j.kjs.2024.100198
8. Hidayah, E., Lee, W.-K., Halik, G., Pradhan, B., Indarto: Assessing coastal flood susceptibility in East Java, Indonesia: comparison of statistical bivariate and machine learning techniques. Water **2314**, 3869 (2022). https://doi.org/10.3390/w14233869
9. Lyu, H.-M., Zhou, W.-H., Shen, S.-L., Zhou, A.-N.: Inundation risk assessment of metro system using AHP and TFN-AHP in Shenzhen. Sustain. Cities Soc. **56**, 102103 (2020)

10. Thanh Son, N., Thi Thu Trang, N., Bui, X.T., Thi, D.C.: Remote sensing and GIS for urbanization and flood risk assessment in Phnom Penh, Cambodia. Geocarto Int. **2021**, 1–18 (2021)
11. Wang, Y., Xu, Y., Tabari, H., Wang, J., Wang, Q., Song, S., Hu, Z.: Innovative trend analysis of annual and seasonal rainfall in the Yangtze River Delta, eastern China. Atmos. Res. **231**, 104673 (2020)
12. Praveen, B., Talukdar, S., Shahfahad Mahato, S., Mondal, J., Sharma, P., Islam, A.R.M.T., Rahman, A.: Analyzing trend and forecasting of rainfall changes in India using non-parametrical and machine learning approaches. Sci. Rep. **10**(1), 10342 (2020)
13. Weday, A.M., Tabor, K.W., Gemeda, D.O.: Flood hazards and risk mapping using geospatial technologies in Jimma City, southwestern Ethiopia. Heliyon **9**(4), e14617 (2023). https://doi.org/10.1016/j.heliyon.2023.e14617
14. Das, A.: Water quality assessment using water quality index (WQI) under GIS framework in Brahmani Basin, Odisha. In: International Conference Innovation in Smart and Sustainable Infrastructure, Aug 2022, pp. 131–150. Springer Nature Singapore, Singapore (2022)
15. Das, A.: Assessment of potability of surface water and its health implication in Mahanadi Basin, Odisha. Mater. Today Proc. (2023)
16. Das, A.: Water criteria evaluation for drinking purposes in Mahanadi River Basin, Odisha. In: International Conference on Trends and Recent Advances in Civil Engineering, Aug 2022, pp. 237–260. Springer Nature Singapore, Singapore (2022)
17. Das, A., Agarwal, V., Kumar, M.: Evaluation of water quality in the Brahmani River Basin, Odisha: a multicriteria decision-making approach for sustainable management. In: River Basin Ecohydrology in the Indian Sub-Continent, pp. 139–165. Elsevier (2024)
18. Costache, R., Tien, B.D.: Identification of areas prone to flash-flood phenomena using multiple-criteria decision-making, bivariate statistics, machine learning and their ensembles. Sci. Total Environ. **712**, 136492 (2020)
19. Census: Population Census 2011. https://www.census2011.co.in/ (2011)
20. Shahiri Tabarestani, E., Afzalimehr, H.: A comparative assessment of multi-criteria decision analysis for flood susceptibility modelling. Geocarto Int. **2021**, 1–24 (2021)
21. Fu, M., Fan, T., Ding, Z., Salih, S.Q., Al-Ansari, N., Yaseen, Z.M.: Deep learning data-intelligence model based on adjusted forecasting window scale: application in daily streamflow simulation. IEEE Access **8**, 32632–32651 (2020). https://ieeexplore.ieee.org/document/9007531/
22. Shamshirband, S., Hashemi, S., Salimi, H., Samadianfard, S., Asadi, E., Shadkani, S., Kargar, K., Mosavi, A., Nabipour, N., Chau, K.-W.: Predicting standardized streamflow index for hydrological drought using machine learning models. Eng. Appl. Comput. Fluid Mech. **14**(1), 339–350 (2020)
23. Das, A.: Multivariate statistical approach for the assessment of water quality of Mahanadi Basin, Odisha. Mater. Today Proc. **65**, A1–A11 (2022)

Source Apportionment for Spatial Variation of Surface Water Quality in Brahmani River Basin, Odisha (India), Utilizing Water Quality Index (WQI), Pollution Index (PI), and Compromise Programming (CP) Techniques

Abhijeet Das

Abstract Surface water pollution is now a major environmental concern on a worldwide scale. To maintain the water quality and to understand the current state of the water resource in Brahmani River, Odisha, surface water monitoring and assessment are crucial. In this study, we utilized three models—Weighted Arithmetic (WA)-Water Quality Index (WQI), Synthetic Pollution Index (SPI) and multi-criteria decision-making (MCDM) approach namely Compromise Programming (CP)—to predict surface water drinking suitability. A total of 15 water samples were collected from 7 monitoring sites and were analyzed using standard methods. All assessing parameters were discovered to be within permissible bounds for agricultural, drinking, and environmental uses. The physicochemical trends indicated a sequence of cations: $Ca^{2+} > Mg^{2+} > Na^{+} > K^{+}$, and the anions are reflected as: $SO_4^{2-} > Cl^{-} > NO_3^{-} > F^{-} > PO_4^{3-}$. Subsequently, several indices, such as WA-WQI and SPI were calculated and indicated that water in the river at Q-(1), (3), (7), and (2) had high levels of Ca^{2+}, SO_4^{2-}, TDS, EC, and Mg^{2+} concentrations, altered pH, and elevated DO values. The WA values revealed that 57.14% sites were the most polluted which rendered them unsuitable for drinking or bathing. Notably, the majority of the sites used for sampling have SPI values > 0.7, indicating high levels of pollution. The mid-stream stations such as: Q-(4), (5), and (6), were categorized under good water zone. The results obtained by CP, demonstrated a value of 0.941 at Q-(1), 0.813 at Q-(2), and 0.649 at Q-(7). These highly polluted areas demonstrated that fertilizer, organic waste, and soil leaching were the main causes of variations in water quality. This knowledge can be helpful to planners, policy makers, and managers of water resources in creating plans that will ensure the river's cultural and aesthetic value is preserved for future generations.

A. Das (✉)
Department of Civil Engineering, C.V. Raman Global University (CGU), Bhubaneswar, Odisha, India
e-mail: das.abhijeetlaltu1999@gmail.com; abhijeetlaltu1994@gmail.com

Y. Zhu (ed.), *Proceedings of the 4th International Symposium in Environmental Science and Industrial Ecology*, https://doi.org/10.1007/978-981-96-1578-0_20

Keywords Surface water · Brahmani · WA-WQI · SPI · CP · Bathing · Leaching · Fertilizer

1 Introduction

Water, as a solvent, is vital for living organisms. It is a limited resource for drinking, agriculture, and industrial purposes. Polluted water is dangerous for plants, animals, and overall biodiversity and can be a vital threat to human health. The quality of surface water can be influenced by various natural factors, including discharge, weathering, hydrological features, runoff, rainfall, erosion of soil, and physiographical characteristics of water bodies [1]. With the rapid economic and social development of recent decades, pollution sources from livestock, poultry, aquaculture, planting, and rural domestic sewage have drawn much attention. Because of this, the dissolved components in water are continually changing in quantity due to both natural processes like rainfall, rock weathering, and soil decomposition, as well as human activities like residential and commercial activities [2]. Consequently, a significant quantity of organisms and inorganic impurities consistently enter the water, leading to water pollution and the presence of suspended solids like mud, algae, detritus, and faecal matter, which contribute to the contamination [3]. However, water resources, including both surface water and groundwater, are extensively utilized natural assets that are currently experiencing significant issues of water pollution and scarcity on a global scale. In previous decades, historical evidences show that change in water quality and quantity due to anthropogenic pressure has led to vanish many settlements of human beings. Mineral extraction, stormwater runoff, industrial effluents, nuclear reactions, and natural processes like rock weathering, mid-ocean ridge spreading, and volcanic activity all contribute toxic elements to the environment [4]. For this cause, preserving the reliability of surface water resources has become a paramount concern. This is primarily due to the scarcity of freshwater and the significant impact of human activities on the degradation of water quality. Thus, a strict and vigilant approach towards monitoring and assessment is needed to ensure its good health. As a part of sustainable development, quantifying pollution levels of harmful elements, identifying potential sources, and mitigating future pollution risks are very essential.

Water quality monitoring serves as a valuable tool to assess the effects of pollution sources and to ensure effective water resource management and the preservation of aquatic ecosystems. An important challenge in monitoring the quality of surface water is managing the intricate datasets that arise from the numerous water quality components measured at various monitoring locations [5]. Therefore, it is necessary to adopt Water Quality Index (WQI) mechanism, in order to create efficient methodologies to assess water resources for the purpose of sustainable development and ensuring human health safety. The WQI, a unitless single number, essentially sums up the overall quality of the water at a certain place and is determined by combining data from several water quality documents that are acquired by matching measured values to regulatory standards [6]. Hence, Weighted Arithmetic (WA) and Synthetic Pollution Index (SPI) approaches are simple to comprehend and help raise public awareness of water pollution. They also support the enforcement of laws requiring the proper management of wastes released from various sources and placing limitations on the extraction of surface water, all of which can contribute to the development of efficient management strategies that safeguard water from contamination. In this regard, both approaches basically convey intricate information about water quality in a way that is understandable to mathematicians and useful to both the general public and decision-makers [7]. Further, this strategy assists in understanding complex data matrices to gain insights into the water quality and ecological condition of the samples under investigation. Therefore, these techniques help identify factors that impact water systems and serve as an essential tool for effective water resource management and prompt problem-solving.

Monitoring river water quality is a useful strategy for making sure that water resources are managed effectively, protecting the environment, and determining whether surface waters are suitable for drinking. However, traditional evaluation of water quality is a technique where the concentration of each water quality measure is compared to the corresponding standard criteria. This conventional method of ensuring the purity of water for irrigation and drinking is expensive, time-consuming, and labour-intensive. Also, this approach necessitates extensive and specific data regarding water quality that the general public and policy makers may find difficult to comprehend [8]. Furthermore, a number of research recommend preparing data for modelling and prediction utilizing new methods. In order to compensate for the limitations of the WQ technique, use of multi-criteria decision-making (MCDM) approaches like Compromise Programming (CP) technique is employed to compare and evaluate the various possibilities in order to choose the best solution. Thus, this approach is highly objective in determining the weight and ultimate ranking since it eliminates human subjectivity [9]. The capacity of the CP approach to examine many parameters at various scales was its main benefit. Additionally, it facilitates the creation of a single output map by scientists by allowing them to combine data from many criteria based on their relative weights, supported by expert knowledge. As a result, this work proposes using an MCDM method such as CP Assessment. This method has also been applied in a number of other domains, including human system reliability, environmental disaster management, water quality research, agriculture, and risk prioritization [10].

The difference between classical statistics and geo-statistics is the spatial dependency assumption. Many research has used various techniques to illustrate the global surface water characteristics' spatial distribution. Thus, Geographical Information System (GIS) can also aid in resolving some of these problems; in particular, using the IDW approach in conjunction with a water quality evaluation method can significantly enhance the visualization of study findings [11]. Additionally, visualizing regions with low to high concentrations that might not be appropriate for exploration is the fundamental function of GIS analysis. Several of the interpolation techniques employed in earlier research, including Inverted Distance Weighted (IDW). It was recognized as a successful technique for calculating cell values using a linearly weighted combination of a number of sample locations inside the research area. Thus, this approach is considered effective in detecting shared patterns and dispersion anomalies, decreasing the original dataset dimension, and enhancing comprehension of the ambient and geogenic sources of soluble ions and metals in water [12]. In case of IDW, the weight is determined by the inverse distance and was allocated to every place according to its distance.

Water quality monitoring serves as a useful instrument for protecting aquatic life and achieving sustainable management of water resources (quantity and quality). Thus, no single research evaluates the quality of drinking water using an integrated method based on WA, SPI, and CP for the selected catchment. To make certain the sustainable use of surface water resources, it is important to characterize the chemistry of surface waters, the state of their quality, and the variables that influence them within a given area. In light of this, the purpose of this study was to evaluate and measure the extent of the Brahmani River Basin's water quality degradation, especially at the regional and temporal scales. In view of this, the investigation seeks to differentiate between all processes associated with water surface evaporation, siltation brought on by sediment deposits, and urban and industrial discharge. For this reason, a study was carried out, aiming for the following goals: (1) to determine the physicochemical characterization and their spatial distribution, (2) source identification of the major ions in the surface water, (3) assessment of surface water quality for drinking purpose using WA, SPI, GIS, and CP. The results of the present study could be helpful in the creation of workable strategies to improve the drinking water networks between rural and urban areas in the chosen monitoring locations.

2 Study Area

The Brahmani is a major seasonal and 2nd largest river in the Odisha State of Eastern India. It travels generally south and south-east passed Bonaigarh and Talcher and then turns east to join northern branches. The basin area in Odisha constitutes 57.63% of the total basin area and 15% of total state geographical area [13]. It consists of total drainage area of 39,628 km^2, out of which, 22,516 km^2 lies in Odisha, and around 15,405 km covers in Jharkhand State, and finally, about 1347 km^2 corresponds to Chhattisgarh State. The region is distinguished by a semi-arid to arid environment with an average of 1400 mm of precipitation per year and a mean annual temperature of 25.7 °C. The Brahmani is created by the Sankh and South Koel rivers meeting at 22° 15′ N and 84° 47′ E, and crosses several areas, that includes, Sundergarh, Kendujhar, Dhenkanal, Cuttack, and Jajpur [14]. In this basin nearby, there are very few treatment facilities on the river. Numerous regions experience high levels of pollution in their water quality. The land uses largely consists of residential, commercial, industrial, mining, livestock, pasture, row crops, forestry, and water. The river's water quality is deteriorating daily as a result of the massive flow of industrial wastes into it. The river is being contaminated by the chemical and steel companies' wastewater in the

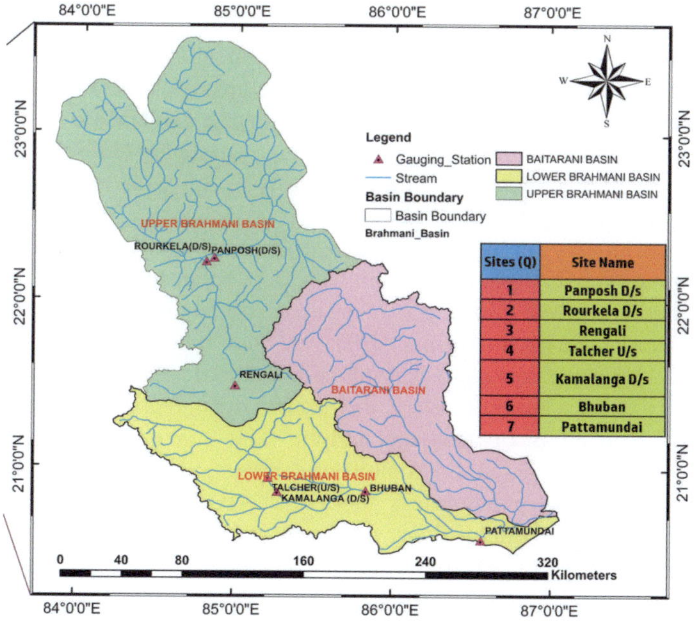

Fig. 1 Index map of the study area

riverbed. Moreover, ArcGIS 10.8 was utilized to create the study area's digital base map, which is displayed in Fig. 1.

3 Sampling and Analysis

Surface water samples were collected from densely populated and agricultural field study areas. The samples were gathered in 1 L plastic bottles made of high-density polyethylene (HDPE), and they were later utilized in physicochemical investigation in the lab. In the study's investigation, the location of the samples that were acquired was identified by Global Positioning System (GPS) tool. A total of 15 surface water samples were collected during the monsoon season before the COVID 19 pandemic period from 7 sampling locations based on the available sources of surface water. The bottled surface water samples were brought to the lab and placed in an icebox. The parameters of water quality were measured and reported in milligrams per litre (mg/L), with the exception of pH and EC (mS/cm). The sampling procedure was followed by the standard Methods for the Examination of Water and Wastewater [15]. These samples were analysed for dissolved oxygen (DO), electrical conductivity (EC), Alkalinity, sodium (Na^+), potassium (K^+), calcium (Ca^{2+}), total dissolved solids (TDS), magnesium (Mg^{2+}), fluoride (F^-), chloride (Cl^-), nitrate (NO_3^-), pH, temperature, sulphate (SO_4^{2-}) and phosphate (PO_4^{3-}), within 48 h of collection.

The titrimetric technique was employed to ascertain Ca^{2+} and Mg^{2+} concentrations. The Cl^- ion was calculated utilizing the titration method with silver nitrate. The temperature, pH, EC, and TDS, were found in the area utilizing a calibrated hand-held conductivity multi-parameter sensor. The soluble K^+ and Na^+ ions were assessed using flame photometer. The concentration of $SO_4{}^{2-}$ and $NO_3{}^-$ were established by means of a spectrophotometer instrument, in association with a visible ultraviolet spectrum. A titration by H_2SO_4 is used to measure the level of alkalinity. DO was recorded using an Oxygen meter, while temperature was analyzed by a digital thermometer. Various quality assurance techniques were applied during the examination of the water sample [16]. Following a second assessment of the experimental findings in the laboratory, ionic balance errors (IBE) were determined, samples were checked three times, and the average value was allotted. The IBE method was used to verify the recorded values in order to assess the accuracy of the analytical findings. The value of IBE is given as a percentage, and the value's range is smaller than 10%. To create maps of the spatial distribution of different chemical elements, ArcGIS software with IDW interpolation was applied. The IDW interpolation approach is useful in determining a value for any unknown place and prioritizes points in the research region that are closest to the forecast locations.

4 Methodology

Reliable assessments of surface water quality and necessary modifications are made possible by the water quality indexes. The Weighted Arithmetic (WA) Water Quality Index (WQI) are important techniques to assess surface water's appropriateness for human consumption. The WHO standards suggested the use of the WQI for drinking water quality standards [16]. Overall, the WA-WQI provides a more precise and thoughtful weight for each parameter when determining the index value for every sample site. Following that, each surface water sample's WQI score was assigned to one of five water quality classes, as shown: < 25 represents excellent water; 25–50, as good; and 50–100, indicates poor drinking water. The key components in this method that represent the impact of different sources of pollution on the overall quality of the water fit for human consumption are weights. Additionally, it distils a sizable amount of data into the thorough and instructive value for every sample site.

For calculations of pollution score, a synthetic pollution index (SPI) approach has been adopted. The SPI was developed to quantify the effect of every single particle and is employed to determine the overall damage brought about by pollution. Thus, this technique aids in capturing the degree of disorderliness and randomness in the data [17]. It is an all-around pollution indicator that considers the cumulative impact of multiple contaminants for a specific application. The deviation brought on by human factors can be minimized, and the index weight is determined based on the degree of fluctuation of each index value. Relevant particles are included in the index values for the creation of an index for any specific function, such as drinking and irrigation [10]. Thus, it is a productive method of presenting probability and uncertainty. The water quality is divided into five grades based on SPI: 0–0.25 (excellent), 0.25–0.50 (good), 0.50–0.70 (poor) and > 0.75 (very poor).

Multi-criteria analysis is one of numerous tools and techniques, particularly CP versions, is frequently employed as a strategy that effectively integrates uncertainty, various sectors, various stakeholders, geographical and temporal concerns in poorly organized problems, and qualitative and quantitative criteria. It is commonly referred as a multi-criteria decision-making (MCDM) method, that uses user-defined criteria and entropy weights to create a weighted normalized matrix [12]. However, it is important to note that by evaluating each parameter's weight independently, the process of constructing WQI for water quality evaluation may be made simpler, which addresses both its advantages and disadvantages [18]. In addition, this mechanism is employed to demonstrate how the number of features used in the indicator division has increased.

5 Results and Discussions

The statistical factors of the water's physicochemical properties are reported and addressed in the sections that follow. Afterwards, using the GIS environment and the analytical data, the numerical spatial distribution of the parameters was generated. The IDW approach was then used to create the maps of the spatial distribution of the water quality parameters. It is significant to remember that techniques using the WA-WQI, SPI, and CP were also identified. The temperature varied from 25 to 30 °C with an average of 28.32 °C. The pH level has an impact on both chemical and biological processes, while temperature influences the concentration of oxygen available in the water. In this work, pH values range from 7.5 to 8.07, indicating slightly alkaline in nature. This range between (6.5 and 8.5) is typical of soil and water basins where human activity and natural processes have an impact. Also, EC helps determine the best irrigation water quality, nitrogen levels, and soil salinity for agricultural production. Therefore, it is essential for monitoring the health of aquatic ecosystems, detecting pollution, and evaluating water quality, especially in aquaculture. The EC of the surface water samples ranged from 152 to 310.6 μS/cm. The ideal value considered is 250 μS/cm. High EC values at Q-(1), (2) and (7), can be indicative of saline conditions or fertilizer application, which are critical factors for soil salinity and water quality management. The inorganic cation calcium is one of the most abundant in drinking water. In surface water that occurs naturally, it is a crucial parameter. In the investigated region, the Calcium (Ca^{2+}) concentrations exists between 29.74 and 66.35 mg/L, that are within the ranges, as reported by WHO standards (75 mg/L) of agricultural soils and water sources. These levels are crucial for preserving plant health and soil structure, and their moderate variability suggests normal nutrient dispersion. Similarly, Mg^{2+} ions are crucial for assessing soil fertility and optimizing plant nutrition, given magnesium's role in chlorophyll synthesis and enzyme activation. Similar studies have reported comparable magnesium levels, emphasizing the importance of maintaining optimal magnesium through soil management practices. With a mean value of 22.93 mg/L, Mg^{2+} varied from 18.80 to 29.98 mg/L. Evidently, 100% of the samples are below the allowable limit of 30 mg/L. Excessive sodium (Na^+) can result in problems with soil salinity, which will impact plant development and water absorption, emphasizing the need for cautious management. The sodium levels ranging from 5.39 to 15.34 mg/L, align with findings from both local and global studies. The acceptable value for Na^+ in this study is 200 mg/L. The decreased quantity of K^+ is a result of the greater durability of potash feldspar in the face of chemical weathering. Again, the increased concentration of both sodium and potassium in freshwaters is caused by sewage from homes. The concentrations ranged from 1.76 to 7.194 mg/L, averaging 3.16 mg/L, are consistent with agricultural studies that emphasize potassium's role in plant growth and stress resistance. In addition, NO_3^- can come from a variety of places, including fertilizers, livestock manure, and other non-point sources of pollution. Samples water had NO_3^- concentrations ranging from 0.65 to 3.53 mg/L. The element fluoride is present in the Earth's crust and is extremely harmful to aquatic life in freshwater environments. The concentration of F^- was varied between 0.30 and 1.16 mg/L. The chloride (Cl^-) in surface water usually comes from a combination of natural and man-made sources, including run-off, animal feed, and irrigation runoff. Indicators of sewage discharge are also crucial. In samples of this region, the concentration of Cl^- ranged from 8.20 to 23.54 mg/L. Almost all-natural water contains SO_4^{2-}, making it a major dissolved component of rain. The concentration of SO_4^{2-} in these 7 water samples, ranged from 7.89 to 40 mg/L. The leaching of gypsum causes sulphate to naturally occur in water. Managing sulphate levels effectively through suitable farming practices is essential to optimizing crop yields and quality while reducing environmental effects. The concentrations of PO_4^{3-} in the water samples varied from 0.07 to 0.29 mg/L. Sites Q-(1), (7) and (2) reflects higher values that exceeds the WHO limit of 0.1 mg/L, which indicates that these levels are caused by anthropogenic activities. Thus, variability in phosphate levels suggests differences in soil fertility and nutrient management practices. The primary contributors to alkalinity are rocks that contain CO_3^{2-} and HCO_3^- ions when dissolved in water. The alkalinity value ranged from 51.72 to 72.89 mg/L. Therefore, it quantifies the ability of water to act as a buffer. Another indicator namely, TDS, is thereby, composed of inorganic salts, trace amounts of organic matter, and dissolved substances in water. The concentration of Total Dissolved Solids (TDS) varied from 86.5 to 182.4 mg/L. Notably, sites such as Q-(1), (2) and (7) exhibits high TDS values, which can be attributed to the pollution of water caused by urban and agricultural runoff. In the natural surface water system, DO is a crucial parameter since it affects the aquatic organisms. Because of its critical function in maintaining aquatic life in moving water, it is vulnerable to even minor changes in its surrounding environment. A range of 6.785–7.67 mg/L was observed for DO. From the results, it is noticed that DO amount is sufficient throughout the study area.

From the physicochemical results, it is noticed that the controlling factor in the order of dominance of cations in the study area is $Ca^{2+} > Mg^{2+} > Na^+ > K^+$, whereas the anions is $SO_4^{2-} > Cl^- > NO_3^- > F^- > PO_4^{3-}$. The concentrations of Mg^{2+},

K^+, Na^+, Mg^{2+}, and Ca^{2+} exhibit an ascending pattern, primarily as a result of efficient leaching from the rock structure, in addition to human activities. Na^+ is derived from rocks containing plagioclase, while potassium is obtained from K^+-feldspars or minerals such as orthoclase, microcline, and biotite found in metasediments. Mg^{2+} is obtained from calcium magnesium silicates, primarily sourced from plagioclase feldspars. Moreover, the study area lacks any naturally occurring sources of sulphate, while it is well-established that $SO_4{}^{2-}$ behaves as a dominant anion, that comes from car emissions, phosphatic fertilizers in urban areas, and industrial effluents. This may lead to lower crop quality and yield, which calls for cautious management techniques. The use of suitable irrigation techniques and water treatment technologies like desalination are examples of mitigation options. A vital component of evaluating risks to human health and agricultural sustainability is surface water quality monitoring. Overall, effective management of Ca^{2+} and $SO_4{}^{2-}$ in surface water is necessary to keep agricultural output high and drinking water safe. The spatial distribution map of all the selected water quality indicators showed that good drinking water quality was distributed in most of the study area (Fig. 2a–n).

The computed WA-WQI value in the study area ranges between 49 and 72, during the chosen period. The highest value of WA-WQI has been found at the site Q-(1) and the lowest value of WA-WQI has been observed at a site Q-(5). It is clear from the calculated WA-WQI, that most of the samples fall into the category of poor drinking water, which covers about 57.145 of the total samples which are not healthy for drinking, whereas, 42.86% of the samples belongs to good. The poor water exists within the upstream and downstream region. And this might be due to excessive TDS, Ca^{2+}, Mg^{2+}, $SO_4{}^{2-}$, Cl^- and $NO_3{}^-$. As seen in Fig. 3a, the geographical and temporal patterns of WA-WQI are shown across the research region, WA-WQI maps demonstrate that the vast majority of the study areas have WA-WQI > 50 mg/L, and are not suitable for drinking. In contrast, the distribution map displays those 4 locations of the present study region comes under the vulnerable zones which includes stations like Q-(2), (1), (3), and (7), respectively. At these locations, some parameters show higher readings. Magnesium (Mg^{2+}) and Calcium (Ca^{2+}) mostly enter water via leaching from rocks; calcium is frequently found in streams that pass-through minerals and rocks high in calcium, such as limestone, gypsum, and dolomite. High Na^+ values will lessen the permeability of the soil, causing salts to concentrate at the surface and preventing plant development [19]. It is noticed that poor water exhibits are because of industrial and domestic waste water, soil erosion and weathering, leaching, organic and natural pollution, were identified as critical sources contribution to pollution of river water.

The results of the calculated water quality indices for all 7 monitoring stations along the Brahmani River Catchment, Odisha, are presented below. A water's suitability for different uses, such as irrigation or drinking, can be determined using chemical parameters and indices. Nonetheless, managers and administrative organizations may find it useful to rate water based on its quality and consumption using specific characteristics when making decisions. The SPI evaluation has noteworthy outcomes, as seen in Fig. 3b. The river water's computed SPI value was discovered to be within the range of 0.31–0.68. According to the interpolated results of SPI

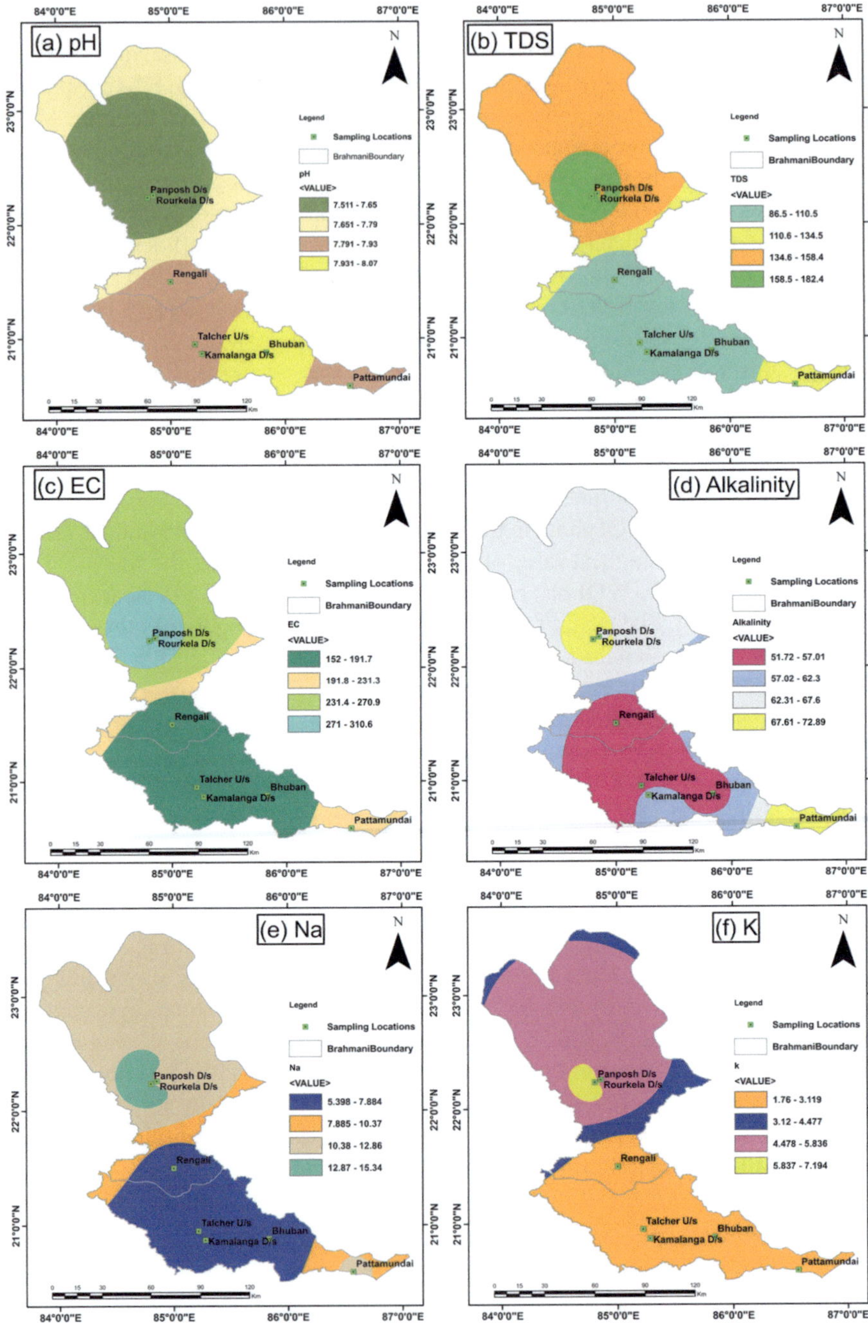

Fig. 2 Spatial distribution of the parameters: **a** pH, **b** TDS, **c** EC, **d** alkalinity, **e** Na^+, **f** K^+, **g** Ca^{2+}, **h** Mg^{2+}, **i** F^-, **j** Cl^-, **k** NO_3^-, **l** SO_4^{2-}, **m** DO and **n** PO_4^{3-}

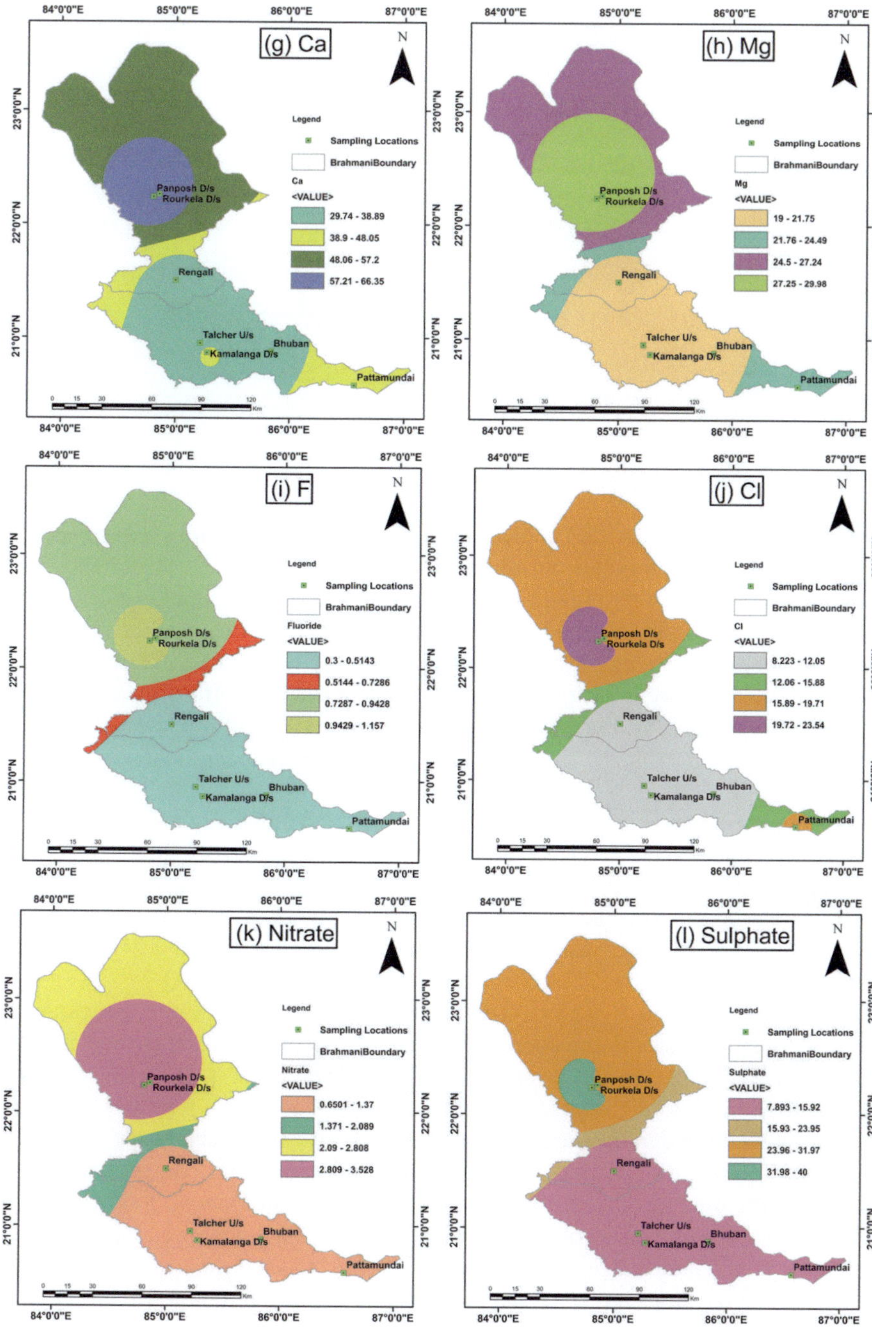

Fig. 2 (continued)

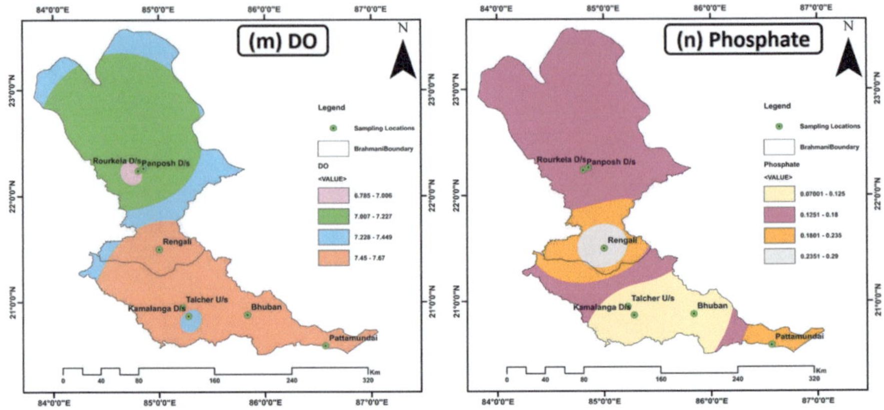

Fig. 2 (continued)

method, 42.9% were classified in the good class and 57.1% locations were placed in the poor class, thus, promotes unsuitability for drinking purposes. The method was graded under the category of good-poor quality scale. The sampling locations Q-(4), (5), and (6) were graded as good, by the quality scale. The quality of the remaining riverside sampling sites was low. The primary cause of elevated EC, TDS, and other water parameters may be attributed to leaching of elements from rocks, particularly gypsum rocks [20]. Furthermore, elevated Na may result in increased sodium adsorption and ion exchange into clay particles, hence reducing soil permeability and drainage.

Developing accurate predictive models is essential to reducing the costs related to evaluating and analyzing surface water quality for drinking reasons. In this work, CP techniques—which have demonstrated notable advantages in water quality estimation—were used to assess the overall performance of this cutting-edge research. The Calculated CP of the river water (Fig. 4), varied between 0.50 and 0.94. The value obtained by the respective CP scores are as follows: Q-1 (0.941), Q-2 (0.813), Q-3 (0.549), Q-4 (0.501), Q-5 (0.559), Q-6 (0.504) and Q-7 (0.649). From the results, it is cleared that the water is considered as polluted at Q-(1), which is termed as most polluted station, and contains a rank of 1, by acquiring highest CP score of 0.941. It is primarily due to the effects of human activity. The rise in Cl^- and SO_4^{2-} at the chosen location demonstrates how evaporation affects the chemistry of surface water. The main cause of the significant evaporation was the rise in water level brought on by prolonged irrigation. The 2nd and 3rd polluted site is Q-(2) and (7). It is evident that the chemistry of surface water in the research area was influenced by both natural and anthropogenic causes simultaneously. According to this study, the water in these locations has poor quality, very bad quality, and unfit for agricultural and human consumption. In addition, WA-WQI > 50 and SPI > 0.70 is also noticed, which indicated that the drinking water's quality has declined. However, CP method gives them their total ranks, which makes the relative pollution level easy to identify. Policymakers from both technical and non-technical backgrounds can make good

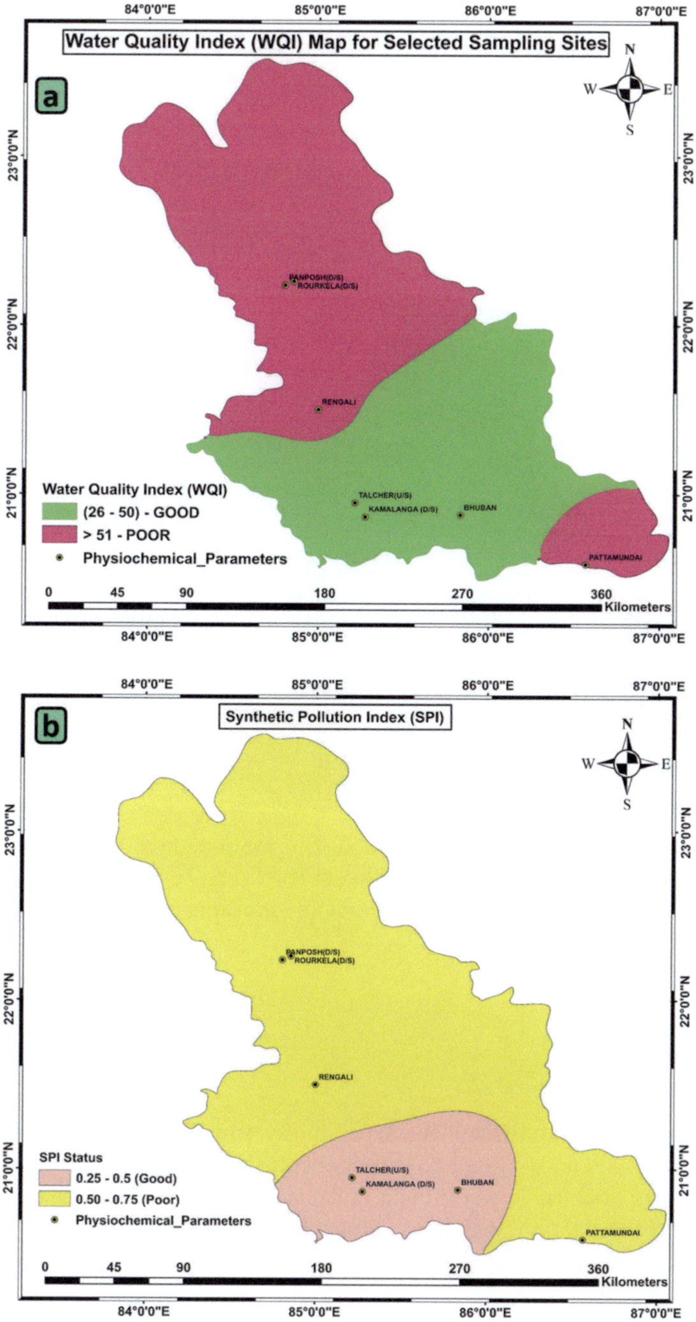

Fig. 3 Layout of interpolated results for **a** WA-WQI, and **b** SPI

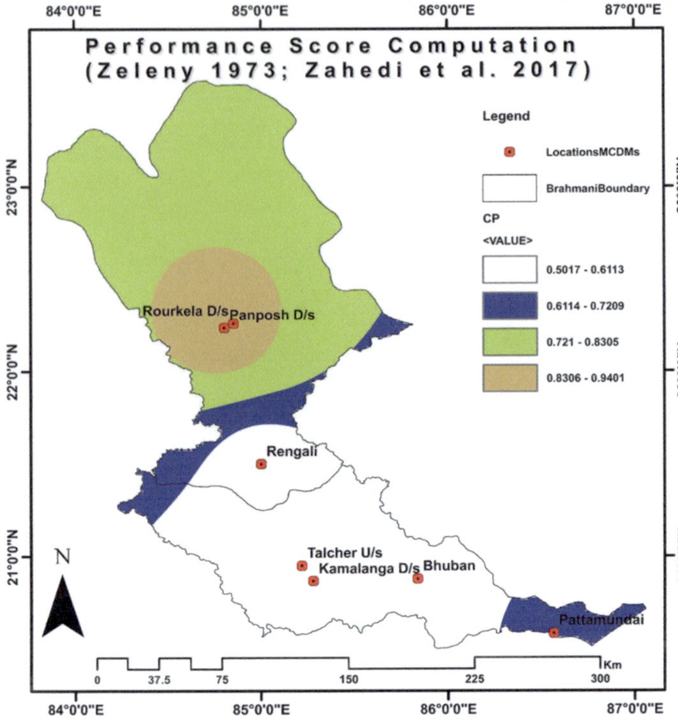

Fig. 4 Layout of interpolated results for CP method

decisions with the help of such a ranking. So, the document's recommended CP version demonstrates that soil water pollutant levels can be accurately assessed, and it also eliminates the need for laborious computations that come with the usual water quality index (WQI).

6 Conclusions

This investigation demonstrated the accessibility and dependability of employing these WA-WQI, SPI and CP methods in conducting research on surface water quality. High quantities of chloride were found in the surface water according to the hydrogeochemical investigation, which was probably caused by human activity, seawater incursion, or salt dissolving. Significant but varied correlations among chemical parameters were observed, with WQ analysis indicating strong inter-parameter relationships. According to WA and SPI, it is observed that high WA-WQI and SPI values at Q-(1) were due to the high value of EC, TDS, Cl^-, SO_4^{2-}, NO_3^- and Ca^{2+}. This could be ascribed to seasonal variations, agricultural practices, and other human-induced activities. Moreover, CP can provide ranks for the calculation and improve the accuracy. It is noticed from the CP results, that the water quality at Q-(9) is considered as the most polluted place, because it is impacted by over-extraction of surface water in the research area as well as industrial, anthropogenic, and agricultural activities. It is recommended that to develop sophisticated modelling methods to forecast future surface water accessibility and detect regions experiencing excessive exploitation or contamination. Investigate alternative water sources, such as treated wastewater, and brackish water, for non-potable purposes. Public awareness campaigns regarding the detrimental effects of subpar water quality on human wellness and agricultural areas. The limitation of this study is: Field sampling limitations, such as accessibility to certain locations or constraints in sampling frequency, could influence the representativeness of water quality assessments across the study area. The recommendation can be described as: Expand the spatial coverage of groundwater quality assessments to encompass regions with varying hydrogeological characteristics. This approach will facilitate comparative analyses and improve the applicability of findings to broader environmental contexts.

Acknowledgements The present research work is supported under the C.V. Raman Global University (CGU), Bhubaneswar, Odisha, India. The author acknowledges Ranu – Prativa Research Group for this idea of study. Constructive reviews by two anonymous reviewers helped in improving the manuscript. The author again expresses their sincere gratitude towards Dr. Bhagirathi Tripathy, IGIT, Sarang, Odisha, for his support and appreciation, ongoing assistance within the research.

References

1. Nong, X., Shao, D., Zhong, H., Liang, J.: Evaluation of water quality in the south-to-north water diversion project of China using the water quality index (WQI) method. Water Res. **178**, 115781 (2020)
2. Haque, M.M., Niloy, N.M., Nayna, O.K., Fatema, K.J., Quraishi, S.B., Park, J.H., Kim, K.W., Tareq, S.M.: Variability of water quality and metal pollution index in the Ganges River, Bangladesh. Environ. Sci. Pollut. Res. **27**, 42582–42599 (2020)
3. Ewaid, S.H., Abed, S.A., Al-Ansari, N., Salih, R.M.: Development and evaluation of a water quality index for the Iraqi rivers. Hydrology **7**(3), 67 (2020)
4. Gupta, D., Shukla, R., Barya, M.P., Singh, G., Mishra, V.K.: Water quality assessment of Narmada River along the different topographical regions of the central India. Water Sci. **34**(1), 202–212 (2020)
5. Matta, G., Kumar, A., Nayak, A., Kumar, P., Pant, G.: Pollution complexity quantification using NPI and HPI of River Ganga system in Himalayan Region. Proc. Indian Natl. Sci. Acad. **88**(4), 651–663 (2022)
6. Nandi, I., Pokharia, C., Chaturvedi, V., Shah, K.: Developing a robust tool for quality health assessment of rivers using optimized weighted arithmetic water quality index method: a study on River Ganges at Varanasi in India: health assessment of River Ganga using WAWQI method. Int. J. Res. Appl. Sci. Biotechnol. **9**(3), 7–20 (2022)
7. Zeleny, M.: Compromise programming. In: Cochrane, J.L., Zeleny, M. (eds.) Multiple Criteria Decision Making. University of South Carolina Press, Columbia, South Carolina (1973)
8. Zahedi, S.: Modification of expected conflicts between drinking water quality index and irrigation water quality index in water quality ranking of shared extraction wells using multi criteria

decision making techniques. Ecol. Indic. **83**, 368–379 (2017). https://doi.org/10.1016/j.ecolind.2017.08.017
9. Sharma, G., Lata, R., Thakur, N., Bajala, V., Kuniyal, J.C., Kumar, K.: Application of multivariate statistical analysis and water quality index for quality characterization of Parbati River, Northwestern Himalaya, India. Discov. Water **1**, 1–20 (2021)
10. Trombadore, O., Nandi, I., Shah, K.: Effective data convergence, mapping, and pollution categorization of ghats at Ganga River Front in Varanasi. Environ. Sci. Pollut. Res. **27**, 15912–15924 (2020)
11. Akhtar, N., Ishak, M.I.S., Ahmad, M.I., Umar, K., Yusuff, M.S.M., Anees, M.T., et al.: Modification of the water quality index (WQI) process for simple calculation using the multi-criteria decision-making (MCDM) method: a review. Water **13**(7), 905 (2021)
12. Wanguba, B., Siriba, D.N., Okumu, B.M.: GIS-based multi-criteria decision analysis model for utility water demand: the case of Lodwar Municipality, Turkana County, Kenya. Heliyon (2024)
13. Kumar, P., Dasgupta, R., Dhyani, S., Kadaverugu, R., Johnson, B.A., Hashimoto, S., et al.: Scenario-based hydrological modeling for designing climate-resilient coastal water resource management measures: lessons from Brahmani River, Odisha, Eastern India. Sustainability **13**(11), 6339 (2021)
14. Swagatika, S., Paul, J.C., Sahoo, B.B., Gupta, S.K., Singh, P.K.: Improving the forecasting accuracy of monthly runoff time series of the Brahmani River in India using a hybrid deep learning model. J. Water Clim. Change **15**(1), 139–156 (2024)
15. APHA (American Public Health Association): Standard Methods for the Examination of Water and Wastewater, 23rd edn. American Public Health Association, Washington, DC (2017)
16. WHO: WHO Guidelines for Drinking-Water Quality, vol. 1. World Health Organization (2008). Available online at: https://www.who.int/water_sanitation_health/dwq/fulltext.pdf
17. Ribas, J.R., Severo, J.C.R., Gonçalves, L.C.: A fuzzy multicriteria approach to estimate the water quality index of hydroelectric reservoirs. Water Resour. Manage. **38**(6), 2155–2170 (2024)
18. Ribeiro, V.H.A., Reynoso-Meza, G.: Multi-criteria decision-making techniques for the selection of Pareto-optimal machine learning models in a drinking-water quality monitoring problem. Int. J. Inf. Technol. Decis. Mak. **23**(01), 447–474 (2024)
19. Das, A.: Multivariate statistical approach for the assessment of water quality of Mahanadi Basin, Odisha. Mater. Today Proc. **65**, A1–A11 (2022)
20. Das, A.: An Integrated Assessment of Spatial Variability Mapping of Surface Water Properties and Its Impacts on Drinking Water in Mahanadi River Basin, Odisha. Green Energy and Environmental Technology, India (2024)

The Issue of Qualitative Water Treatment in Eastern Donbas Region

Vladimir Dmitrienko, Mikhail Lemeshko, Evgenia Kostromina, and Victoria Pushkina

Abstract The report highlights a consistent rise in both the volume of water consumed and the amount of wastewater discharged. The contamination of surface water sources for the population by anthropogenic factors has reached critical levels, making it difficult to implement effective water treatment. The majority of reservoirs in the European part of Russia have been classified as "polluted", "very polluted" or "dirty" due to the presence of pollutants. Following the cessation of lengthy coal mining operations in the Eastern Donbass, the flooding of the techno gene space with underground waters has created a need for the organization of a program for the removal of polluted mineralized waters. The current technological infrastructure of treatment facilities is unable to achieve the necessary reduction in pollution levels due to the high volumes of pumped mine water and the significant contamination with dissolved substances. This study demonstrates the potential for effective post-treatment of mine waters using a floating aquatic plant, Eichhornia, as a natural filtration solution. One of the potential solutions to this issue is the utilization of the heat generated by pumped underground waters. This is demonstrated by the example of the initial pond at the Deep mine treatment facilities in the town of Shakhty. A regression equation was developed to assess the decrease in the temperature of the settling pond based on a series of experimental studies conducted over several years and their subsequent statistical evaluation. The equation was used to predict the conditions of Eichhornia vegetation in winter.

Keywords Mine water · Waste water · Pollution · Chemical composition · Waste water post-treatment · Eichhornia · Vegetation

V. Dmitrienko (✉) · M. Lemeshko · E. Kostromina
Institute of Service and Business (Branch) of DSTU in Shakhty, Shakhty, Rostov Region, Russia
e-mail: vadmitrienko@rambler.ru

V. Pushkina
Automobile Road Institute (Branch) of SRSPU (NPI) named after M.I. Platov, Shakhty, Rostov Region, Russia

Y. Zhu (ed.), *Proceedings of the 4th International Symposium in Environmental Science and Industrial Ecology*, https://doi.org/10.1007/978-981-96-1578-0_21

1 Introduction

Water is a highly valuable natural resource in the present day. The amount of water consumed by the population for various household and industrial purposes is significant, and it is increasing annually [1]. The volume of fresh water on Earth contained in surface sources, primarily represented by rivers, constitutes a mere 0.01% of the total volume [2]. In many regions of the world, the degree of sanitary suitability and availability of sources may vary due to the significant influence of natural and man-made factors on the most important hygienic characteristics [3].

The ongoing urbanization of cities, growth in industrial production and intensification of agricultural activity are accompanied by a constant increase in both consumption and wastewater discharge. Furthermore, the contamination of both surface and groundwater by human activities is increasing at such a rate that current water treatment methods cannot guarantee the production of water of the required quality [4].

In the Rostov region, which has historically been relatively successful in providing the population with water, there has been a significant deterioration in water quality over the past 40–50 years. In the 1960s, water from even small rivers was used in certain cases for cooking purposes. The banks of these rivers were home to numerous springs with water of exceptional quality. Currently, the concentration of pollutants in many of these water sources exceeds the maximum permissible concentration (MPC).

The reduction of anthropogenic pollution will only be achieved through the rational use of water resources, an increase in the volume of closed-loop industrial enterprises and an improvement in the quality of wastewater treatment [3].

2 Relevance of Small Rivers Pollution

The quantity of clean drinking water remaining on Earth is diminishing, with the number of such sources declining annually. In the majority of cases, surface and underground waters are used to provide potable water to the population [4].

Russia is one of the most water-rich countries in the world. The Volga, Don, Ob, Yenisei, Lena and Amur River basins provide a reliable water supply to a significant portion of the country [4]. Nevertheless, if the degree of contamination in Siberian rivers has not yet reached a critical threshold, then in the European part of Russia the majority of rivers are classified within the categories "polluted", "very polluted" and "dirty". Furthermore, the majority of pollution is anthropogenic in origin. This indicates that human activity is the primary cause of damage to the country's ecological systems and the deterioration of human health [5].

The majority of domestic wastewater, following treatment, and in some cases without it, is discharged into small rivers, which convey their waters to larger rivers and ultimately discharge into riverbeds that provide water to the population.

The annual growth in water consumption is in excess of 5%. In consequence, the discharge of wastewater, mine, quarry and collector-drainage water into surface water bodies is also increasing [5].

The majority of wastewater treatment technologies currently in use are designed to remove hazardous substances. However, there is limited attention paid to reducing salt discharges, as the removal technologies, such as reverse osmosis, require significant capital and operational costs. As a result, the quality of water from water supply sources in the Rostov Region is declining.

The proportion of water samples from surface sources of centralized water supply that fail to meet the requisite sanitary standards for sanitary and chemical indicators over the past three years has reached 20% for surface water samples and 70% for groundwater samples [5].

The Rostov region is distinguished by a high level of long-term coal mining, as well as the liquidation and conservation of exhausted and unprofitable mines. This has resulted in a significant and active impact on the state of the subsoil of the Eastern Donbass. The extensive closure of coal mines has resulted in a significant exacerbation of environmental issues within the region, both during the liquidation of the mines and following their closure [6].

The liquidation of mines in the Rostov region using the "wet" method has resulted in the flooding of 95% of the techno gene space of the Eastern Donbass with underground acidic waters (pH 5–6) and mineralization of up to 20 g/dm^3 with a high content of sulfates and iron [5, 7]. The maintenance of the design level of flooding of mines is ensured by drainage systems, which include treatment facilities. However, the efficiency of these treatment facilities does not ensure compliance with the regulatory parameters of water quality before discharge into water bodies.

The total inundation of the mine excavation site, which resulted in the cessation of underground operations, gave rise to significant environmental concerns for both the local population residing in the mining region and the surrounding ecosystem. This was due to the environmental consequences of the flooding, which had a considerable impact on the surrounding ecosystem [8]. In particular, the rise in the level of mine waters represents a novel and significant anthropogenic factor in the contamination of the regional hydrographic network and fertile soils [9, 10]. In this context, the enhancement of wastewater treatment quality is becoming a matter of significant importance.

It is widely acknowledged that the introduction of additional purification processes, facilitated by the growth of higher aquatic plants, represents an effective strategy for the reduction of wastewater pollution [11], the most straightforward method for addressing this issue is to utilize flora indigenous to the region. However, the disposal of these materials presents a significant challenge due to the accumulation of harmful substances [12]. Furthermore, the efficiency of the process of removing salts from water is not particularly high [13].

It is our contention that the ecological safety of small rivers in the Shakhty area can be enhanced through the implementation of phyto purification techniques, specifically the utilization of the floating plant Eichhornia (water hyacinth), for the remediation of waste and groundwater [14]. This assertion is predicated on the premise that water hyacinth is capable of degrading the majority of pollutants rather than accumulating them, which significantly streamlines the disposal of biomass. Nevertheless, the specific requirements of Eichhornia with regard to vegetative conditions preclude the possibility of its preservation at low temperatures [15, 16]. This necessitates the formulation and substantiation of technical solutions for the preservation of plants during the winter period.

3 Methods and Means of Research

The preceding laboratory studies [17] permitted the determination of the minimum temperature requisite for the preservation of water hyacinth during the winter season. It can therefore be concluded that in order to ensure the reliable preservation of Eichhornia in the conditions of the Rostov region, it is necessary to carefully substantiate the temperature regime of reservoirs intended for the storage of plants at low positive water temperatures.

During the investigation of the geographical area surrounding the treatment facilities, a series of measurements were taken to ascertain the prevailing wind speed, temperature and air humidity. These were recorded using an ATE-1038 electronic anemometer, which also had the functionality of storing the data internally. The data was subsequently processed and analyzed in the laboratory. The aforementioned indicators were periodically monitored utilizing an MV-4M aspiration psycho meter and a mercury thermometer with a division value of 0.1 °C.

The temperature of the water in the settling tanks was gauged at a depth of 3–4 cm using UT71C multi meters with DTPL011-0.3/1.5 temperature converters, with the results recorded in the memory of the multi meters. The temperature converters were suspended from a line at the end of a 4.5-m-long telescopic rod. The thermocouples were situated at distances of 0.03 m and 0.10 m, respectively, from the float. The thermocouples were connected to the multi meters via a temperature compensation cable with an 8-m length. The configuration of the sensors permitted alterations to be implemented at disparate locations along the surface of the mirror as the apparatus traversed the dam between the ponds.

A special stationary complex with remote DS1820 sensors was employed for the continuous monitoring of air temperature. This allowed readings to be taken at intervals of 1–60 s, stored in a database, and visualized on a computer screen located in the institute's laboratory. The analysis of the daily thermo grams (86,400 s) (Fig. 1) enabled the precise determination of the time of measurement at the treatment facilities. In other words, the temperature measurements taken in the settling ponds were conducted during periods with comparable values of air temperature and humidity, wind speed, and cloud cover.

This is essential to mitigate the impact of diurnal temperature fluctuations and the thermal inertia of the ponds, which can otherwise lead to errors in data interpretation. Given that solar radiation contributes to an increase in the temperature of the settling tanks, and the objective is to maintain a minimum permissible temperature

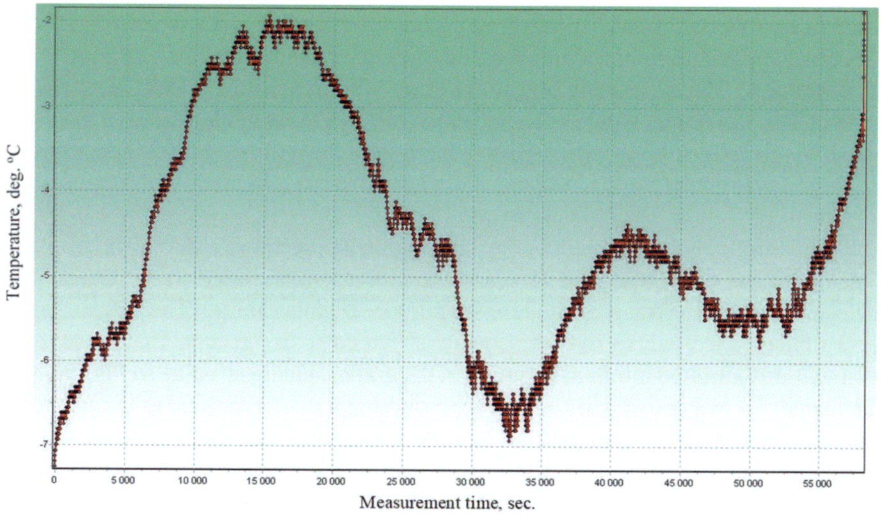

Fig. 1 Thermogram of outside air on January 26–27, 2021

for Eichhornia, it is essential to ascertain the extent of the potential decrease in water temperature. It was thus resolved that the research should be conducted exclusively in conditions of cloud cover. This approach enables the minimization of the impact of abrupt temperature fluctuations (Fig. 2) during the process of clearing. It was thus resolved that the research should be conducted exclusively in cloudy weather.

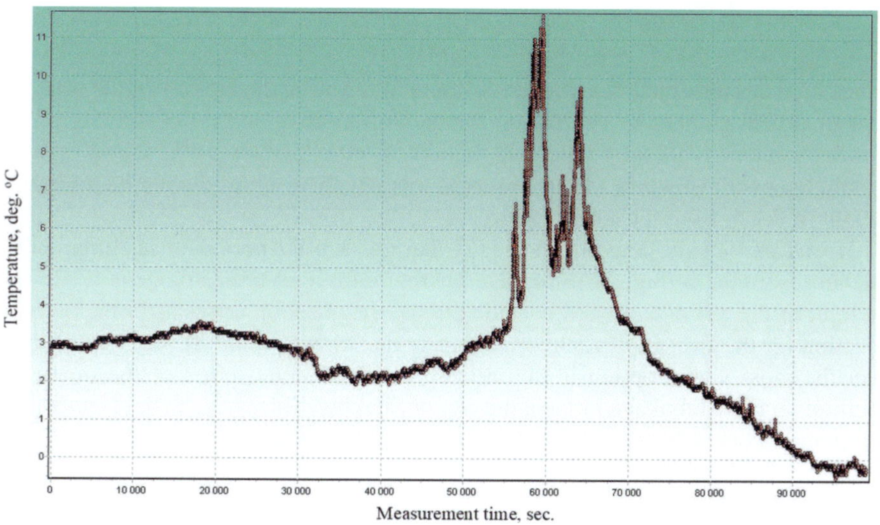

Fig. 2 Thermogram of outside air on December 21–22, 2020

Prior to undertaking any measurements, all electronic devices were subjected to a rigorous examination in accordance with the established protocols, utilizing reference samples. The measuring and monitoring complex with DS1820 temperature sensors was calibrated using a reference mercury thermometer with a division value of 0.1 °C, given that the specified sensors have an accuracy of ± 0.5 °C. The resulting correction was entered into a bespoke software program designed for the calculation of temperatures in a spreadsheet environment.

The methodology employed for the measurement process was as follows: In accordance with the meteorological forecast, the measurement range was selected. The measuring apparatus was activated for the purpose of data capture. The thermo grams were subjected to analysis over the course of two days. In the event that the average temperature exhibited minimal variation in the graphs, the mean value for the day was calculated. To this end, a daily array of measurements was recorded in the database, typically from 8:00 the previous day to 8:00 the current day, and the mean temperature was calculated in a spreadsheet. The time of measurement at the treatment facilities was approximately determined based on the rate of temperature change.

This sequence of research was adopted on the basis of preliminary studies and analysis of the influence of various factors on the objectivity of measurements.

The values obtained for air and water characteristics as a result of measurements were recorded in a table. Once the measurements were complete, this data was processed in a spreadsheet to assess the influence of various factors on temperature in settling ponds or in the Statistics program when a regression equation was being compiled.

4 Research Results

It has been demonstrated that the temperature of the air in the vicinity of the wastewater treatment plant is susceptible to fluctuations contingent on the velocity of the wind, the intensity of solar radiation, and the specific location of the measurements within the pond. However, it was observed that negative values did not manifest at a height of 0.1 m from the surface mirror.

It was previously demonstrated [17] that for a brief period of acclimation of Eichhornia in the spring, the optimal temperature for long-term storage is 17–18 °C. In light of the aforementioned considerations, the question arises as to the optimal location for the storage of Eichhornia during the winter period. It is our contention that the most viable option is the utilization of the initial pond of the treatment facilities at the Glubokaya mine in the city of Shakhty. This is due to the fact that the temperature of the water extracted from the mine workings is, on average, 19.1 °C throughout the year.

However, the results of the chemical analysis of mine waters indicated a notable excess of certain substances in the initial three ponds. In particular, the concentration of iron was found to be significantly higher than the permissible level for water hyacinth. Concurrently, the mine water is devoid of essential nutrients, including nitrogen, phosphorus, potassium and sodium. It is therefore necessary to plant the plants in containers that are isolated from the water of the settling pond.

Notwithstanding the considerable volume of water being pumped out at a rate exceeding 900 m^3 per hour during the coldest periods of January, the temperature at the outlet of the treatment facilities was recorded at + 5.2 °C. It is important to note that the long-term storage of water hyacinth, even at temperatures between 12 and 13 °C, has a significant inhibitory effect on its growth, with an adaptation period that can extend beyond one month [17]. It is therefore recommended that water hyacinth be planted in sealed floating containers with protection from air currents, and placed in the initial pond. This approach allows for an extension of the active growth period of the plant, with the planting of mature specimens in the post-treatment facilities during the spring at a relatively low financial cost.

In order to prevent plant death, it is essential to evaluate the impact of fluctuating water temperatures in the first pond under a range of meteorological conditions. This will facilitate the reliable storage of Eichhornia in floating pools during the winter period. The objective of this research is to develop a model for assessing the temperature reduction (Δt) of the initial settling pond of a treatment facility, based on experimental studies.

Given that the temperature at the entrance to the first settling pond of the Deep mine, even at temperatures below − 15 °C, decreases from 19.1 to 18.9 °C, it can be assumed to be constant at an average of 19.0 °C.

The level of the bypass pipes into the second settling tank remains constant, thus the volume of water and the surface area of the pond remain constant. Consequently, the primary variables influencing the convective heat exchange with the surrounding environment and the associated evaporation heat costs will be wind speed (V), temperature (T) and air humidity (W).

The weather in our region is characterized by significant instability from November to March. It is challenging to identify periods with comparable meteorological conditions and to incorporate the potential for on-site observations. Accordingly, the sample for analysis was formed over a period of four years. The temperature and humidity speed measurements were conducted using a single device at four points located at the corners of the pond. Additionally, measurements of the water and air temperature at the surface mirror were taken at five points. A further point was included in the centre of the pond.

The mean values of the air environment characteristics and water temperature fluctuations were calculated based on the data obtained. The resulting calculations were then entered into the sample array. The statistical indicators of the array were evaluated using the Statistics program and are presented in Table 1.

The results of the correlation analysis indicate that there is a statistically significant relationship between the ambient air temperature and wind speed and the change in temperature in the pond. However, the correlation level for humidity is relatively low at 0.042. It can therefore be surmised that this factor can be excluded when creating the model.

Table 1 Results of the correlation analysis of variables

Parameters	T (°C)	V (m/s)	W (%)	Δt (°C)
T (°C)	1.000000	– 0.260312	– 0.440174	– 0.706023
V (m/s)	– 0.260312	1.000000	– 0.218916	0.686740
W (%)	– 0.440174	– 0.218916	1.000000	0.041898
Δt (°C)	– 0.706023	0.686740	0.041898	1.000000

The influence of air temperature (T, °C) and wind speed (V, m/s) on the temperature differential between the water entering and leaving the pond (Δt, °C) can be evaluated through the utilization of the response surface depicted in Fig. 3.

It can thus be concluded that both variables (air temperature and wind speed) are significant and exhibit a weak correlation with one another.

A regression equation was derived from the calculated results using a linear formulation, with an approximation reliability level of $R^2 = 0.7698$.

$$\Delta t = 1.410 - 0.079 \cdot T + 0.240 \cdot V \tag{1}$$

The analysis of the utilization of non-linear dependencies does not result in a notable enhancement of reliability; however, it does introduce complexities with regard to the potential for transformation of the model in relation to other variables. Therefore, the derived Eq. (1) accurately represents the initial data set with a high

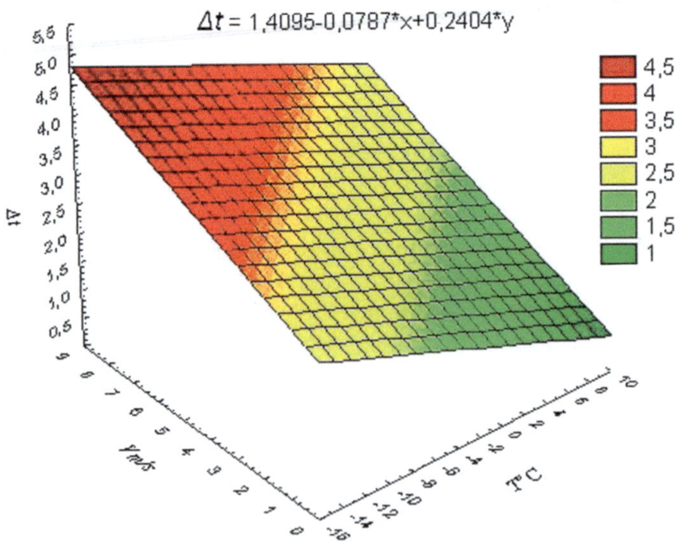

Fig. 3 Graph of the dependence of the decrease in water temperature at the outlet of the first pond on the temperature and air velocity

degree of reliability, and can therefore be used to predict the thermal regime of the settling pond.

The provided equation enables the forecasting of a decline in pond water temperature based on meteorological data, thereby facilitating the implementation of measures to avert the demise of plants. Furthermore, it is possible to ascertain critical values of air temperature or wind speed for the safe storage of plants.

5 Discussion of Research Results

Notwithstanding the considerable construction of new treatment facilities and the reconstruction of existing ones in the Eastern Donbass, the pollution of small rivers remains at a high level. The results of the chemical analysis indicate that the majority of the purified household and groundwater discharged into the environment is classified as "insufficiently purified". The total discharge of mine waters in the Shakhty area is 1500 m^3/h. This equates to a potential pollution load of over 100 tons per day.

One potential solution to this problem is phyto purification. The topography of the majority of treatment facilities allows for the construction of ponds for supplementary purification at a minimal cost.

In Russia and globally, a substantial corpus of evidence has been amassed on the enhancement of wastewater treatment efficacy through the deployment of higher aquatic plants [15]. In particular, the positive effect of using Eichhornia has been observed; however, there is a paucity of information regarding its storage in natural reservoirs during the winter months. The primary issue is the potential for plant mortality when the water temperature declines. It is evident that in the majority of instances, the temperature within the primary purification tanks of both domestic and mine waters aligns with the optimal conditions for the normal growth of water hyacinth. Nevertheless, at this stage of purification, the concentration of contaminants is markedly in excess of the permissible value of [16]. Concurrently, the subsequent stages of purification are characterized by a reduction in water temperature to levels that are detrimental to Eichhornia [17].

The conducted studies and the proposed equation for estimating the decrease in the temperature of the water of the settling pond of the "Glubokaya" mine make it possible to predict the storage conditions of Eichhornia in winter.

The principal disadvantage and limiting factor of utilizing analogous studies is the restricted applicability, confined to a singular water body. Attempting to estimate the temperature reduction of the entire complex of treatment facilities, even within a single site, is accompanied by an increase in the deviations between the calculated and actual values of heat loss due to the influence of a multitude of factors.

At the time of writing, the level of pollution of water sources in our region has not yet reached critical values. However, the environmental situation is deteriorating year on year. In making pivotal decisions, the findings of the aforementioned studies will enable us to justify the potential risks associated with investing in the construction of wastewater phyto purification facilities for the removal of salts.

6 Conclusions

The eastern region of Donbass was distinguished by the prevalence of minefields with high water content. Following the cessation of coal mining operations and the inundation of the mountainous region, there was a risk of flooding in certain areas. This necessitated the implementation of a dewatering strategy.

The mineralization of the mine waters is notably elevated, accompanied by a considerable quantity of dry residue even following purification. The technological schemes employed for the remediation of groundwater in the region are designed to facilitate the removal of a range of potentially harmful pollutants. A considerable quantity of dissolved substances is discharged into minor watercourses.

It is possible to reduce the anthropogenic load by subjecting the wastewater to additional purification with the use of higher aquatic plants. However, the climatic conditions of the Rostov region preclude the full implementation of phyto purification. This issue can be solved by utilizing the heat of pumped groundwater.

In order to substantiate the possibility of storing water hyacinth in winter in the territory of the Eastern Donbass, an Eq. (1) was developed based on experimental studies to estimate the decrease in temperature of the settling pond of the former mine "Glubokaya". This will facilitate the prediction of conditions conducive to the survival of plants during the winter months and the reduction of the risk of mortality.

References

1. Larionov, V., Sheremetyeva, E.: The current state of world water resources and main directions for increasing their availability. Bull. Irkutsk State Univ. Econ. **25**(4), 590–596 (2015)
2. Afanasyev, V.: "Clean water" and dirty "water resources". Clean Water Probl. Solut. (1), 13–18 (2009)
3. Venitsianov, E.V., Pryazhinskaya, V.G.: Water for life—in the world and in Russia. Rationalization of water resources management for sustainable development. Ecol. Life (12), 72–76 (2012)
4. Melnikov, O., Larionov, V., Alabuzhev, D., Romanteev, P.: Water resources as an object of modern economy. Water Chem. Ecol. (4), 23–27 (2009)
5. Ecological Bulletin of the Don: On the State of the Environment and Natural Resources of the Rostov Region in 2023. Under the General Editorship of M.V. Fishkin, 369 p. Ministry of Natural Resources and Ecology of the Rostov Region, Rostov-on-Don (2024)
6. Kurakov, Yu., Kukhtin, V., Suvorov, V.: Impact of mine closure on the environment of mining towns and villages. GIAB (6), 123–123 (2004)
7. Molev, M., Maslennikov, S., Zanina, I.: Environmental Safety of Coal Mining Regions, 113 p. ISOiP (Branch) of DSTU in Shakhty, Shakhty (2018)
8. Kulikova, A., Sergeeva, Yu., Ovchinnikova, T., Khabarova, E.: Formation of mine waters and analysis of their purification methods. Min. Inf. Anal. Bull. (7), 135–145 (2020)
9. Khramtsova, T., Stom, D., Benefit, V.: Use of macrophytes for additional treatment of municipal wastewater. Probl. Ecol. (2), 260–262 (1995)
10. Demkin, V., Kharionovskiy, A., Gusev, N., Schastlivtsev, E., Kaplunov, V., Kirsh, Y.: On the problem of complex processing of mine waters. Kuzbass-1. Min. Inf. Anal. Bull. (7), 209–219 (2009)

11. Kaplunov, Yu., Limansky, A., Gusev, N.: Analysis of the state of the problem of protection and rational use of water resources in the context of restructuring of the coal industry. Monit. Sci. Technol. **1**(6), 27–34 (2011)
12. Krot, Yu.: Use of higher aquatic plants in biotechnologies for surface and waste water treatment. Hydrobiol. J. **42**(1), 76–91 (2006)
13. Stepanova, S., Simonova, G.: Water hyacinth—a natural water purifier. Bull. SGUGiT **24**(1), 264–267 (2019)
14. Kholodova, S., Rudikov, D.: On the possibility of using water hyacinth to purify polluted water. Water Ecol. Probl. Solut. **3**(79), 70–76 (2019)
15. Abuova, G., Kharlamova, A., Sardina, A.: Efficiency of using water hyacinth (*Eichhornia crassipes*) in wastewater treatment. Eng. Constr. Bull. Casp. Reg. Sci. Tech. J. **1**(39), 33–37 (2022). GAOU AO VO "AGASU", Astrakhan
16. Zhukov, A., Rogacheva, S., Gubina, T.: Study of the possibility of desalination of the cooling pond of the Balakovo NPP using higher aquatic plants. Bull. Samara Sci. Cent. Russ. Acad. Sci. **12**(8), 2125–2128 (2010)
17. Dmitrienko, V., Merenkova, N., Pashkova, O., Zanina, I.: Justification of the winter temperature regime of water hyacinth content for urban wastewater treatment. Environ. Sci. Eng. 153–162 (2024)

Integrating Satellite Imagery with In-situ Data to Monitor Groundwater Storage: A Case Study of the Dibdibba Aquifer in Iraq

Hanan Kaduim Mohammed, Imzahim A. Alwan, and Mahmoud Saleh Al-Khafaji

Abstract In recent decades, population expansion and technological advancements have led to an increased demand for water supplies. The accurate evaluation of fluctuations in groundwater storage in Iraq is critical for sustainable water resource management. The crucial part of water management involves gathering water-table data in order to create a time series of fluctuations in groundwater storage. Nonetheless, in many regions, there is a lack of continuous, reliable data, or access to such data necessitates authorization from the pertinent governments. In this paper, the capacity of the GRACE satellites to enhance the investigation of spatiotemporal sustainability of groundwater resource was assessed through investigating the Dibdibba Aquifer in Iraq. Furthermore, groundwater depletion was evaluated by analyzing fluctuations in groundwater levels from Jan 2009 to Dec 2013 with GRACE satellite data. Results indicate that the groundwater elevation can estimate with a coefficient of determination and Pearson's correlation of 0.10, 0.10 and 0.15, and − 0.31, − 0.34, and 0.38 for each of GSFC, JPL_D, and Giovanni, respectively, when compared with observed data. The findings indicated the continued depletion of groundwater in the Dibdibba Aquifer, and that there is a match between the results based on GRACE satellite and monitoring well data. These findings contribute to highlighting and contemplating the effectiveness of using satellite data in monitoring groundwater in light of the rapid development of these technologies.

Keywords GWS · Dibdibba Aquifer · GRACE satellite · Monitoring wells · Time series analysis

---

H. K. Mohammed · I. A. Alwan (✉)
Civil Engineering Department, University of Technology, Baghdad, Iraq
e-mail: imzahim.a.alwan@uotechnology.edu.iq

M. S. Al-Khafaji
Department of Water Resources Engineering, University of Baghdad, Baghdad, Iraq

Y. Zhu (ed.), *Proceedings of the 4th International Symposium in Environmental Science and Industrial Ecology*, https://doi.org/10.1007/978-981-96-1578-0_22

1 Introduction

Iraq is experiencing a critical water issue because of the too fast population expansion and economic development, with a reduction of inflow water into Iraq through the Tigris and Euphrates rivers resulting from dam building in upstream countries and accelerating evaporation and declined precipitation due to the climate change [1, 2]. This issue has intensified in recent years owing to recurrent drought and population growth [3]. This crisis has resulted in the degradation of the land, a scarcity of drinkable water, and an exacerbation of conflicts over water resources. Climate change is a significant worldwide challenge in the management of water resources, as drought becomes an escalating worldwide concern [4].

In the present investigation, four monitoring wells were utilized for the assessment of groundwater fluctuation. Studying variations in equivalent water thickness using conventional techniques is difficult since it requires hydrologic data and deals with extensive geographical scales [5]. Remote Sensing is a powerful technique for obtaining information about an object by analyzing data collected from a distance [6, 7].

A time-series evaluation was conducted in the United Arab Emirates to analyze changes in groundwater storage. This analysis utilized remote sensing data from the latest release (RL06) of the GRACE satellite observations and the GLDAS products. Utilizing the in-situ data collected from wells between 2010 and 2016. The comparative results demonstrated strong concurrence, and the validity of the findings was confirmed by the use of Sen's slope trend and Kendall–Mann's analysis. A former study conducted by [8] assessed alterations in regional groundwater storage in Illinois utilizing monthly and seasonal GRACE-TWS data alongside in-situ soil moisture measurements. Compared calculated Terrestrial Water Storage (TWS) variations from GRACE to well level readings. Investigations revealed that the seasonal pattern and magnitude of GRACE-TWS variations correspond well with in-situ data, nevertheless substantial differences were seen in month-to-month variability. In accordance [9], the Saq Aquifer in Saudi Arabia and Jordan is seeing a reduction in Groundwater Storage (GWS) at a rate of -6.52 ± 0.29 mm per year. The noted reduction in GWL is significantly correlated with the increasing exploitation of GW for agricultural irrigation, alongside the reported decrease in water levels in surrounding wells that function as local sources of water.

The methodology applied to the present paper may be outlined as follows: Comparing the time series of groundwater storage for the four in-situ monitoring wells with available data from 2009 to 2014 and with groundwater storage data derived from several platforms, including Giovanni, Earth Data NASA, and Google Earth Engine. Finally, the statistical analysis is utilized to ensure the reliability and preciseness of the data.

Analyzing variations in water reserves is crucial as it helps understand their sensitivity and vulnerability. In order to address those issues and cross the gaps left by previous investigations, this paper concentrated on analyzing the spatiotemporal groundwater storage fluctuations derived from the satellite imagery in the Dibdibba Aquifer, in Iraq, from Jan 2009 to Dec 2013.

2 Study Area

The area of study, Dibdibba Aquifer, is lying in the eastern region of the Karbala-Najaf Plateau in Iraq, positioned between the sacred towns of Karbala and Najaf. It extended over a total area of around 1209 km^2. The ground surface elevation varies between 16 and 93 m above mean sea level (m.a.s.l), having a mean altitude of 73 m.a.s.l. There are four groundwater monitoring wells, as illustrated in Fig. 1. The formation has a thickness varies between 45 and 60 m, representing the highest layer in the exposed series, forming the foundation of the desert plain stretching between Najaf and Karbala [10].

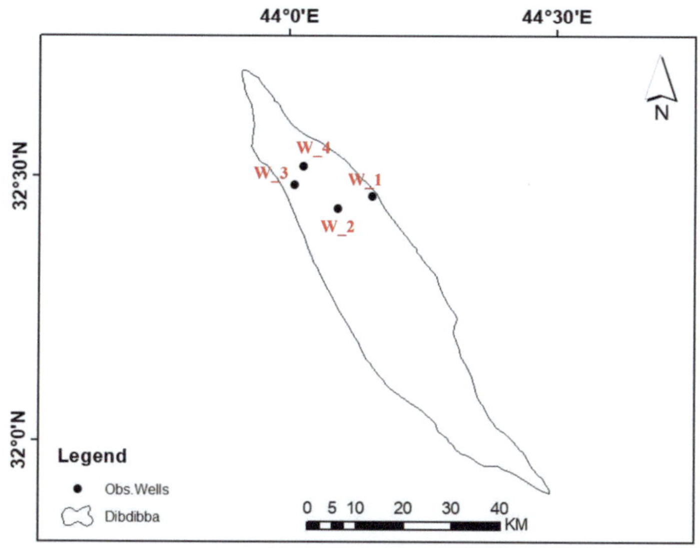

Fig. 1 Location of the observation wells in the Dibdibba Aquifer

3 Dataset

3.1 In Situ Data

The monitoring wells are W_1, W_2, W_3, and W_4 with depths of 20, 40, 60, and 40 m, respectively, in the study area, and these are unpublished data to verify GRACE results. The specific yield (Sy) values are typically determined by the analysis of pumping tests done over a period of hours in situ; the average (Sy) is 0.07 unit-free value [11].

3.2 The GRACE Mission Data

The twin satellites in the GRACE mission orbit around the Earth in an elliptical trajectory, maintained at an approximate distance of around 200 km from one another [12]. In this paper, the following two solutions were employed: the GSFC-RL06 v02 (Jan 2009–Dec 2013) mascon solution, $0.5 \times 0.5°$, these data were downloaded from https://earth.gsfc.nasa.gov/geo/data/grace-mascons, accessed on 16 April 2024. In addition, the GRACE (Jan 2009–Dec 2013) was utilized. For calculating GWS, each equal area $3° \times 3°$ spherical cap mascon block in the JPL-RL06.1 v3 mascon [13]. These data were downloaded from https://podaac.jpl.nasa.gov/dataset/TELLUS_GRAC-GRFO_MASCON_GRID_RL06.1_V3, accessed on 16 April 2024. The (JPL) is the dataset that will be downscaled.

3.3 Google Earth Engine

GEE is an open-source platform that encompasses a vast and varied collection of geographical data. It incorporates satellite images with a high resolution, which can be accurately analyzed using a wide range of advanced spatial analytical tools [14]. In the context of this research, GRACE monthly version 03—mascon was used, the height and water storage anomalies are expressed in centimeters, which is the corresponding unit of water thickness and resolution of 55 km.

3.4 Global Land Data Assimilation System

In this paper, the NOAH data with a spatial resolution of $0.25° \times 0.25°$ and a temporal resolution of monthly for the period Jan 2009–Dec 2013 was utilized, keeping consistent with GLDAS and GRACE. GLDAS model NOAH025_Mv2.1 was used for each Soil Moisture (SM), Canopy Water Storage (CWS), Surface Water (SW), and Snow

Water Equivalent (SWE). These data were downloaded from the NASA data server https://giovanni.gsfc.nasa.gov/giovanni, accessed on 16 April 2024. The average of GWS from GLDAS model (GLDAS_CLM025_DA1_Dv2.2) mm unites with a spatial resolution of $0.25° \times 0.25°$ and a temporal resolution of daily for the period Jan 2009–Dec 2013 was utilized; these data were downloaded from the NASA data server https://giovanni.gsfc.nasa.gov/giovanni, accessed on 1 Aug 2024.

4 GWS Estimation

The approach utilized to estimate GWS in this research performed subtraction of GRACE mascon data from GLDAS Noah model data related explicitly to SM. Importantly, in arid and semi-arid regions, where fluctuations of gravitational in SW are negligible [15], in additional, ΔCWS and ΔSWE have minimal impact and are negligible. Consequently, Eq. (1) becomes:

$$\Delta GWS = \Delta TWS - \Delta SM \tag{1}$$

where ΔGWS, ΔTWS, and ΔSM are groundwater storage change, Terrestrial water storage change, and soil moisture change, respectively.

5 GWS from Monitoring Wells

Monthly measurements are taken to determine the GWL data by the General Authority for Groundwater, Iraq. Ensuring temporal consistency in data while selecting a study period. The average groundwater level was calculated and then multiplied by a given yield to get the GWS, as shown in Eq. (2) [16].

$$GWS = GWL * Sy \tag{2}$$

where GWS, GWL, and Sy are groundwater storage, groundwater level, and average specific yield respectively.

6 Result

Figure 2 illustrates data regarding fluctuations in groundwater levels obtained from four monitoring wells for the period from Jan 2009 to Dec 2013. The continuous lines represent monthly fluctuations in groundwater levels, while the red dashed lines reflect the general trend of these fluctuations, allowing any long-term patterns to be detected.

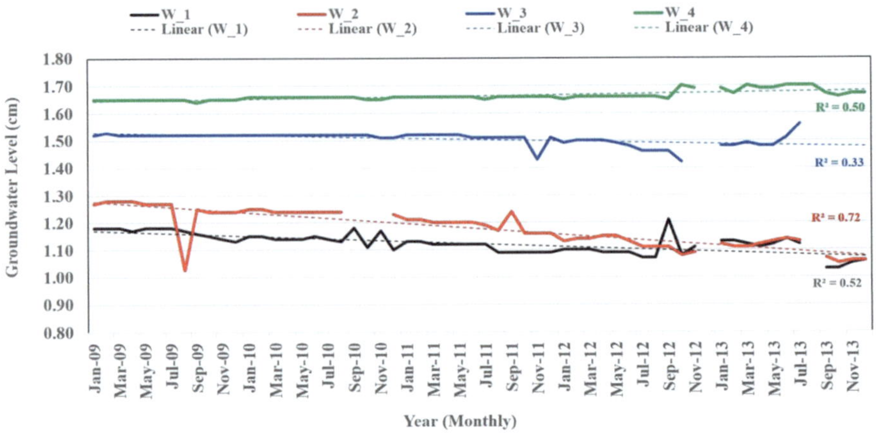

Fig. 2 Monthly fluctuations in groundwater levels in the Dibdibba Aquifer for the period Jan 2009–Dec 2013

The data revealed a continued depletion of the groundwater aquifer, where all wells showed a decline in the water level. The depletion rates of the W_1 and W_2 wells are greater in comparison to the W_3 and W_4 wells, as shown depletion trend of -5.41 ± 6.95, -1.11 ± 9.52, -2.96 ± 5.96, and 2.02 ± 2.69 cm/yr, respectively. Consequently, the depletion occurs gradually, on the other hand, the discharge rate is higher than the recharge. The correlation coefficients (R^2) in the four wells reflect a significant decrease in the GWL, as shown in Fig. 2 and Table 1.

Figure 3 shows distinct seasonal fluctuations in water levels across all models, yet the monitoring wells data remain impervious to these seasonal fluctuations. This seasonal fluctuation mostly indicates the effect of rainfall and evaporation on the replenishment of groundwater reserves. The Pearson's correlation (r) monthly GRACE-based results reach -0.31, -0.34, and 0.38 for the GSFC, JPL_D, and Giovanni, respectively. While, the R^2 was 0.10, 0.10 and 0.15 for each of GSFC, JPL_D, and Giovanni, respectively.

Figure 4 illustrates the monthly variations in stored groundwater in the Dibdibba Aquifer from 2009 to 2013. It can be noted that the results based on GRACE and the data based on monitoring wells are consistent for the period between Jul and Oct 2009

Table 1 Statistical analysis of GWL data in the Dibdibba Aquifer

Parameters	W. No.			
	W_1	W_2	W_3	W_4
Slope	-5.41 ± 6.95	-1.11 ± 9.52	-2.96 ± 5.96	2.02 ± 2.69
Pearson's correlation (r)	-0.72	-0.85	-0.57	0.71
R-square (R^2)	0.52	0.72	0.33	0.50
Adj. R-square	0.51	0.72	0.31	0.49

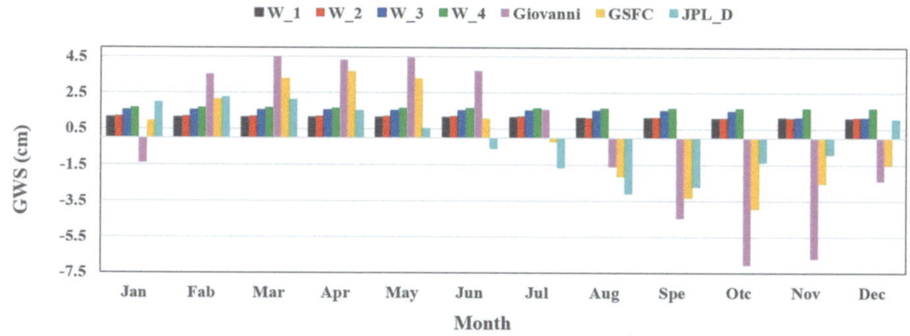

Fig. 3 Comparison of the monthly GWS in the Dibdibba Aquifer

for the W_2 and W_4 wells, while the results showed a clear match for all monitoring wells except the W_2 well due to missing data for the period Oct 2010. After Oct 2011, the groundwater storage trend for the W_3 well aligns with the groundwater storage trends determined from remote sensing. Furthermore, throughout the period from Oct 2012 to Jan 2013, despite the lack of data in the monitoring well time series, there is a noticeable overall decline in GWS.

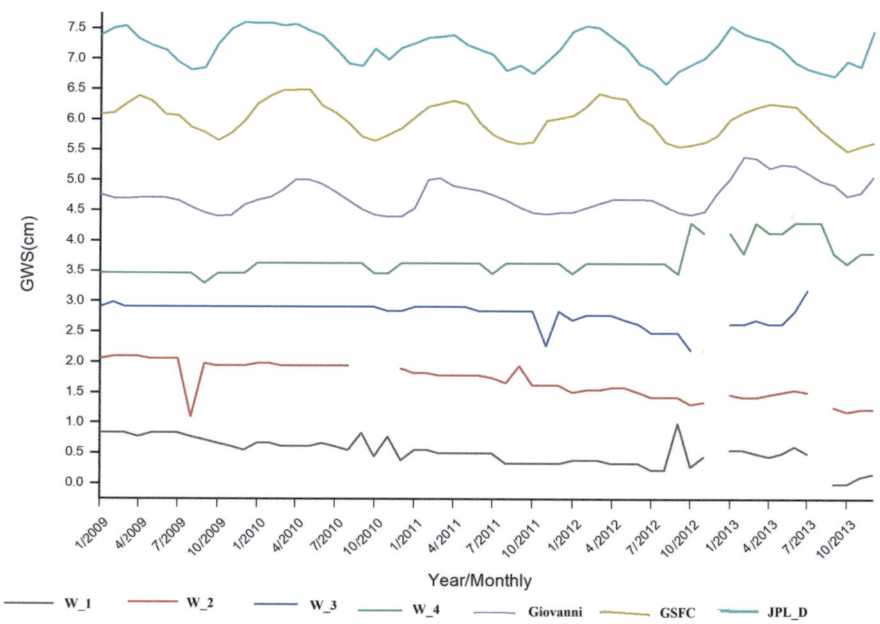

Fig. 4 Groundwater storage in the Dibdibba Aquifer from Jan 2009 to Dec 2013

7 Conclusion

The overall evaluation of the condition of the Dibdibba Aquifer in Iraq, the statistical analysis reflects that W_4 had 0.71 Pearson's correlation (r), while the W_1, W_2, and W_3 had − 0.72, − 0.85, and − 0.57, respectively. Also, the Pearson's correlation (r) monthly GRACE-based results reach − 0.31, − 0.34, and 0.38 for the GSFC, JPL_D, and Giovanni, respectively. This means that the depletion rate is higher than the recharge rate. Nevertheless, the curves derived from GRACE-based results exhibit more pronounced changes over months compared to point-wise (well) readings.

The data indicates that the monitoring wells remain unaffected by seasonal variations, in contrast to the GRACE data, which exhibited evident GWS fluctuations throughout the research period.

It is possible to infer from the results of this research the feasibility and effectiveness of using satellite data in monitoring groundwater currently. It should be noted that the rapid development in obtaining data using satellites will increase the effectiveness and reliability of using this data in monitoring groundwater, which contributes to providing the necessary data to ensure sustainable management of groundwater.

References

1. AlJumaili, K.K., Al-Khafaji, M.S., Al-Awadi, A.T.: Assessment of evapotranspiration estimation models for irrigation projects in Karbala, Iraq. Eng. Technol. J. **32**(5), 1149–1157 (2014)
2. Saeed, F.H., Al-Khafaji, M.S., Al-Faraj, F.: Hydrologic response of arid and semi-arid river basins in Iraq under a changing climate. J. Water Clim. Change **13**(3), 1225–1240 (2022). https://doi.org/10.2166/wcc.2022.418
3. Saeed, F.H., Al-Khafaji, M.S., Al-Faraj, F.A.M., Uzomah, V.: Sustainable adaptation plan in response to climate change and population growth in the Iraqi part of Tigris River Basin. Sustainability **16**(7), 2676 (2024)
4. Hatem, I., Alwan, I.A., Ziboon, A.R.T., Kuriqi, A.: Assessment of agricultural drought in Iraq employing Landsat and MODIS imagery. Open Eng. **14**(1), 20220583 (2024). https://doi.org/10.1515/eng-2022-0583
5. Yin, W., Hu, L., Zhang, M., Wang, J., Han, S.: Statistical downscaling of GRACE-derived groundwater storage using ET data in the North China plain. J. Geophys. Res. Atmos. **123**(11), 5973–5987 (2018)
6. Aziz, N.A., Alwan, I.A.: An accuracy analysis comparison of supervised classification methods for mapping land cover using sentinel 2 images in the Al-Hawizeh marsh area, southern Iraq. Geomat. Environ. Eng. **15**(1), 5–21 (2021)
7. Alwan, I.A., Karim, H.H., Aziz, N.A.: Investigate the optimum agricultural crops production seasons in Salah Al-Din Governorate utilizing climate remote sensing data and agro-climatic zoning. Iraqi J. Sci. 2087–2094 (2019)
8. Yeh, P.J., Swenson, S.C., Famiglietti, J.S., Rodell, M.: Remote sensing of groundwater storage changes in Illinois using the Gravity Recovery and Climate Experiment (GRACE). Water Resour. Res. **42**(12) (2006). https://doi.org/10.1029/2006WR005374

9. Fallatah, O.A., Ahmed, M., Save, H., Akanda, A.S.: Quantifying temporal variations in water resources of a vulnerable middle eastern transboundary aquifer system. Hydrol. Process. **31**(23), 4081–4091 (2017). https://doi.org/10.1002/hyp.11285
10. Hassan, W.H., Nile, B.K., Mahdi, K., Wesseling, J., Ritsema, C.: A feasibility assessment of potential artificial recharge for increasing agricultural areas in the Kerbala desert in Iraq using numerical groundwater modeling. Water **13**(22), 3167 (2021). https://doi.org/10.3390/w13223167
11. Al-Kubaisi, Q.Y., Al-Abadi, A.M., Al-Ghanimy, M.A.: Estimation of groundwater recharge by groundwater level fluctuation method of Dibdibba aquifer at Karbala-Najaf plateau, central of Iraq. Iraqi J. Sci. 1899–1909 (2018)
12. Tapley, B.D., Bettadpur, S., Ries, J.C., Thompson, P.F., Watkins, M.M.: GRACE measurements of mass variability in the earth system. Science **305**(5683), 503–505 (2004). https://doi.org/10.1126/science.1099192
13. Wiese, D.N., Landerer, F.W., Watkins, M.M.: Quantifying and reducing leakage errors in the JPL RL05M GRACE mascon solution. Water Resour. Res. **52**(9), 7490–7502 (2016)
14. Kazemi Garajeh, M., Haji, F., Tohidfar, M., Sadeqi, A., Ahmadi, R., Kariminejad, N.: Spatiotemporal monitoring of climate change impacts on water resources using an integrated approach of remote sensing and Google Earth Engine. Sci. Rep. **14**(1), 5469 (2024)
15. Seraphin, P., Gonçalvès, J., Hamelin, B., Stieglitz, T., Deschamps, P.: Influence of intensive agriculture and geological heterogeneity on the recharge of an arid aquifer system (Saq–Ram, Arabian Peninsula) inferred from GRACE data. Hydrol. Earth Syst. Sci. **26**(22), 5757–5771 (2022). https://doi.org/10.5194/hess-26-5757-2022
16. Strassberg, G., Scanlon, B.R., Rodell, M.: Comparison of seasonal terrestrial water storage variations from GRACE with groundwater-level measurements from the High Plains Aquifer (USA). Geophys. Res. Lett. **34**(14) (2007). https://doi.org/10.1029/2007GL030139

Investigating of Land Cover Changes and Their Effects on Urban Heat Using Remote Sensing Data in Baghdad, Iraq

Raghad H. Hasan, Zaidoon T. Abdulrazzaq, and Nadia A. Aziz

Abstract Recognizing, sensing, and reducing the urban heat island (UHI) effect have become a fundamental subject in many countries, regions, and big cities during the last two decades. This study aims to investigate the changes in land cover and how they affect urban heat in Baghdad, Iraq by using time-series remote sensing dataset. The specific objectives of this study are to provide comprehensive documentation of land cover changes in Baghdad for the years 2000, 2014, and 2024 using Landsat datasets. The Land Surface Temperature registered a continuous increase in the temperature values due to the global warming during the study period (2000–2024). The maximum land surface temperature increased about 12 °C during examination time period. In addition, the UHI intensity of Baghdad showed higher values over the 20 years. The results reveal a clear, significant, and positive correlation between the LST and the locations of the neighborhoods (residential buildings), indicating that the LST increases in proportion to the neighborhood locations, where R2 values about 0.7 for the study period. This suggests that urban behavior tends to control local microclimates.

Keywords LST · NDBI · Urban · Climate · GIS

R. H. Hasan
Surveying Engineering Department, College of Engineering, University of Baghdad, Baghdad, Iraq

Z. T. Abdulrazzaq (✉)
Space Research and Technology Center, Scientific Research Commission, Baghdad, Iraq
e-mail: zaidoon.taha@live.com

N. A. Aziz
Civil Engineering Department, University of Technology, Baghdad, Iraq

Y. Zhu (ed.), *Proceedings of the 4th International Symposium in Environmental Science and Industrial Ecology*, https://doi.org/10.1007/978-981-96-1578-0_23

1 Introduction

Investigating land cover changes and their effects on urban heat is an essential component of understanding the urban environment system, particularly in urban regions where rapid expansions, increased population, and intensive transformations of urban surfaces have taken place [1, 2]. Recognizing, sensing, and reducing the UHI effect have become a fundamental subject in many countries, regions, and big cities during the last two decades [3, 4]. Reasons for an increase in urban temperature are broadly categorized into two: natural and urban-related elements [5, 6]. In recent decades, and with the increase in worldwide energy requirements, both natural and manmade urban processes have led to a significant alteration in regional climates [7, 8]. Baghdad, the largest city in Iraq, is an example of one of these urban regions and was chosen as the study area of this research [9]. The objectives of this study are therefore to update land use/land cover (LULC) in Baghdad using remote sensing (RS), determine urban heat, and identify factors that cause urban heat increase based on an analysis of LULC changes [10].

Iraq, especially Baghdad, is a hot zone as it is characterized by exceedingly high temperatures in summer. The combination of air and surface temperature depends on land cover and the vegetation type. Vegetation has a bigger capacity to alter the major factors that lead to surface temperature. Urban heat islands (UHIs) are one of the important urban climatic phenomena taking place in cities globally [11]. The issue of land cover and land surface temperature (LST) changes in tropical areas has been studied for many years [12]. Iraq, particularly Baghdad, is part of this phenomenon, having already witnessed a significant escalation in its temperature of around 3 °C in the last 30 years. As a result, people and communities in urban areas are confronting some new challenges posed by elevated city temperatures. This study aims to assess the effects of changing land cover on urban heat, focusing on Baghdad, Iraq's capital city. In view of the scarcity of up-to-date LULC maps and RS spatial models that offer an in-depth insight into urban heating in Baghdad due to their application in the past, this research adds novel contributions to the currently accessible literature, in addition to the potential change policies [13, 14].

For many years, the remote sensing approach has been increasingly used to detect LULC changes [15–17]. Multispectral satellite-based sensors play a crucial role in the normal analysis of land surface changes or urban development's [18, 19]. The researchers used remote sensing images, which were important as a valuable source to distinguish land use/land cover classifications over the past years. They also used different time-ordered satellite data sets to monitor LULC changes, thus being able to manage the natural resources of the area [20–22]. Many researchers have used remote sensing data to identify changes in LULC and to estimate the effects of surface temperature or urban heat, as shown by many research studies in other countries in the world [23–26].

This study aims to investigate the changes in land cover and how they affect urban heat in Baghdad, Iraq by using time-series remote sensing dataset. The specific objectives of this study were: (1) to provide comprehensive documentation of land cover changes in Baghdad for the years 2000, 2014, and 2024 using Landsat datasets, (2) to identify the areas with the most urban expansion in the city of Baghdad, (3) to assess the effect of land cover changes on LST variation in the city of Baghdad. The study considered the City of Baghdad, Capital of Iraq.

2 Study Area

The study area represented in Baghdad located in the Midwestern part of Iraq in a series of plains on both sides of the Tigris. Baghdad is the largest and most populous city in Iraq, as well as the capital and administrative center of the country. The city is located between latitude 33° 15′ 06″ to 33° 42′ 36″ and longitude 43° 90′ to 45° 00′, with a total area of 2,107,625 ha, as shown in Fig. 1. The spatial and temporal land cover changes, such as conversion from green and dense forestland to no built-up, an empty town, or conversion from the entire agricultural farm land to an urban developed area and city centers, have diverse effects. Up to date, limited studies have examined LST-land cover relation in Baghdad.

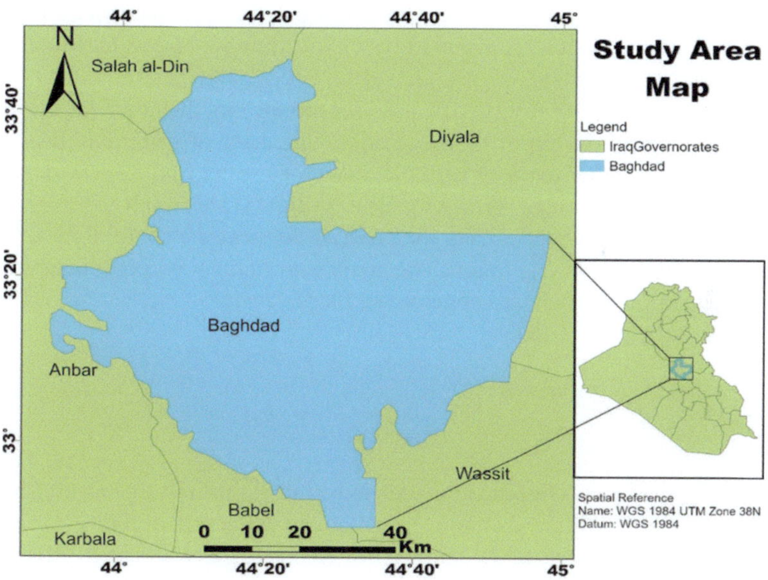

Fig. 1 Location of the study area

3 Methodology

In this study Landsat images was used to estimation and monitoring the changes in land cover and how they affect urban heat in Baghdad, Iraq. The images were downloaded from USGS website for April month, and prepressed using ArcGIS pro. Then spectra indicator (NDVI, and NDBI) was computed, the LST and UHI was estimated. Finally, Pearson correlation was identified to estimate the relationship between land cover and UHI.

3.1 NDVI

NDVI is calculated as a ratio between the red and near infrared bands values using the formula [27]:

$$(\mathrm{NIR} - \mathrm{R})/(\mathrm{NIR} + \mathrm{R}) \quad (1)$$

Typically, NDVI range of values between − 1 and 1. NDVI value above 0.5 indicates the increase vegetation cover, while through the urbanized areas, it appears from the lack of vegetation represented by the declining of the values to zero and below.

The application and the ability to be used for monitoring the urban heat condition were due to the relationship of the Vegetation/Character of soil/urban heat effect. On the other hand, other parts of land cover such as water bodies provide a lower temperature called Urban cool. Where loneless sunlight is absorbed by the water body in the day and cools down occurring the coast, it does offer a cooler temperature at night. The second level is vegetation. The green environment is cooler than the areas with less vegetation, asphalt and concrete surfaces.

3.2 NDBI

The NDBI is used for the classification and detection of the built-up and non-built-up areas [28]. The built-up area is characterized by the extraction of light wavelengths, while the water body shows a low ratio value. The NDBI uses the band NIR and the Short Wavelength Infrared (SWIR) to extract the built-up areas, which depend on the materials' absorption and reflection characteristics. The NDBI equation is:

$$\mathrm{NDBI} = (\mathrm{SWIR} - \mathrm{NIR})/(\mathrm{SWIR} + \mathrm{NIR}) \quad (2)$$

The NDBI index, which depends on classifying buildings from other land cover through different criteria, was utilized in previous studies for the automatic detection and classification of land use land cover from satellite images. The ascending value of SWIR values and the non-inclusion in the calculation of land cover classes leads to obtaining the unique signature of any built-up area in the imagery. The low built-up class values lead to indicating the similarity in the spectral width among the built-up class and any other classes within the studied area with slight differences. The SWIR band is known as the highest reflection band in the case of impervious surfaces.

3.3 LST

The use of LST is one of the key indicators for assessing surface energy conditions and its correlation with vegetation cover extremes in temperature distribution also contributes to the assessment of land cover changes. LST is also considered a significant factor in the assessment of UHIs. Several studies have involved the use of LST to investigate UHIs in different case studies. For instance, Landry and Choi used the potential LST concept for the assessment of UHIs. Sun and Zhang applied LST in natural oases and newly developed urban areas. Sobrino et al. assessed and explained UHIs and their effects in detail. Dousset et al. investigated the effects of UHIs on the health of inhabitants in urban areas downwind.

LST is an effective parameter used to detect urban heat effects. However, in our study, Landsat satellite data were used for the extraction of LST from the thermal band (TIRS, 10–12 μm) of images were selected for the assessment. The LST was automatically extracted from the radiance temperature in Celsius degree from the product LANDSAT/LC08. For the computation of LST firstly calculate the proportion of vegetation index (PV) from Eq. (3), and emissivity (4) and finally the LST derived from the following [29]:

$$PV = \text{Square}\ (\text{NDVI} - \text{NDVI min})/(\text{NDVI max} - \text{NDVI min}) \tag{3}$$

$$\varepsilon = 0.004 * PV + 0.986 \tag{4}$$

$$LST = (BT/(1 + (0.00115 * BT/1.4388) * \ln(\varepsilon))) \tag{5}$$

3.4 UHI

The UHI effect leads to a smoggy climate. The UHI has been extracted using the following method [30]:

$$UHI = \mu + \sigma/2 \tag{6}$$

in which μ is the mean LST value of the study area, and σ is the standard deviation of the LST.

4 Results and Discussion

This study examines land cover changes between 2000 and 2014 and 2024 to assesses their effects on UHI intensity in Baghdad, Iraq. The NDVI results showed a significant change in the land cover classes with a decrease in the vegetation cover and an increase in impervious sites and bare land over the investigation period, as shown in Fig. 2. It can be notice that the vegetation cover was decrease in 2024 than 2000.

Very few maps are available for Baghdad, Iraq, to analyze the urbanization and the increasing of impervious surfaces during the last 20 years. In this context, the development and urban changes were evaluated using the Normalized Difference Build-up Area (NDBI) through the remote sensing database and the highlighted changes in Baghdad, Iraq. In addition, an examination of the size and location of urban escalating regions showed how these communities were embraced by the urban encases in Baghdad. The NDBI map shown in Fig. 3.

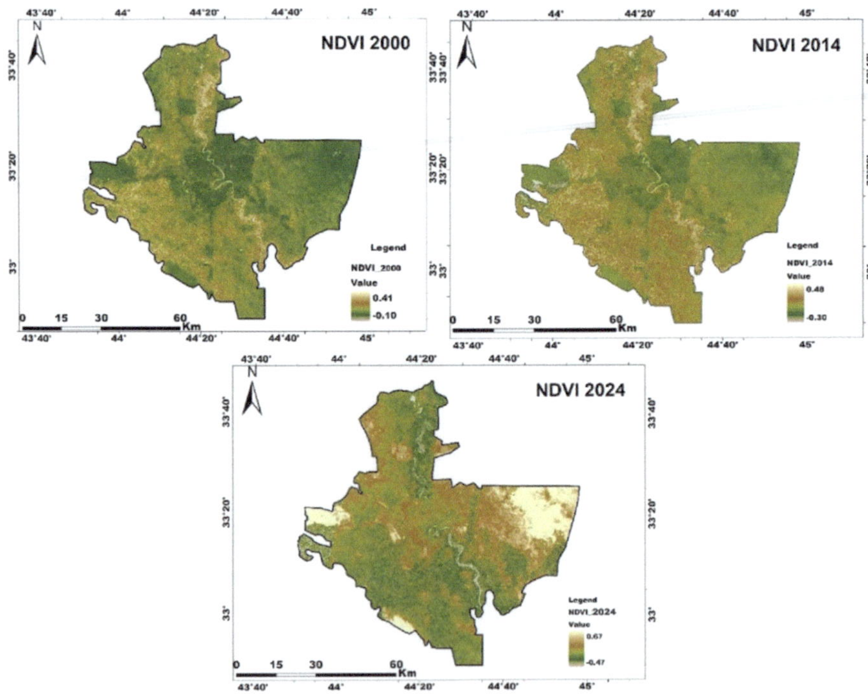

Fig. 2 NDVI for the years 2000, 2014, and 2024

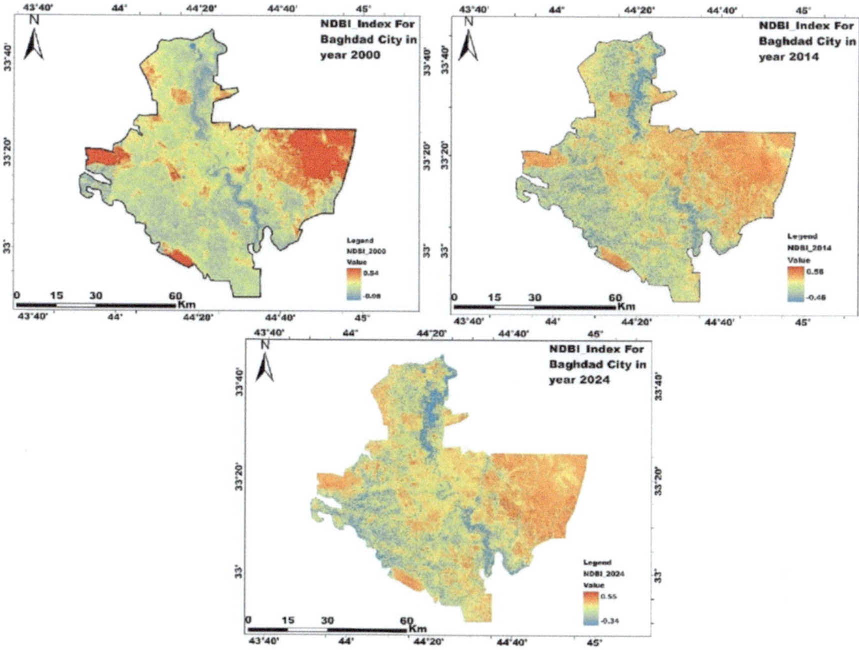

Fig. 3 NDBI for the years 2000, 2014, and 2024

The LST registered a continuous increase in the temperature values due to the global warming during the study period (2000–2024). The maximum LST increased about 12 °C during examination time period. From Fig. 4, it can be notice that the LST in 2000 ranges between 46.66 and 18.77 °C. While, in 2014 the LST ranges from 48.39 and 21.9 °C, and in 2024 ranges between 59.38 and 23.17 °C. Which indicate that the maximum temperature increased about 12 °C and minimum temperature increased about 6 °C.

UHI is one of the considerable climatic features of urban analysis. UHI occurs when there is a disparity between the urban and rural or metropolitan core and rural temperatures, where the city center is higher with enclosed edges.

UHI is investigated through assessing the LST. LST of Baghdad is extracted from the LST data of April for the years 2000, 2014, and 2024. The extracted LST data is cut to the same extent as Baghdad's boundary to ensure the analysis considers the urban areas (Fig. 5).

UHI is analyzed involves modeling the LST by fitting a simple linear model with the Pearson's correlation coefficient. By computing the LST coefficients, this method reveals a clear, significant, and positive correlation between the LST and the locations of the neighborhoods (residential buildings), indicating that the LST increases in proportion to the neighborhood locations, the R2 about 0.7. This suggests that

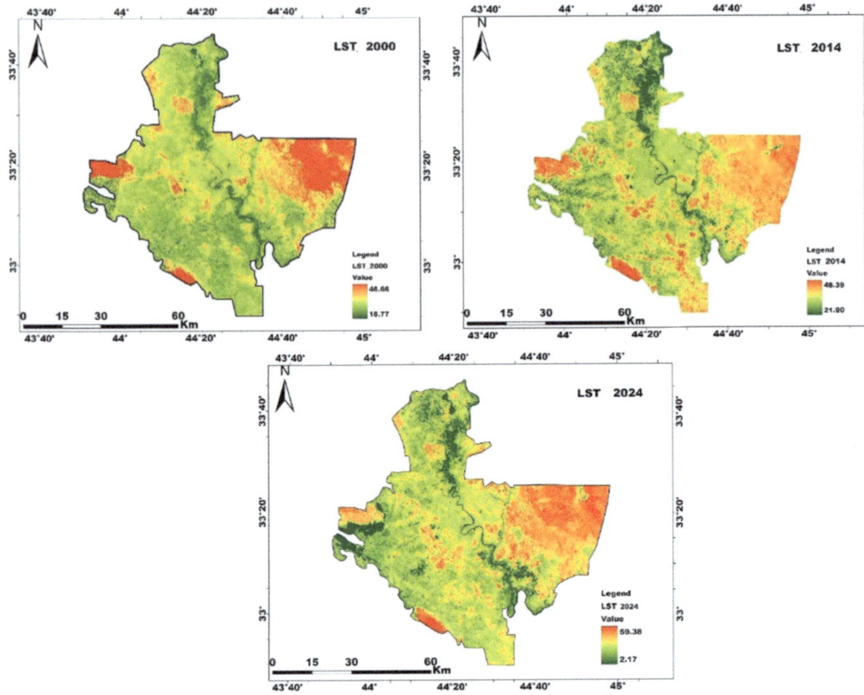

Fig. 4 LST for the years 2000, 2014, and 2024

urban behavior tends to control local microclimates. Figure 6a shows the correlation between UHI and NDBI for the study period. In addition to that, the UHI intensity of Baghdad showed higher values over the 20 years. The results of the Pearson correlation indicated that the LST was strongly correlated with the land cover changes in terms of the built-up area and the vegetation cover with the UHI intensity. Also, to assess the urbanization impact on UHI intensity and investigate the contribution of changes in built-up areas in Baghdad, Iraq, on variations in the annual Normalized Difference Vegetation Index (NDVI) and LST during the study period. Figure 6b shows the correlation between UHI and NDBI for the study period, where R2 values about 0.2 this indicated a negative correlation.

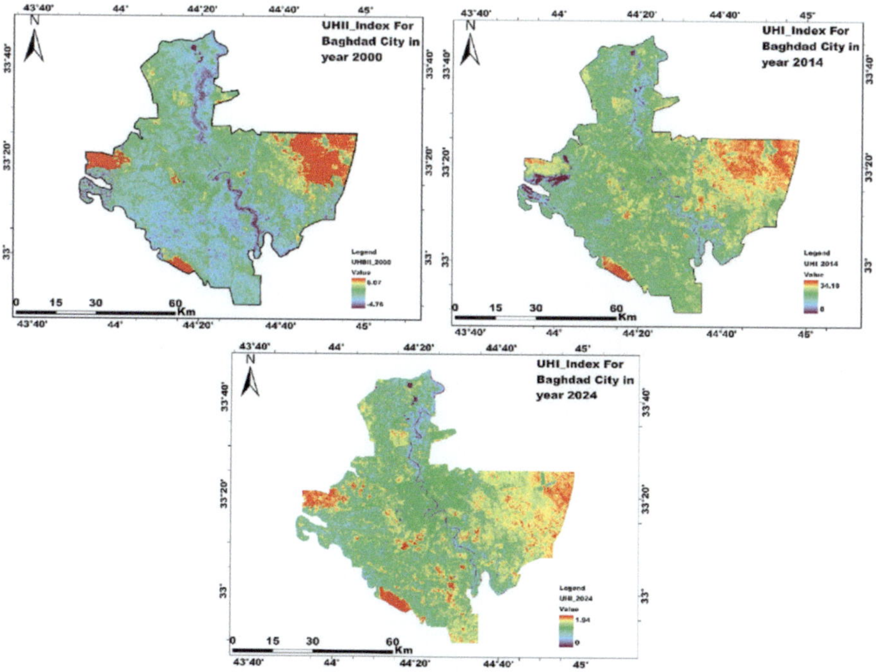

Fig. 5 UHI for the years 2000, 2014, and 2024

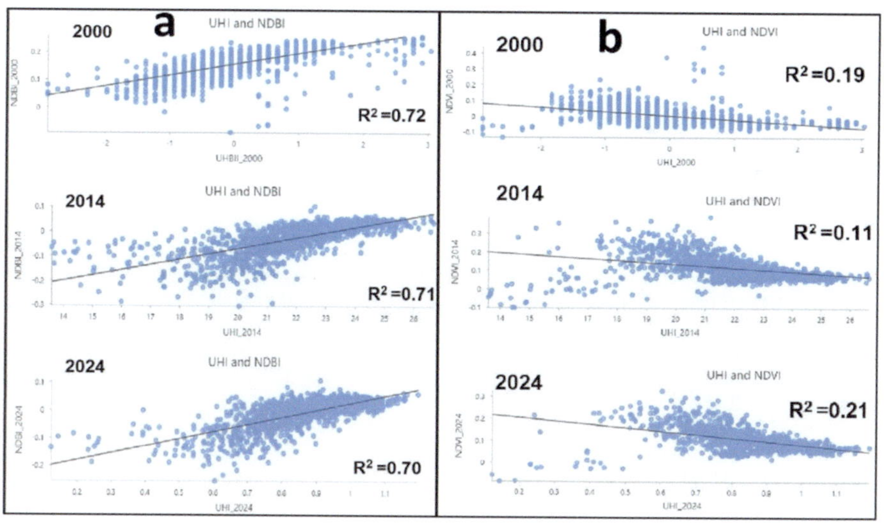

Fig. 6 Correlation **a** UHI and NDBI, **b** UHI and NDBI for the study period

5 Conclusions

Baghdad is a big city with a notably rising population. Urban growth in Baghdad is oriented to accommodate this increased population. This has led to land cover changes that have made the urban heat effect more pronounced. In this study UHI was analyzed using remote sensing data for the years 2000, 2014 and 2024. This study had two main objectives. The first objective was examining the changes in land cover in Baghdad and the effect of these changes on urban heat. The second objective was examining how urban heat has spread over the years. The results showed that, LST registered a continuous increase in the temperature values due during the study period (2000–2024). The maximum LST increased about 12 °C during examination time period. In addition to that, the UHI intensity of Baghdad showed higher values over the 20 years. The results reveal a clear, significant, and positive correlation between the LST and the locations of the neighborhoods (residential buildings), indicating that the LST increases in proportion to the neighborhood locations, where R2 values about 0.7 for the study period. This suggests that urban behavior tends to control local microclimates. In conclusion, the temperature will continue to rise in Baghdad. The integration between urban planners and the local, national, and international experts is needed, and it must include sub-fields such as engineering, social scientists, environmentalists, doctors, economists, and so on, among other environmental and sustainable public sectors, in order to develop sustainable solutions.

References

1. Al-Alem, T.: Investigating land cover changes and their effects on urban heat using remote sensing data in Baghdad, Iraq. Interciencia **46**, 638 (2021)
2. Said, L.W., Aziz, N.A., Hussein, L.Y.: Spatiotemporal mapping of agricultural and meteorological drought in Wasit Province based on GIS and remote sensing data. Ecol. Eng. Environ. Technol. (EEET) **25**(6), 115–122 (2024)
3. Chavez, A.D.: Image-based atmospheric corrections: revisited and improved. Photogramm. Eng. Remote Sens. **62**, 1025–1036 (1996)
4. Agbasi, O.E., Aziz, N.A., Abdulrazzaq, Z.T., Etuk, S.E.: Integrated geophysical data and GIS technique to forecast the potential groundwater locations in part of South Eastern Nigeria. Iraqi J. Sci. **60**(5), 1013–1022 (2019)
5. Beklan, M.A., Muktash, S.A., Gadal, I.M.: The ability to predict land uses in South Al-Madaen City using remote sensing data and geographic information system (GIS). Diyala J. Agric. Sci. **10**, 85–90 (2018)
6. Difar, H.F., Abed, F.M.: Automatic extraction of unmanned aerial vehicles (UAV)-based cadastral map: case study in Al-Shatrah district-Iraq. Iraqi J. Sci. **63**(2), 877–896 (2022)
7. Faraj, S.M., Al-Alem, T.A.: Investigation of the surface temperature of urban areas using the conversion of the image from the Landsat-8 image in Diyala Governorate of Iraq. J. Univ. Anbar Pure Sci. (UAPS) **14**, 95–107 (2020)
8. Aziz, N.A., Alwan, I.A., Agbasi, O.E.: Geospatial selection of rainwater harvesting in Wadi Sarkhar: an analytical hierarchy process-multi-criteria evaluation approach. J. Indian Soc. Remote Sens. **52**, 1501–1515 (2024)
9. Jaber, A.S., Abed, F.M.: The potential of hybrid data integration approaches based UAV-SfM photogrammetry in identifying new archaeological features in the archaeological UR site. Iraqi J. Sci. **65**(8), 4677–4694 (2024)

10. Abed, F.M., Noori, A.M., Al-Saedi, A.S.J.: Optimizing application of UAV-based SfM photogrammetric 3D mapping in urban areas. Iraqi J. Sci. **65**(5), 2958–2975 (2024)
11. Aziz, N.A., Alwan, I.A., Agbasi, O.E.: Integrating remote sensing and GIS techniques for effective watershed management: a case study of Wadi Al-Naft Basins in Diyala Governorate, Iraq, using ALOS PALSAR digital elevation model. Appl. Geomat. **16**(1), 67–76 (2023)
12. Abdulrazzaq, Z.T., Ahmad, M.M., Salih, S.A., Asfahani, J.: Hydro-geophysical parameters estimation of porous aquifer using geoelectrical technique-case study taken from Baiji-Tikrit sub-basin, Iraq. ANAS Trans. Earth Sci. **1**, 3–16 (2022)
13. FAO: State of the World's Forest Genetic Resources. Food and Agriculture Organization (FAO) of the United Nations (2013)
14. FAO: The impacts of disasters and crises on agriculture and food security. In: Strengthening the Evidence Base on Disaster Preparedness, Response and Resilience in the Agriculture Sectors. Food and Agriculture Organization of the United Nations (FAO) (2018)
15. Alwan, I.A., Aziz, N.A.: Monitoring of surface ecological change using remote sensing technique over Al-Hawizeh Marsh, Southern Iraq. Remote Sens. Appl. Soc. Environ. **27**, 100784 (2022)
16. Abdulrazzaq, Z.T., Agbasi, O.E., Alnaib, A., Asfahani, J.: Determining the optimum drilling sites for groundwater wells based on the hydro-geoelectrical parameters and weighted overlay approach via GIS in Salah Al-Din area, central Iraq. Iran J. Geophys. **16**(4), 153–163 (2023)
17. Aretouyap, Z., Asfahani, J., Abdulrazzaq, Z.T., Tchato, S.C.: Contribution of the fuzzy algebraic model to the sustainable management of groundwater resources in the Adamawa watershed. J. Hydrol. Reg. Stud. **43**, 101198 (2022)
18. Tijani, A., Abduljabbar, M.K.J.: A comparison of different land cover changes detection methods for Baghdad. Iraq Int. J. Remote Sens. Appl. **7**, 20–25 (2017)
19. Birkmann, J., Garsuch, A.: Health: A Key for Urban Resilience. UNU-EHS, Bonn, Germany (2017)
20. Brasfield, J., Moseley, W., Lure, M.: Urban Heat Impacts. Environmental Justice Atlas (2016)
21. Hardy, C.: Urban Development in Baghdad: A Publicly Disputed Process. Chatham House, London, UK (2018)
22. Khudhur, M.H., Alwan, I.A., Aziz, N.A.: Comparison of the accuracies of different spectral indices for mapping the vegetation covers in Al-Hawija district, Iraq. AIP Conf. Proc. **2775**(1), 050005–050010 (2023)
23. Abduljabbar, M.K.J.: Spatial configuration of urban growth and expansion in Baghdad, Iraq. Dissertation, Bonn University, Bonn, Germany, 2017
24. Al-Kishie, A., Naser, T., Abdulqahar, K.: Monitoring urban growth in Baghdad city from 2003 to 2015. Tikrit J. Agric. Sci. **19**, 17–29 (2019)
25. Alwan, I.A., Karim, H.H., Aziz, N.A.: Investigate the optimum agricultural crops production seasons in Salah Al-Din Governorate utilizing climate remote sensing data and agro-climatic zoning. Iraqi J. Sci. **60**(9), 2087–2094 (2019)
26. Khudhur, M.H., Aziz, N.A., Alwan, I.A.: Comparative study of supervised classification methods of land cover mapping using remote sensing data: a case study in Al-Hawija district/ Iraq. AIP Conf. Proc. **3105**(1), 050070–050080 (2024)
27. Rouse, J.W., Haas, R.H., Scheel, J.A., Deering, D.W.: Monitoring vegetation systems in the great plains with ERTS. In: Proceedings, 3rd Earth Resource Technology Satellite (ERTS) Symposium, vol. 1, pp. 48–62 (1974)
28. Zha, Y., Gao, J., Ni, S.: Use of normalized difference built-up index in automatically mapping urban areas from TM imagery. Int. J. Remote Sens. **24**(3), 583–594 (2003)
29. Orimoloye, I.R., Ololade, O.O., Mazinyo, S.P., Kalumba, A.M., Ekundayo, O.Y., Busayo, E.T., Nel, W.: Spatial assessment of drought severity in Cape Town area, South Africa. Heliyon **5**(7), e02148 (2019)
30. Thapa, P.: Urban heat island analysis using Landsat 8 satellite data. In: Conference: 11th International Geographic Information Science, pp. 1–16 (2021)

Assessment Land Suitability for Solar Energy Farm Using GIS-Based Multi Criteria Analysis in Iraq

Sarah Ali Mustafa, Nadia A. Aziz, and Imzahim A. Alwan

Abstract Natural resources are thought to be the primary source of economic activity in every nation on the planet, and one of the most important worldwide indicators of development is energy. Because of reasons like population increase, scientific and technological advancements, and other considerations, the need for energy is rising daily on a global scale. This study aims to assess the land suitability for solar energy farm in Iraq using GIS-Based Multi Criteria Analysis (MCA) based on various meteorological, economic, and environmental parameters. Five Criteria was used namely; solar energy, slope, Roads, electricity lines, and water. Also, the AHP method was used to determine the weight for each criteria layer according to its importance. Then all the criteria layers combined together using weighted overlay function in ArcGIS 10. The final map classify the study area into three zones; high suitability, moderate suitability and unsuitable, and the area for each class was computed. the result could be useful for the planers and policy maker for Sustainable development.

Keywords GIS · Solar · Sustainable · Renewable energy

1 Introductions

Natural resources are thought to be the primary source of economic activity in every nation on the planet [1]. One of the most important worldwide indicators of development is energy, because of reasons like population increase, scientific and technological advancements, and other considerations, the need for energy is rising daily on a global scale [2]. For many years, the goal has been to meet 80% of the primary energy demands of humans using fossil fuels [3]. But the supply of fossil fuels is finite and the cost of producing them will rise in the future. Fossil fuel energy production is also the main contributor to greenhouse gas emissions into the atmosphere, which is what has led to global climate change [4–6]. The need for dependable and safe

S. A. Mustafa · N. A. Aziz (✉) · I. A. Alwan
Civil Engineering Department, University of Technology, Baghdad, Iraq
e-mail: nadia.a.aziz@uotechnology.edu.iq

Y. Zhu (ed.), *Proceedings of the 4th International Symposium in Environmental Science and Industrial Ecology*, https://doi.org/10.1007/978-981-96-1578-0_24

energy sources has led to a sharp rise in the investigation of alternatives to fossil fuels, that provide the energy and the reduce greenhouse gas emissions [7]. Due to renewable energy ability to address a wide range of environmental and economic issues, it is currently seeing a global trend of increased use, particularly in relation to solar energy. The International Energy Agency projects that by 2050, solar energy will generate 11% of the world's electricity. A prime example of year-round, naturally occurring renewable energy that doesn't negatively impact the environment is solar farms [8]. Thousands of photovoltaic panels are erected on vast tracts of carefully chosen land to create solar farms, which store energy and supply the main electrical grid. Solar farms can be anywhere in size from a few megawatts to hundreds of megawatts [9]. Solar energy is considered as a feasible renewable energy source since it is regarded as a stable energy source that is not significantly impacted by variations in seasonal weather patterns [10]. Also, solar energy technologies have become more attractive to use because of their increased manufacturing efficiency in recent years [11].

When establishing solar projects, selecting a suitable location is essential since the initial goal is to produce a high-performing, reasonably priced project [12] used multi-criteria decision analysis approach and estimated the weight using the Analytical Hierarchy Process (AHP) to find and map the best solar PV locations for each characteristic using GIS technology with good results evaluated of solar energy generation by conducting technical and theoretical assessments to determine the quantity of solar energy that can be employed in creating electric power [13].

The purpose of this study is to use GIS-Based MCA to assess the land suitability for solar energy farm in Iraq based on various meteorological, economic, and environmental parameters.

2 Study Area

Iraq is the study region, situated in the southwestern region of Asia. The total area about 438,320 km^2. Iraq located between 29 00′ and 37 20′ North latitude and 38 45′ and 48 45′ East longitude. Iraq's terrain is diverse, with alluvial plains in the center and south, plateaus in the west, and mountain ranges in the north. Iraq's four distinct seasons and diverse geography contribute to the country's diverse climate. Iraq is renowned for its scorching, dry summers and chilly, wet winters [14, 15]. Iraq has a lot of solar energy potential since it is situated in the global solar belt (Fig. 1) and receives plenty of sunshine all year round. The research region is highly suitable for investment in solar farms due to its varying periods of solar radiation, which range from 2800 to 3300 h annually, and an average daily quantity of 6.5–7 kWh/m^2 [16, 17].

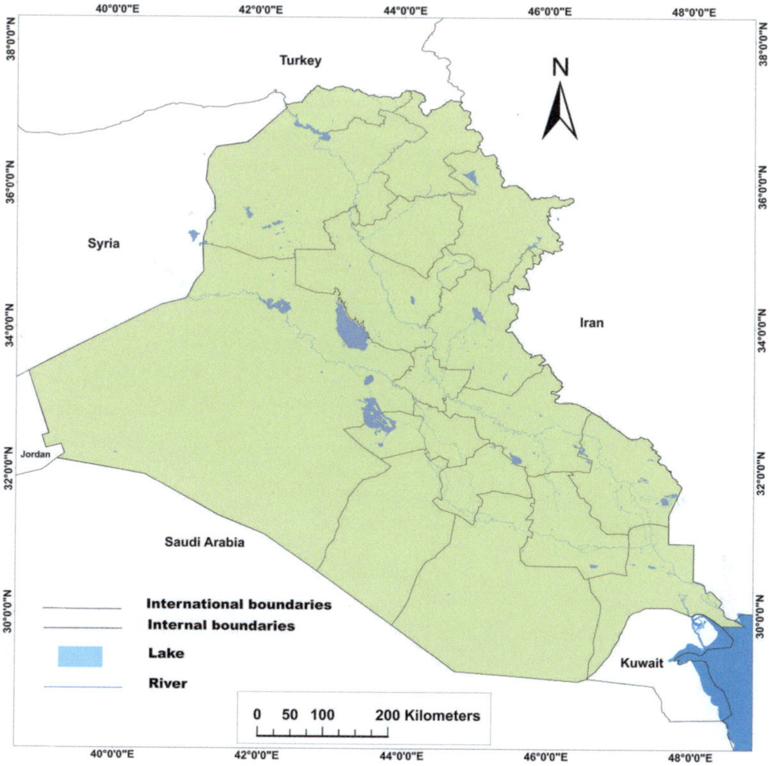

Fig. 1 Location of the study area

3 Methodology

Because of their increased manufacturing efficiency in recent years. Based on a number of criteria, a methodical approach was used in this study to identify the most suitable locations for solar PV installations. The slope is a crucial variable that offers essential information about the topography of a certain location [10]. Slope criterion layer was driven from SRTM DEM data. In order to place solar energy harvesting concept into practice, the data gathering and analysis phase is essential. To perform this procedure, the raw data must be formatted such that it may be utilized for more complex studies. In Fig. 2, the study approach flowchart is displayed. Weights must be chosen in order to integrate the criterion layers. AHP, which considers the weights allocated to each dataset, was utilized in this investigation to establish the weights.

Based on a range of important geographical criteria that influence solar power plant's viability, the current study is conducted. This study included many datasets, such as digital elevation model (DEM), average annual solar radiation for the period 2000–2024 from NASA website, shapefiles and statistical data for roadways and

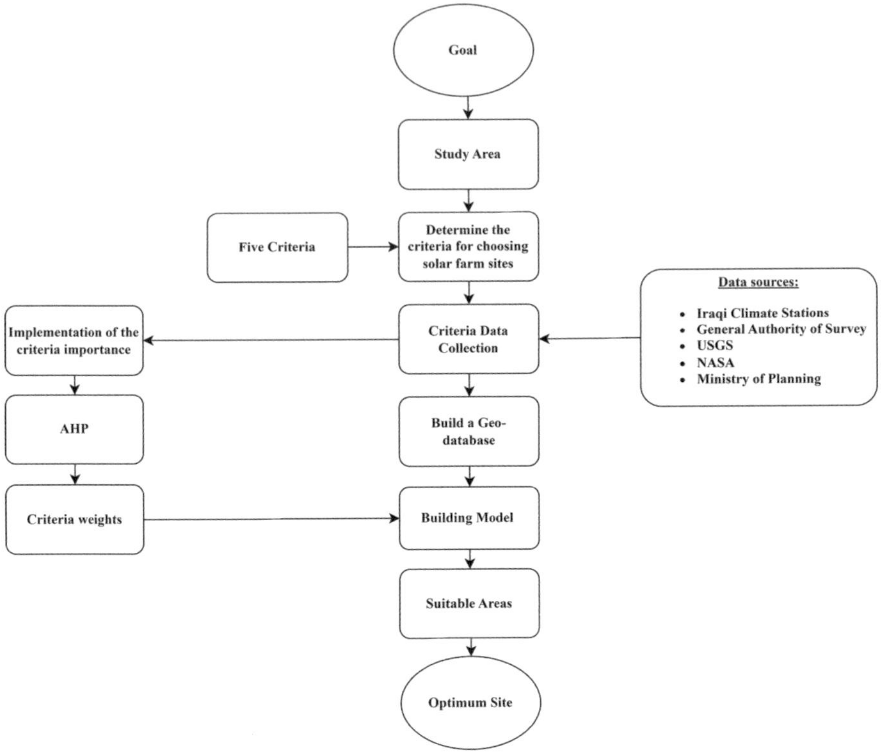

Fig. 2 Flowchart schematic for methodology

electricity lines from the Ministry of Planning's. ArcGIS 10.8.2 software was used for all data analysis and processing, and to produce the final site suitability map.

The slope map was derived from DEM data by degree. The Euclidean distance compute the distance based on the straight-line distance, also it changes the layer's formats to a raster form for the criteria of power lines, and roads. Then, after prepare the five criteria layers (solar, slope, Electricity, roads, and water), all these layers were classified into uniform scale from 1 to 5, where 1 indicated unsuitable and 5 indicated very suitable. The weight of each layers assigned using AHP approach. Finally, all these classified layers combined using Weighted overlay function in ArcGIS.

3.1 Selection Criteria for Suitable Solar Sites

The goal of this study is to identify the suitable solar panel installation sites in Iraq. Five criteria were used solar radiation, slope, electricity lines, Roads lines, and water bodies to provide significant insights on the optimal site of solar panels, as shown in Table 1. The amount of solar radiation present has high impact on solar panels

Table 1 The proposed criteria for this study

Criteria	Layer source	Spatial resolution	Condition	References
Solar radiation	R.S data	0.25° × 0.25°	Annual solar radiation for the years (2000–2024)	–
Slope	STRM DEM	30 m	Less than 5% slope	[20]
Electricity lines	Shapefile	Vector	The distance should be not more than 0.5 km to the main electric power lines	[25]
Roads	Shapefile	Vector	The distance should be at least 60 m from roads	[25]
Water	Shapefile	Vector	The distance should be at least 150 m from water bodies	[25]

generate power. Data on solar radiation have been gathered from NASA website in NetCDF form then interpolated using Kriging method. Another important factor to consider while setting up solar farms is slope [18, 19]. Solar doesn't work well at sites with a lot of elevation gain or slope. Medium elevation sites with a slope of less than 5% are ideal for solar power [20], because steep slopes are thought to result in greater building costs, slope and elevation are significant factors. In a similar vein, low-lying areas will save money [21]. Open, level sites are typically the best choices for solar photovoltaic installations since they get more sunshine [22]. When choosing the best locations for renewable energy projects, economic considerations like proximity to electricity transmission lines are crucial. Solar energy transmission and lower construction and maintenance costs depend on the distance to power transmission lines. Thus, the best locations are those that are near electricity transmission lines. The distance between the solar power plant and the transmission line is an additional economic aspect to take into account. It can be exceedingly costly and time-consuming to build new power lines to connect to the solar power plant [23]. Accordingly, the study concluded that the best locations for solar power plants are 60 m away from infrastructure, such as main and minor highways, railroads, and airfields. The presence of neighboring bodies of water may have an impact on the efficiency of solar panel installations [24].

3.2 *Analytical Hierarchy Process (AHP)*

GIS-based AHP-MCE approach consider an essential technique for making decisions. The best locations for a certain activity may be chosen using a variety of factors [26, 27]. Some of the factors or considerations mentioned above are used to assess if a specific location is appropriate for a given application. Standardization, or converting the criterion data into a consistent numerical measure, is the first and most important stage in the review process. The AHP-MCE technique was employed in this study to find appropriate locations for the installation of solar panels. To accomplish this, five target layers were used: Solar radiation, slope, power lines, transmission lines, and water bodies. Determining whether a site is appropriate for solar panel installations requires consideration of each of these factors. A standardized scale ranging from 1 to 5 was used to standardize these objective layers. It is essential to take into account every factor when assessing if solar energy collecting is practical at a certain location. The weighted overlay approach yields the final appropriateness and facilitates the construction of the weighted hierarchy [28]. The following procedure was followed. For every factor [28], of which five were included in this investigation, a pairwise matrix was generated.

$$p = \frac{p11\ p12\ p13}{pn1\ pn2\ pnn} \tag{1}$$

where p represents the matrix's factors. Next, the normalized weights for the factors are calculated in the second step:

$$wn = \frac{GMn}{\sum_{i=1}^{n} GMn} \tag{2}$$

where G.M. is the average of all column weights and w is the column accumulative weight of factors. The computation of the C.R. determines the judgment coherence:

$$CR = \frac{CI}{RI} \tag{3}$$

where Consistency Index (CI) is:

$$CI = \frac{\lambda - n}{n - 1} \tag{4}$$

where λ is the judgment matrix's eigenvalue, and it may be computed as follows:

$$\lambda = \frac{\sum_{i=1}^{n} piw}{N \cdot w} \tag{5}$$

where N is the total number of strata and p is the factor weight. A satisfactory C.R. value is less than 0.10. The randomness index (R.I.) was derived using the standard tables used by [29].

The consistency ratio (CR) must be determined in order to properly evaluate the consistency of a pairwise comparison. Recalculating the outcome and modifying the research will assist in resolving this issue [28]. Consideration is given to the Pairwise Comparison Matrix, which is made possible by the AHP approach, when choosing weights. As shown in Table 2, the matrix created by the individual components offers a comparative analysis of the relative importance of each criterion in relation to the

Table 2 The pairwise comparison matrix of the criteria layers

Criteria	Solar	Slope	Electricity	Roads	Water	Weight (%)
Solar	1	2	4	7	8	44.3
Slope	1/2	1	3	6	7	29.6
Electricity	1/4	1/3	1	5	6	16.8
Roads	1/7	1/6	1/5	1	2	5.5
Water	1/8	1/7	1/6	1/2	1	3.8
Consistency ratio (C.R.) = 8%						

others. Each criterion is given a priority or weight throughout the pairwise comparison process to indicate its significance within the larger framework for decision-making. The elements are ranked in order of importance using past experience and expert opinion in this field [29, 30].

4 Results and Discussion

The amount of solar radiation that reaches the planet is the first requirement for gathering solar energy; the earth's surface slope comes next. Five criteria layers were used in this study, then, all the criteria were reclassified into uniform scale from 1 to 5 to facilitate analysis and presentation, as shown in Fig. 3. The South eastern part of the study area has high sun radiation values, whereas the Northern sections have lower sun radiation values, as shown in Fig. 3a. These results are consistent with the global average horizontal radiation, which means that solar farms would be better suited for installation in radiation-rich areas. The slope map shown in Fig. 3b, it can be observed that the majority of the research area has high fitting scores, with the exception of the Northern and Northeastern area, which have steep slopes and should be avoided for harvesting. Another significant issue is the distance from the power line network; the farther away from the network, the more energy is wasted during transmission and the higher the cost of energy transmission (Fig. 3c). The requirement distance for the electrical distribution network, noting that the best locations are those that are no more than 500 m from the network's distribution lines. While selecting the best locations for solar energy harvesting, other key considerations include proximity to major cities and roadways. The road network almost encircles the whole study region (Fig. 3d). The distance to the water sources, rivers and lakes, are seen in Fig. 3e, it supposes as there is a 150-m minimum distance between the solar panels and water sources. The final map classify the study area into three zones; high suitability, moderate suitability and unsuitable, with percentages ranging from 54% for unsuitable regions to 26% for low suitability areas, 20% for high suitability areas. The overall size of the high suitability sites is 87154 km^2, which is a solid indication of Iraq's potential for solar energy harvesting, it has been found (Fig. 4).

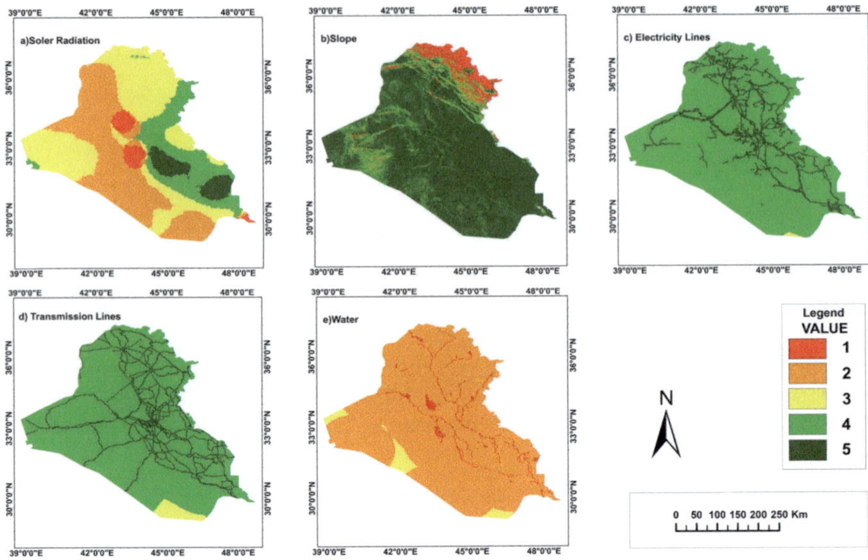

Fig. 3 The reclassified raster for the five criteria

Then the area for each class was computed as shown in Fig. 5 and Table 3. The result could be useful for the planers and policy maker for sustainable development.

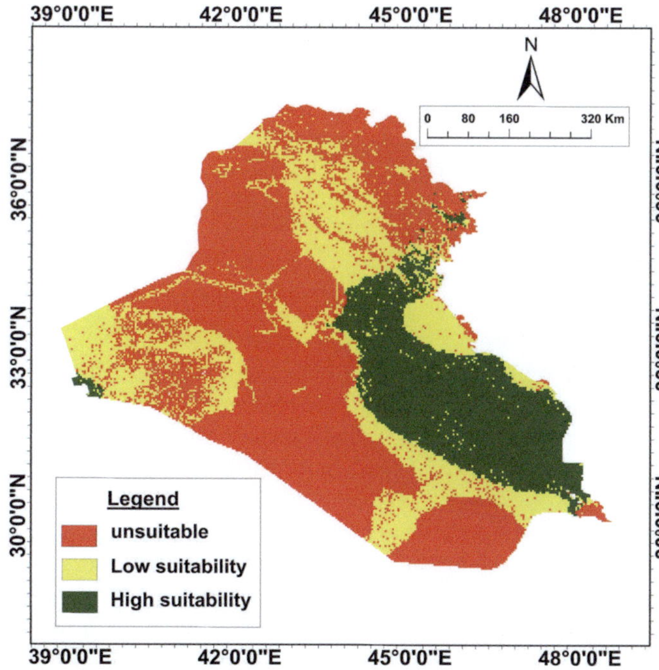

Fig. 4 Final suitability model

Fig. 5 The percentage of each appropriateness model class

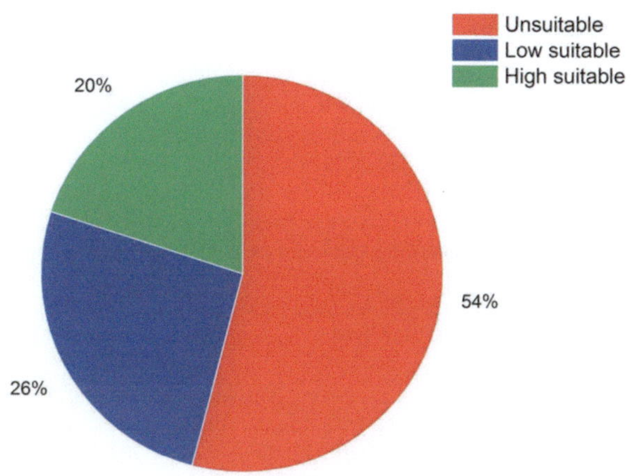

Table 3 The area of each suitability class

Class	Area (km^2)
Unsuitable	235,459.9
Low suitability	112,766.1
High suitability	87,154.1
Total area	435,380.2

5 Conclusion

Iraq's unique position allows to receive enormous amounts of annual solar radiation, adopting solar energy can effectively help to alleviating the problem of energy shortage in Iraq. A move to solar energy would also improve long-term growth, invest local energy resources, diversify the nation's energy mix, and lessen its impact on the environment. The main goal of this study was to determine the best places in Iraq for solar energy collecting utilizing a multi-criteria GIS technique. The study's findings demonstrated the significance of the factors that influence solar farm placement decisions, including solar radiation, slope, proximity to power lines, proximity to major roadways, and distance from primary water sources. The findings showed that 20% of the study area's total land was extremely appropriate for solar farms, 26% was lowly suited, and 54% was unsuitable. The areas with the highest suitability were mostly in the southeast and a small portion of the center region of Iraq.

References

1. Said, L.W., Aziz, N.A., Hussein, L.Y.: Spatiotemporal mapping of agricultural and meteorological drought in Wasit Province based on GIS and remote sensing data. Ecol. Eng. Environ. Technol. (EEET) **25**(6) (2024)
2. Zhou, J., Wu, Y., Wu, C., He, F., Zhang, B., Liu, F.: A geographical information system based multi-criteria decision-making approach for location analysis and evaluation of urban photovoltaic charging station: a case study in Beijing. Energy Convers. Manage. **205**, 112340 (2020)
3. Alwan, I.A., Aziz, N.A.: Monitoring of surface ecological change using remote sensing technique over Al-Hawizeh Marsh, Southern Iraq. Remote Sens. Appl. Soc. Environ. **27**, 100784 (2022)
4. Khudhur, M.H., Aziz, N.A., Alwan, I.A.: Comparative study of supervised classification methods of land cover mapping using remote sensing data: a case study in Al-Hawija district/ Iraq. AIP Conf. Proc. **3105**(1), 050070–050080 (2024)
5. Saeed, I.M., Ramli, A.T., Saleh, M.A.: Assessment of sustainability in energy of Iraq, and achievable opportunities in the long run. Renew. Sustain. Energy Rev. **58**, 1207–1215 (2016)
6. Gusatu, L.F., Yamu, C., Zuidema, C., Faaij, A.: A spatial analysis of the potentials for offshore wind farm locations in the North Sea region: challenges and opportunities. ISPRS Int. J. Geo Inf. **9**(2), 96 (2020)

7. Alwan, I.A., Karim, H.H., Aziz, N.A.: Investigate the optimum agricultural crops production seasons in Salah Al-Din Governorate utilizing climate remote sensing data and agro-climatic zoning. Iraqi J. Sci. **60**(9), 2087–2094 (2019)
8. Bosch, S., Rathmann, J.: Deployment of renewable energies in Germany: spatial principles and their practical implications based on a GIS-tool. Adv. Geosci. **45**, 115–123 (2018)
9. Khudhur, M.H., Alwan, I.A., Aziz, N.A.: Comparison of the accuracies of different spectral indices for mapping the vegetation covers in Al-Hawija district, Iraq. AIP Conf. Proc. **2775**(1), 050005–050010 (2023)
10. Aziz, N.A., Alwan, I.A., Agbasi, O.E.: Geospatial selection of rainwater harvesting in Wadi Sarkhar: an analytical hierarchy process-multi-criteria evaluation approach. J. Indian Soc. Remote Sens. **52**, 1501–1515 (2024)
11. Devabhaktuni, V., Alam, M., Shekara Sreenadh Reddy Depuru, S., Green, R.C., Nims, D., Near, C.: Solar energy: trends and enabling technologies. Renew. Sustain. Energy Rev. **19**, 555–564 (2013)
12. Alhammad, A., Sun, Q., Tao, Y.: Optimal solar plant site identification using GIS and remote sensing: framework and case study. Energies **15**(1), 312 (2022)
13. Ghasemi, G., Noorollahi, Y., Alavi, H., Marzband, M., Shahbazi, M.: Theoretical and technical potential evaluation of solar power generation in Iran. Renew. Energy **138**, 1250–1261 (2019)
14. Al-Kayiem, H.H., Mohammad, S.T.: Potential of renewable energy resources with an emphasis on solar power in Iraq: an outlook. Resources **8**(1), 42 (2019)
15. Al-Umar, M.H., Satchet, M.S., Al-Zaidi, B.M., Abood, A.R.: Spatial study of causes and effects of the sandstorms using meteorological data and GIS: the case of Nasiriyah city, Iraq. Period. Eng. Nat. Sci. **7**(4), 2012–2021 (2019)
16. Mahdi, E.J.: Assessment of solar energy potential for photovoltaic (PV) systems applications in Iraq. Ph.D. thesis, College of Science, University of Baghdad, Baghdad, Iraq, 2018
17. Alasady, A.M.A.: Solar energy the suitable energy alternative for Iraq beyond oil. In: Proceedings of the International Conference on Petroleum and Sustainable Development (IPCBEE), Singapore, vol. 26 (2011)
18. Şener, Ş, Şener, E., Nas, B., Karagüzel, R.: Combining AHP with GIS for landfill site selection: a case study in the Lake Beyşehir catchment area (Konya, Turkey). Waste Manage. **30**(11), 2037–2046 (2010)
19. Akkas, O.P., Erten, M.Y., Cam, E., Inanc, N.: Optimal site selection for a solar power plant in the Central Anatolian Region of Turkey. Int. J. Photoenergy **2017**(1), 7452715 (2017)
20. Turney, D., Fthenakis, V.: Environmental impacts from the installation and operation of large-scale solar power plants. Renew. Sustain. Energy Rev. **15**(6), 3261–3270 (2011)
21. Ding, Z., Zhu, M., Wu, Z., Fu, Y., Liu, X.: Combining AHP-entropy approach with GIS for construction waste landfill selection—a case study of Shenzhen. Int. J. Environ. Res. Public Health **15**(10), 2254 (2018)
22. Kiatreungwattana, K.: Best Practices for Siting Solar Photovoltaics on Municipal Solid Waste Landfills. US Environmental Protection Agency (2013)
23. Ignizio, D.: Suitability modeling and the location of utility-scale solar power plants in the southwestern United States (2010)
24. Aly, A., Jensen, S.S., Pedersen, A.B.: Solar power potential of Tanzania: identifying CSP and PV hot spots through a GIS multicriteria decision making analysis. Renew. Energy **113**, 159–175 (2017)
25. Hassan, Q., Viktor, P., Al-Musawi, T.J., Ali, B.M., Algburi, S., Alzoubi, H.M., Al-Jiboory, A.K., Sameen, A.Z., Salman, H.M., Jaszczur, M.: The renewable energy role in the global energy Transformations. Renewable Energy Focus **48**, 100545 (2024)
26. Saxena, A., Jat, M.K., Kumar, S.: Application of GIS and MCE techniques for optimum site selection for water harvesting structures. J. Rural Dev. 195–206 (2018)
27. Sayl, K., Adham, A., Ritsema, C.J.A: GIS-based multicriteria analysis in modeling optimum sites for rainwater harvesting. Hydrology **7**(3), 51 (2020)
28. Saaty, T.L.: Decision making with the analytic hierarchy process. Int. J. Serv. Sci. **1**(1), 83–98 (2008)

29. Saaty, T.L., Vargas, L.G.: Why is the principal eigenvector necessary? In: Saaty, T.L., Vargas, L.G. (eds.) Models, Methods, Concepts & Applications of the Analytic Hierarchy Process, pp. 63–70 (2012)
30. Aziz, N.A., Alwan, I.A., Agbasi, O.E.: Integrating remote sensing and GIS techniques for effective watershed management: a case study of Wadi Al-Naft Basins in Diyala Governorate, Iraq, using ALOS PALSAR digital elevation model. Appl. Geomat. **16**(1), 67–76 (2024)

MIX
Papier aus verantwortungsvollen Quell
Paper from responsible sources
FSC® C105338

If you have any concerns about our pro
you can contact us on
ProductSafety@springernature.c

In case Publisher is established outside
the EU authorized representative
Springer Nature Customer Service Cent
Europaplatz 3, 69115 Heidelberg, Ger

Printed by Libri Plureos GmbH
in Hamburg, Germany